Michael Aranda
600 B Carr Ave.
Santa Rosa, CA 95404
(707) 525-1085

A Logger's Lexicon

A view of the operations of Deep River Logging Company, out of Deep River, Washington on the Lower Columbia River. Their No. 5, a Lima Shay built in 1900, eases a string of loaded disconnected trucks across a cribbed bridge. Typical of the logger's ingenuity, the cribbed bridge provided a means of spanning canyons if a pile driver wasn't available. The use of so many fine logs wasn't as wasteful as might at first appear, as most bridges of this type were dismanted after an area was cut out, and the logs sent to the mill to be turned into lumber. *Courtesy of Willard Evenson*

A LOGGER'S LEXICON

An Illustrated Reference for Logging Terms and Technology

by John T. Labbe and Lynwood Carranco

TimberTimes ❖ Hillsboro, Oregon

A Logger's Lexicon

An Illustrated Reference for Logging Terms and Technology

By John T. Labbe and
Lynwood Carranco

First Edition, 2001

Published by:
TimberTimes

P.O. Box 219
Hillsboro, OR 97123

Edited by Philip Schnell and Steven R. Gatke
Designed by Philip Schnell
Manufactured in the United States of America

ISBN: 0-9650213-6-X
Library of Congress Control Number: 2001131978

Cover painting The Camp 45 reload was the last place to use a steam donkey on the operations of the narrow gauge West Side Lumber Company, in the Sierra high country out of Tuolumne, California. At this location, logs were transferred from truck to railroad skeleton cars for the trip to the mill by means of a crotch line loading system and a Willamette three-drum loader. *Painting by John Hugh Coker*

Endpapers Cathlamet Logging Company's Shay No. 3 stands by at the landing while loggers use a gin pole with loading tongs to load logs onto disconnected trucks. Location is near Cathlamet, on the Washington side of the Lower Columbia River. *Labbe collection, courtesy of Marshall Morgan*

Loggers prepare to roll a redwood log onto an ox-drawn wagon near Eureka, California. Redwood logging by this means was somewhat unusual in the Coastal Redwood Region, due to the rough ground and constant moisture. In this region, ground skidding with oxen was much more common in the early days. Later, the oxen were replaced with donkey engines. *Freese and Fetrow photo, courtesy of Michael Koch*

With his tools of the trade, an ax and a saw, hanging below him, high climber W.E. "Windy" Illman comes down after topping a Douglas fir spar treee somewhere in Washington State's Cascade Mountains. Next comes the job of rigging the tree with the tons of blocks and line required for high lead logging. This was the highest paying and perhaps most dangerous job in the woods. *Pickett photo, courtesy of Michael Koch*

Introduction

by Philip Schnell

The work of the logger has always been more than just a job. For over two centuries this dramatic, strenuous (and often dangerous) occupation has been carried out from one end of the North American continent to the other, usually in remote places far removed from civilization. The kind of men attracted to logging were thereby tough, hard-working, and sometimes outcasts of society. They were forced to deal with hardships and isolation on a daily basis. The living and working conditions of these men became fertile ground for the development of a language as uniquely descriptive as, for example, that of the cowboy and sailor. So unique is this lingo, that to the uninitiated it can be bewildering at best. Stewart Holbrook aptly demonstrated this in his introduction to Walter McCullah's classic logger's dictionary, *Woods Words*:

"Once upon a time, so legend has it, an injured logger was brought to a hospital, which may have been in New Westminster, or maybe Snohomish, or perhaps it was Tillamook. In any case, the head nurse asked him how the accident had happened. "Nurse" said he, "It was this way: I was setting chokers on the candy side and was just hooking onto a big blue butt when the rigging slinger says to let her go. The hooker yelled Hi! to the punk, the punk jerked the wire, the puncher opened her wide, and . . . well, Nurse, here I am." This wasn't too clear to the nurse. "But," she said, "I don't understand." The logger sat up in his cot. "Damned if I do, either," he replied, "unless that haywire rigging slinger was crazy."

Over the years, the evolution of the language of the North American logger, or lumberjack, has been influenced by many factors. Among these are region, national origin, geography, time frame, mechanization of logging methods, and type of trees sought, to name a few.

In the early years, before the advent of mechanization, the great lumbering centers were in the northeastern part of the continent. As these forests were obliterated, the loggers began a western and southern migration. With each change of venue came different conditions. Winter logging, with its attendant sleighs and ice roads gave way to different methods as the logging frontier expanded. Sometime during the shift from east to west, steam power was introduced into the woods, changing forever the nature of logging. Soon the plodding oxen would find their way into the loggers' stew pots, as railroads and steam donkeys proved their efficiency and capacity. The steam era lasted a mere fifty years or so, and the "lokies" and steam donkeys were to suffer the same fate as their cloven-hoofed predecessors when distillate-powered donkeys, tractors and motor trucks began the process of edging them from their positions of superiority. These changes in locale, methodology and geography had a profound effect on the logger's vocabulary; new logging conditions, tools and methods required new names. Many terms survived the migration, some intact, and others corrupted almost beyond recognition. In fact, it could be said that there exist loggers' *languages*, because many terms used in the industry were unique to certain eras and regions.

A surprising number of terms survive today in common usage, although many of their original meanings have been virtually forgotten. The term "haywire," indicating anything and everything that is either poorly operated or poorly put together, can be traced to the iron wire used to bind bales of hay in the days of animal logging. This handy material was used to cobble repairs on just about every type of equipment used in the woods. To get "in a jam" meant the worst thing that could happen on a river drive—a log jam. The "skidroad," (unfortunately corrupted into "skidrow" in modern usage) was originally a roadway for skidding logs on the ground, either with oxen or steam donkeys. The term now merely indicates the seedy waterfront areas of Pacific Northwest cities, where in earlier times out-of-work loggers sought employment, or at least as often, pleasures of the flesh. Another rich source of logger lingo came from the Native Americans. Terms such as "skookum" and "siwash" come from Pacific Northwest Chinook jargon.

Not all logging terms are entirely original and unique to logging. Quite a number of them were borrowed from the maritime and railroad industries. The "sail guy" on a sailing ship was a line which supported the fore-and-aft sails. In logging, the "sail guy" supported the outer end of the boom on a hayrack or heel boom loading system. Even the term "donkey" may have maritime roots. On steamships, a "donkey" was a small steam-powered hoist of less than one horsepower. Although not generally acknowledged, John Dolbeer, a longtime seafarer, and later lumberman, did not in fact invent the steam-powered hoist—his contribution rested mainly with his adaptation of existing steam hoist technology (resulting in the Dolbeer donkey) and its subsequent introduction into the woods.

Much of the lingo created by loggers was a product of camp life. Poor (or often mediocre) food, inadequate sleeping quarters, abysmal sanitary conditions, the relentless presence of lice, bedbugs, flies, and other pests, and even boredom, inspired descriptively humorous terms—and perhaps through this humor, men were better able to cope with their miseries. If a logger couldn't be comfortable, at least he could joke about it, maybe by turning his shirt inside out to "walk the bugs to death!" Alcohol and fighting seem to have been ever-present elements of camp life, giving, for example, graphic color to the term "logger's smallpox," a condition brought on by the application of a caulked boot to an opponent's face. When loggers weren't in town seeking women, booze, gambling and fights, they often spent their spare time spinning countless bunkhouse tales. Usually, the goal was to top the tales of heroic logging feats told by the previous liar, and through this folklore many legendary characters were either created or glorified. Horror stories involving mythical creatures were also popular, usually intended to amaze and/or terrify newcomers to the logging woods. Among those described in stories were the "agropelter," "hugag," and "side-hill gouger."

Another common reason for the creation of a new term was the logger's mistrust and disdain for his employer. Terms such as "monthly insult," for his paycheck, and "cheat stick," for a scaler's log rule attest to the fact that loggers often felt that they were not being treated squarely, and often rightly so. This sentiment was finally brought to the attention of the outside world by the advent of the I.W.W., or "wobblies," during the first World War. Terms such as "wobbly horror," "strike on the job" and "scissor bill" are examples that illustrate the deep rift between the logger and his employer, especially during the rise of the unions.

As would be expected, most equipment was called by its purpose or function, such as "yarder," "loader," "skidder," "roader," etc., often preceded by the name of its manufacturer or inventor, such as a "Willamette roader," "Lombard steam hauler," "McGiffert loader," etc. Rigging components and tools tended to be named by their function, such as "skyline carriage," or "haulback line."

New words or terms were usually applied to anything newly introduced into the woods. If a logger was unfamiliar with a new idea or questionable piece of equipment, he was often sceptical of its merits, and might give it a disparaging name, such as "abortion," or "fakealoo." Similarly, loggers frequently created descriptive nicknames for any type of machine or tool that possessed certain peculiar characteristics, such as the "sledge hammer two-speed," a type of two-speed yarding donkey with a sliding gear-change pinion which was often difficult to move on its shaft.

Virtually all previously-published dictionaries and glossaries of logging terms have been subject to the publishing taboos of their times. The authors felt that the world was finally ready for an unbowdlerized record of what loggers really called things. Therefore, modest readers should be warned: loggers often used language that was, to say the least, brash, and often profane. In order to offer a more complete listing, and to preserve actual usage, this book is uncensored.

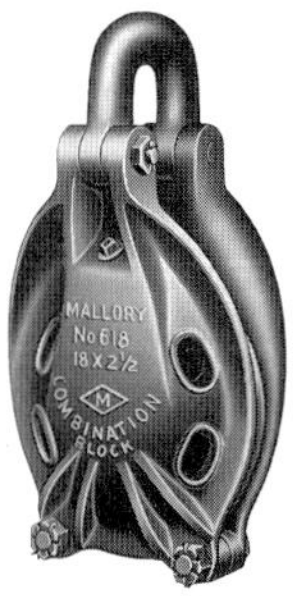

About this Book

Letters in parentheses following definitions indicate the forest regions in which the terms as defined are used. See map.

(App.)	Appalachian Forest
(C.H.F.)	Central Hardwood Forest
(L.S.)	Lakes States Forest
(N.F.)	Northern Forest
(N.W.)	North Woods
(P.C.F.)	Pacific Coast Forest
(P.N.F.)	Pacific Northwest Forest
(Red.)	Redwood Forest
(R.M.F.)	Rocky Mountain Forest
(S.F.)	Southern Forest

If no designation is used, the term is used in all forest regions of the United States and Canada. In a few instances, very local terms are ascribed to a state or country.

The letters "LL" indicate that a definition is quoted, with permission, directly from the book *Lumberjack Lingo: A Dictionary of the Logging Era.* by L.G. Sorden and Jacque Vallier, NorthWord, Inc.,1986.

In the illustration and photo captions, a word or phrase in **bold type** indicates that the term is listed in the body text.

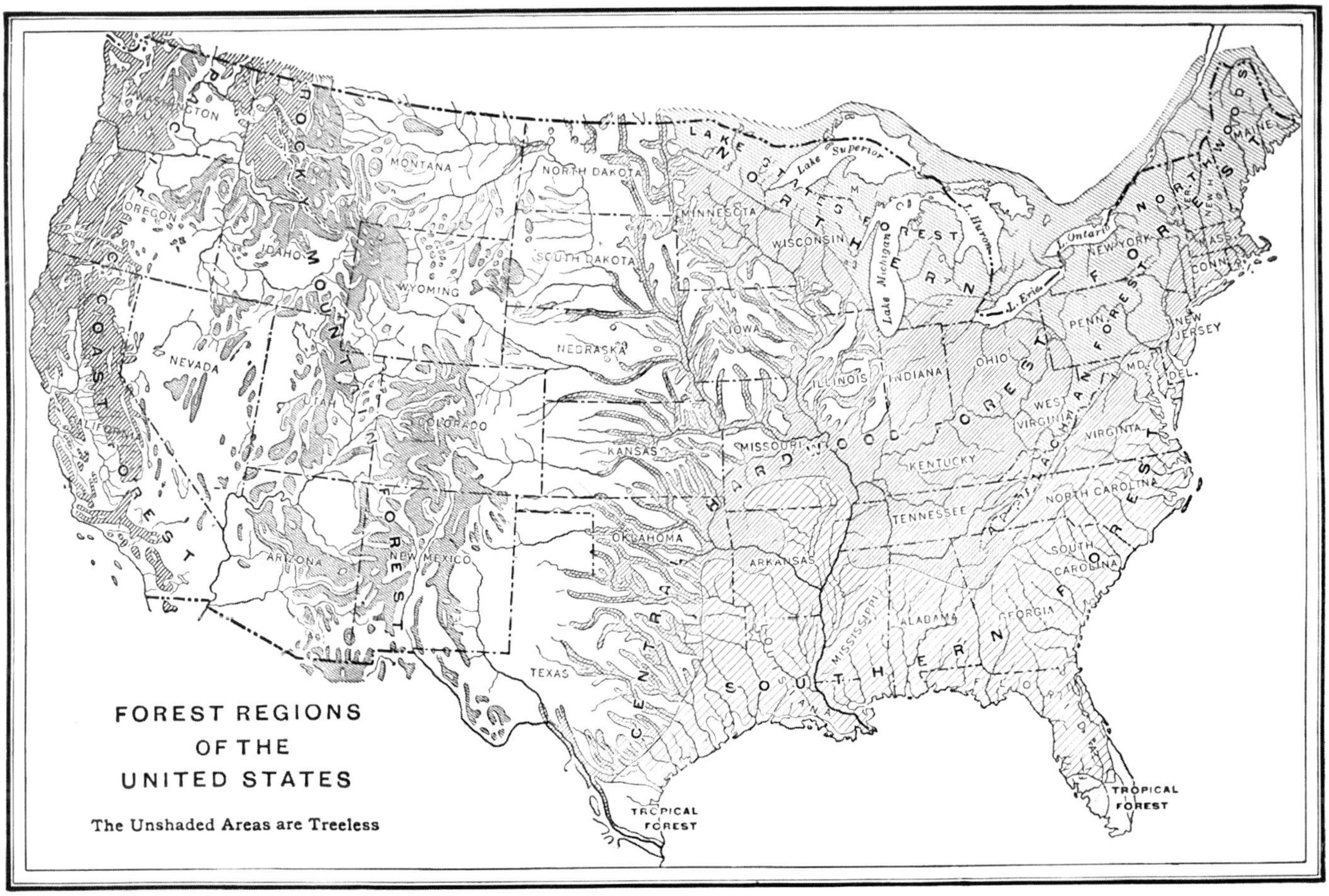

From "Logging," R. Bryant, Wiley & Son, 1923, Philip Schnell collection

Smith Powers Lumber Company ran an extensive logging operation out of Powers, Oregon, and owned a giant mill at Coos Bay. This view of one of their camps, located in a narrow canyon, shows skyline yarding and gin pole loading being carried out right in camp, a not-too-typical practice. There are five donkeys visible, and two high climbers. One climber is in the gin pole loading tree, and the other is near the top of the skyline tree, next to the carriage. It's difficult to determine the function of each machine in this complicated setup. *Courtesy of John and George Powers, Labbe collection*

A

AB That portion of the tree (when it falls and breaks) from the break to the top. AB refers to "after the break." (P.C.F.)
A.F.L. A lumber industry union. (P.C.F.) See L.S.W.U.
abortion 1. A homemade machine or a strange piece of machinery. 2. A haywire piece of machinery. 3. A swear word used on any machine that breaks down.
abutment A rock or timber structure built along the bank of a stream to keep the logs in the channel. (N.F.) See wing jam. LL
according to Hoyle in The Book of Snags This is the authority used by old-timers in proving to a new man how a job should be done. (P.C.F.)
accordion A bucked log with a series of shallow shelves across the

This home-built locomotive, operating on a **wooden railroad**, or **tram,** and **trailing** logs in the woods of Southwestern Washington, might well have been termed an **abortion** by local loggers. She was built by Wilson & Olsen of La Center. *Charlie Bross collection*

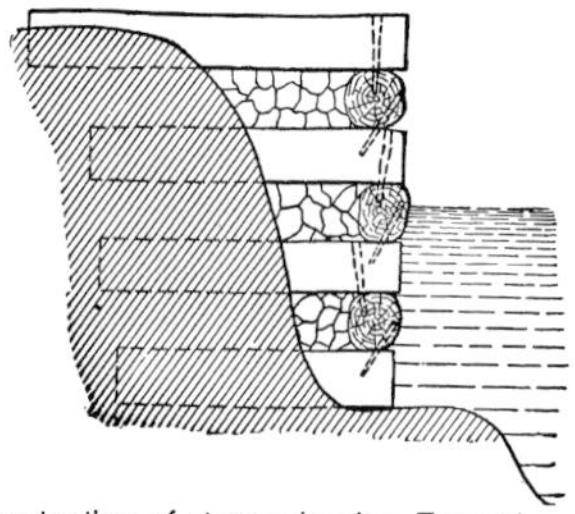

An **abutment** for the protection of stream banks. *From "Logging," R.C. Bryant, Wiley & Sons, 1923, Philip Schnell collection*

end, caused by poor bucking. (P.C.F.)
ace in the hole 1. The timber that is down and bucked. 2. A hidden surprise. (P.C.F.)
acid wood Wood suitable for the manufacture of wood alcohol and other distillation products. LL
acts like a cat in a gale of wind Said of one who moves very briskly. (Maine)
Adams spark arrestor A type of steam donkey spark arrestor. (P.C.F.)

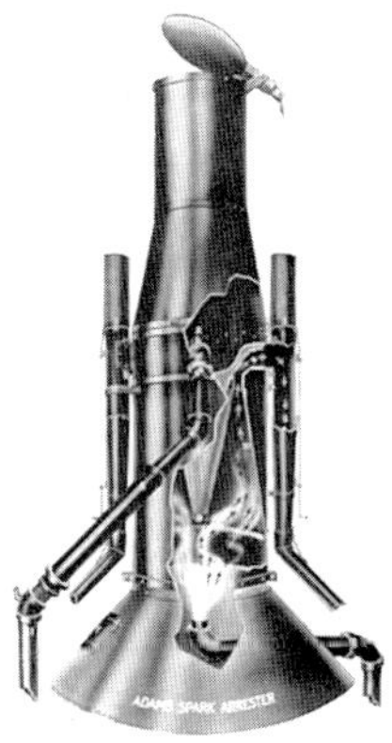

An **Adams spark arrestor**. *Willamette Iron & Steel catalog,1920, Glen Comstock collection*

addle To earn money by working. (L.S.)
admiralty shackle A heavy shackle at the tail tree connecting the skyline to the stub line. See skyline stub.
advanced decay The older stage of decay in which disintegration is easily recognized.
adverse A grade on a logging road or railroad that is against the movement of loaded logs. See against the log.

Milton Creek Logging Company, of Saint Helens, Oregon, here pulls loaded **disconnected trucks** up an **adverse** incline. The hauling line has been secured around the load, rather than to the disconnected truck, to avoid pulling the truck from under the load. *Courtesy of William Beagle*

adz A tool used to hew logs in order to smooth them on one side. The long handle enabled a worker to use it while standing. Also shin hoe and shin-cutter. LL
aerial line See skidder.
aerial roading 1. Hauling logs by aerial tramway. 2. Hauling by a skyline.
aerial skidder Specifically a name for a skidder made by the Washington Iron Works, Seattle, Washington (P.C.F.)
aerial photogrammetry The art and science of interpreting useful information from aerial photographs.
afforestation The planting of trees on a treeless area.
A-frame 1. An A-shaped framework of timbers used to support a block for loading logs. 2. The same type of framework mounted on a raft and used for yarding logs to salt water along the British Columbia and Alaskan coasts. 3. An early name for a logging arch.

Loading logs with an **A-frame** along the Manistee & Grand Rapids RR in Michigan. *Mrs. Ross Wood collection, courtesy of W. Casler*

A-frame show 1. A stand of timber near the shore so that logs can be pulled into the water with an A-frame. 2. A logging operation using an A-frame. (P.N.F.)
aftercut Another term for the back cut or final cut made in felling a tree. (P.N.F.)
after the turn of night Any time after midnight. (L.S.) LL
against the lay At a right angle to the lay of the wire rope.
against the log A logging road grade where the loads have to be hauled uphill. See adverse.
age class One of the intervals, commonly 10 or 20 years, into which the age range of tree crops is divided for classification or use.
agitator A guy who is always stirred up, or stirring up other people, usually for no good reason. Also outlaw.
agropelter An imaginary animal that lived in hollow trees. Any lumberjack who walked by would be killed by a falling limb. (L.S.) LL
air artist Specifically a skilled railroad engineer who knows how to handle air brakes when taking heavy trains down severe grades. Careless use of the air caused wrecks, the loss of logs or the breaking of the train in two.
air bunk On railroad cars, a patent log bunk with inside air pistons which raised or lowered the stakes on the ends of the bunks.
air tongs Loading tongs on the end of a boom or shovel loader which are operated by air, also called earwig from resemblance to that insect.
Alabama wool Cotton underwear. (P.N.F.)
Alaska pine Western hemlock. (P.N.F.)
Alco See American (locomotive).
alder grab The stem of an alder, or other small tree, which is bent over and plugged into a hole bored in a boom stick, or secured in some other way, to hold a boom of logs inshore. (N.F.) LL
alfarky, alforky A horse or mule pack (from the Spanish "alforja"). In the 1800s and early 1900s, supplies were packed into some camps by animals.

A 3/4-yard Marion **shovel loader** equipped with **air tongs** loading logs. *Loggers Handbook, Pacific Logging Congress, 1951, Philip Schnell collection*

alibi day Payday when loggers suddenly developed a toothache or thought of some other alibi to go into town. (P.C.F.)
alkali Any white powdery substance found in camp or in the woods. (P.N.F.)
alky Drinkable alcohol, source usually unknown, but sworn to be of the very highest quality. (P.C.F.)
Aladdin A trade name for a gas mantle lamp used in the late 1800s and early 1900s. (P.C.F.)
all-age (D) management The management of a forest to include trees of diverse ages and, therefore, of diverse sizes.
alley 1. The access lane in a drying yard. 2. The space between two pockets in a booming ground. 3. The long narrow float on which the swifter machine is placed. (P.N.F.) 4. The roofed-over space between the kitchen and sleeping quarters where meat and wood were stored. (N.F.) See dingle.
all hands and the cook The call or cry that a log jam had occurred and that everyone of the river crew was needed to help break it up. (L.S) LL
all hands and the cook and the whiskey-jacks This expression meant for everyone in camp to come running because of some extreme emergency. (P.C.F.)
all heart A descriptive term for a lumber product produced entirely from the heartwood of the log.
all his horsepower Said of a man working as hard as he can at a job. (P.C.F.)
alligator 1. A flat-bottomed, amphibious barge or boat, invented by John West of Simcoe, Ontario, which could be moved overland from one body of water to another. (N.F.) (Canada) 2. A device, usually made from the fork of a tree, on which the front end of a log was placed to facilitate skidding on swampy ground. (N.F.) (Canada) 3. The headworks on a raft in towing logs through the lakes. (L.S.)

Boom men here are working to salvage **sinkers** and **deadheads** from their small boats, sometimes called **alligators**. Note the **ice guards** on the **piers** in the river. This was the site of Camden Lumber Company, of Camden on Gauley, West Virginia. *Courtesy of W. Casler*

alligator mouth A split in a log caused by poor bucking. (Red.) Also bucker's break.
allowable cut The amount of wood fiber that may be harvested annually or periodically from a specified area over a stated period in accordance with the objectives of management. Also annual allowable harvest. Also allowable cut.
alphabet song (shanty-boy's alphabet) Another bit of popular rhyme that centered in the life of the logger and lumberjack. Whether all loggers or lumberjacks knew the alphabet is questionable, but this poem no doubt was a factor in teaching it to them:
A is for axes you very well know,
And B is for boys who can use them also.
C is for chopping we now do begin,
And D is for danger we sometimes get in.
E is for echoes that through the woods rang,
And F is for foreman that headed our gang.
G is for grindstone that swiftly goes round,
And H is the handles so smoothly are worn.
I is for iron that marketh the pine,
And J is for jovial that's never behind.
K is for keen edge our axes do keep,
And L is for lice that keep us from sleep.
M is for the moss we stuffed in our camps,
And N is for needle that mendeth old pants.
O is for owl that hoots at night,
And P is for pine that we always spell right.
Q is for quarreling they do not allow,
And R is for river our logs that do plough.
S is for sleigh so big and so strong,
And T is the teams that haul them along.
U is the use we put our logs to,
And V is the valley we haul them through.
W's the woods we left in the spring,
And now you have heard all I'm going to sing.
(Roland Palmer Gray, Songs and Ballads of the Maine Lumberjacks)
American 1. A type of steam-powered railroad log loader built by the American Hoist & Derrick Company. 2. Locomotives built by the American Locomotive Company. Also Alco.

One type of **American** loader at work near Crosby, Pennsylvania, for N.W. Heineman. *McKean County Historical Society, courtesy of W. Casler*

anchorage A name given to the various lines and holds used to anchor down the end of a skyline. A stump is generally used if one can be found in the proper location; if not, a deadman is used.
anchor brand strawberries Prunes. (L.S.) The term originated on the lumber ships on the Great Lakes. LL
anchor cable A short line that is used to tie down the ends of a donkey sled to prevent tipping up on a heavy pull or to prevent the donkey from climbing the tree.
anchor ice 1. Ice which remained along the banks of streams after the ice in the center of the stream had broken up. 2. The ice which

McCloud River Railroad No. 25 was an **American**, or **Alco** 2-8-2, a product of Schenectady, New York, a subsidiary of the American Locomotive Company. Turned out in 1925, she handled trains for the McCloud River Lumber Company of McCloud, in Northern California, near Mount Shasta. *D.S. Richter photo, Labbe collection*

forms on the bottom of rapids in a fast-flowing stream. 3. Ice that forms around water wheels, booms, anchors, etc. (N.F.)

anchor line A line attached to a small buoy and one fluke of an anchor used in towing a raft of logs. The fluke, which is that part of the anchor which fastens in the ground, is used to free the anchor when fast to rocks or snags. (N.F.) LL Also holdown line.

angle bar A steel plate with a flange base, having from four to six holes through which bolts may be inserted and used to tie rails together on the railroad. LL

An **angle bar** and a number of **Dutchmen** used to connect two lengths of rail. *Labbe photo*

angledozer A tractor with a centrally-mounted bulldozer blade which can be angled to shove dirt to either side.

An International diesel **angledozer** working in Montana. *Loggers Handbook, Pacific Logging Congress, 1950, Philip Schnell collection*

animal 1. An ox, horse or mule used in yarding or skidding. 2. A green hand in the woods. (P.C.F.)

Annie and Mary About 1885, this name was given to the oldest common carrier railroad in California, The Arcata & Mad River Railroad, later operated by the Simpson Timber Company in the Redwood Country.

annual allowable harvest (aah) The scheduled quantity of timber to be removed from a particular management unit in a year.

annual ring A concentric growth layer of wood produced by a tree in a single year.

antifreeze Any alcoholic drink. (P.C.F.)

antigodlin', antigoglin' 1. Said of one wandering from the straight path. 2. At an angle to the correct line. Same as catty-corner. (P.C.F.)

antsy The state of being overanxious.

ape 1. A rigging man. 2. A logger. Also brush ape.

apple knocker 1. Usually the migrant farm workers who tried their luck in the woods. (P.C.F.) 2. A man who sucked up to management.

apron 1. A deck of piling built to hold a scow. 2. The leather covering that workers used to protect their clothing in the lumber mill. 3. A log deck directly below the gate of a splash dam on the downstream side. 4. A deck extending beyond the brow log at the dump to shear logs into deep water. 5. A platform built of timbers at the foot of a chute that guides the logs in a desired direction as they leave the chute.

The **log dump** of the Meadow River Lumber Company, at Rainelle, West Virginia. The **unloader** dumps the logs by the use of an **A-frame;** the logs are carried to the **log pond** over the **apron**. *Courtesy of W. Casler*

Arab A workman who wanders around on the job and is always absent when there is heavy or dirty work to be done. (P.C.F.)

arbuckle Coffee. (App.) (S.F.) LL

arch A piece of equipment used to hold up one end of the logs being yarded by a tractor.

arch cat A tractor used to yard logs with an arch.

arch hook See cat hook, butt hook.

arch line The drum line from the cat through the arch's sheave and to which the chokers are attached. Also bull line.

area regulation A method of controlling the annual or periodic acreage cut from a forest disregarding fluctuations in volumes of fiber yielded. Leads to a type of "regulated forest."

ark A bunkhouse, specifically one on a raft in a floating camp. (N.F.) (P.N.F.) (App.) See wanigan.

Arkansas wedding cake Cornbread. (P.C.F.)

Arkie 1. A green logger in the western woods who is a native of Arkansas or neighboring states. 2. A term of disparagement. (P.C.F.)

armored tit Canned milk. (P.C.F.)

armstrong method Work done by hand and not by machine.

A **cat** with an early **arch**, designed and built by the Red River Lumber Company, of Westwood California. The arch is mounted on a set of **Athey wheels**, which were pivoted on a center mount. If an obstacle, such as a low stump, were encountered, the wheels were capable of flopping over it without interrupting the operation. *From "The Timberman" collection*

armstrong plow A shovel.
Aroostooker An inhabitant of Aroostook County, Maine.
around the horn! A cry used by hookers and hoisters to warn the men around the landing that a log is being turned end-for-end in the air while being loaded. (P.C.F.)
arsneau See jammer.
artificial regeneration Renewal of the forest by direct seeding or planting; establishing a new stand of trees by planting seed or seedlings by hand or machine.
artificial spar tree A raised spar tree used in place of a natural tree.
artist A man who is an expert at any one particular job. (P.C.F.)
as full of ideas as a dog has fleas Said of a very wise man. (P.C.F.)
ashdown hook An open butt hook with a link on the bottom to which the haulback was attached. Used chiefly in clearing and chunking out. (P.C.F.)
ask for time To quit a job.
asparagus A bundle of small logs strapped together to make for easier handling between woods and the pulp mill. (P.C.F.)
asphalt logger 1. A person who would rather be in town than in the woods. 2. A logger who talks big in town but who does little on the job. (P.C.F.)
asshole See wahoo.
association Any organization, such as the Pacific Northwest Loggers or the California Redwood Association, that combine their efforts to help one another.
association fire A forest fire on lands protected by the association, outside of, and not caused by, a logging operation. The association pays the costs rather than the individual logger.
Athey wheels Logging trailers using endless treads which are similar to tractors. These treads are useful in soft or rough ground. They were also used on arches. Also track wheels.

An Athey Tractor Wagon equipped with **Athey wheels**. *Courtesy of Wm. Carson*

at the base An expression used in timber contracts when referring to the diameter of standing timber, meaning at the ground, contrasted with the usual custom of measuring at the stump.
at the boom The downstream price for logs. (N.F.)
at the pond A quotation price on logs delivered at the mill or loading pond.
at the stump In a timber contract, the point at which the bole is severed from the stump; refers to logs cut, not yarded.
auction The action of operators who often bid up publicly-owned stumpage higher than its real worth in order to stick the other fellow. (P.C.F.)
automatic wood harp A chain saw. (P.C.F.)
ax See boy scout ax, branding ax, broad ax, cedar ax, cruiser's ax, falling ax, Hudson's Bay ax, marking ax, surveyor's ax, swamping ax, wire ax.
ax'er To quickly cut something loose in order to avoid an accident. (P.C.F.)
ax handle A unit of measure (the width of a school marm's rear) used facetiously.
ax handle party A fight using ax handles or other wooden weapons. (P.C.F.)
axle grease Butter. (P.C.F.)
axman 1. A name for a faller in the late 1880s. (P.C.F.) 2. A swamper on a survey crew. 3. A chute builder. 4. A sled builder. 5. A man unusually good with an ax.

B

B.B.s Bridge builders. (S.F.)
B.L.M. The Bureau of Land Management, a division of the U.S. Department of the Interior which owns a lot of timber in Western Oregon, the remnant of an old land grant to the Oregon and California Railroad.
B.C. fir See Douglas fir. (P.N.F.)
B.C. Rule The official log rule of the Province of British Columbia which was established in 1909.

A **B.C. Rule**. *From Kloepfer Hardware Co. Catalog, Glen Comstock collection*

babbitt Metal alloy that is poured around the ends of wire rope, setting them in ferrules to make chokers. Also used in bearings and boxings of all types.
Babe Paul Bunyan's blue ox.
baby's born Said when a job is finished. (P.C.F.)
bachelor camp See stag camp.
back The upper or convex part of a saw tooth.
back-around In the early days of truck logging, a Y for turning trailers before trailers were carried back to the woods on a truck. (P.C.F.)
back block An early term for the tail block. (P.C.F.)
backbone 1. The top of a ridge. 2. The ridge pole of a camp building. 3. The bottom log or plank in the V of a log chute. 4. Guts.
back-box In turpentine orcharding, to cut an additional box or boxes, or to place additional cups on a bled tree during the third or fourth year it is worked. (S.F.)
back-breaker 1. Said of a difficult job. 2. A rough-riding crummy, gas speeder or rail motor car. 3. A bough bed carelessly made with big branches.
back cant To reverse the use of a peavey to hold a log back instead of the usual use to roll it forward. (N.F.)
back corner 1. The distant end of a logging operation. 2. The angle where a tail block turns the haulback line around a corner.
back country 1. The extreme back part of a logging show. 2. Uncut timber in the distant hills. 3. Any distant place that is difficult to reach. 4. A ridge which is partly down into a logging show and which must be logged over.
back cut In felling timber, the final cut put in a tree after the undercut has been completed. Also pay cut, felling cut.
back drum The donkey drum nearest to the boiler on a steam rig, or next to the engine on a gas or diesel rig.
back up, back 'er up To cut the back side of a tree after the undercut has been completed. (P.C.F.)

Fallers on **springboards** are **backing up** a **yellow fir** tree. The **falling wedges** have been driven into the cut and an **oil bottle** hangs within easy reach of the **head faller**. Scene is at Camp 9 of the Big Creek Logging Company, near Knappa, Oregon. *Weister photo, Labbe collection*

backfire To set a fire in front of an onrushing main fire to create a fire break. LL

back forty 1. The back end of an operation and the area beyond. 2. Any distant place. 3. The ideal logging show which you never find. (P.C.F.)
back guy The guy line which is back of the spar tree, opposite to the lead of the mainline or skyline. It takes most of the pull when yarding logs.
back his play 1. To back up a man. 2. To back up a claim or statement. (P.C.F.)
backing chain 1. When logging trucks were first used, a chain to prevent the trucks from sliding under the logs. 2. A chain to prevent light loads from pulling apart.
backin' the breeze Said of a talkative man. (P.C.F.)
back landing A tail hold which consisted of horizontal logs attached to the tops of vertical logs driven into the round at intervals along the opposite side of a loading area. Blocks were chained at intervals along the horizontal logs. A line from a donkey or gypsy locomotive was run through one of the blocks and directly across to a log. The logs were pulled onto railroad cars, which were pushed, between the back landing and the raised skidroad. (Red.)
backs up to the window for his paycheck Said of a man who does so little work he's ashamed to receive his pay.
back the tree up To set the wedges before the back cut is completed. (P.C.F.)
back to camp To go back to camp after a shutdown. (P.C.F.)
Badger 1. A brand of smoking or chewing tobacco used by the lumberjacks. (L.S.) LL See Wilder. 2. A make of rut cutter.
bad-order Said of anything which is broken or not in shape to work, men or machines. Originally a railroad car in need of repairs. (P.C.F.)
bag boom An open flexible boom used to catch logs at the mouth of a stream emptying into a lake or similar body of water. The ends are made fast to the shore below the mouth of a stream, and when the boom is filled, the ends are brought together and closed, forming a round boom. (N.F.) (Canada)
Bagley grader A scraper pulled by a donkey engine which was used on railroad cuts and fills before the days of bulldozers; designed by John Bagley who was manager of the Tacoma & Eastern Railroad. Also scoop scraper.
bagnio A whore house. (N.F.)

A small **gypsy** locomotive switches a **landing** for the John Vance Mill & Lumber Company near Arcata, California, in the 1880s. The **back landing**, which was used as a **tailhold** for the loading lines, was common in the Coastal Redwood Country. *A.W. Ericson photo, Carranco collection*

back line See haulback.
back rigger 1. A logger who rigs tail trees by hand. 2. One who helps the second rigger rig up tail trees on a skidder or slackline.
back rigging Special rigging on the tail tree of some skidder systems, specifically to improve the tail hold.
back-scale To deduct something from the scale or measurement of piece work done by a man or crew.
back spar A tail tree.
back-spiker One of the members of a crew who completes the spiking of rails to crossties after the track has been laid by the steel gang. LL
back strap See molasses.
bait 1. Good chow. 2. A block used for leverage in prying. (P.N.F.)
bait can A lunch bucket. (P.C.F.)
bake a cake To get a good fire going in a locomotive or donkey engine. (P.N.F.)
bakehead A railroad fireman who shoveled coal to the fire box, thus "baking his head." (P.C.F.)
bake oven A large stove with a series of ovens placed one above the other which was used for baking. (N.F.)
Baker raft A flat, ocean-going raft with each log chained to a cross log which had a total of six logs chained behind it. A permanent frame enclosed the group of cross logs and their tows. See ocean-going raft.

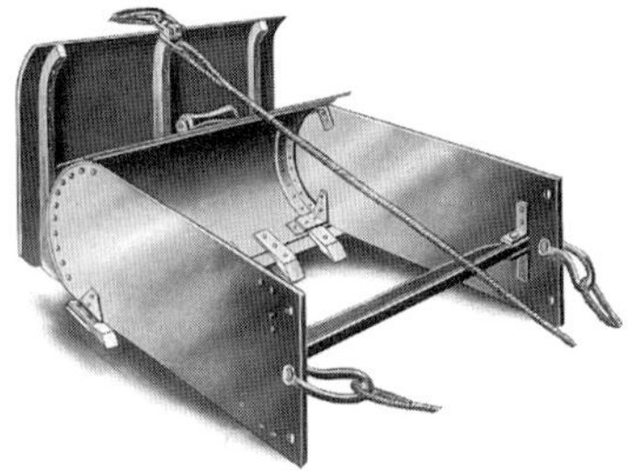

A **Bagley grader**. *Mallory Catalog No. 6, Glen Comstock collection*

bald-facin' logs Sniping logs. (P.N.F.)
bald-headed 1. Said of the butt rigging sent back to the woods without chokers. 2. Said of smooth tires. 3. Said of one who tackles a job by strength and ignorance, having no real knowledge of the specific job. (P.C.F.)
Baldwin The Baldwin Locomotive Works of Philadelphia, Pennsylvania was the most popular supplier of rod engines used in the woods.
bale of hay A crude splice in a wire rope which usually had jaggers.
ball The running surface of a rail.
ball and chain 1. A job that a man cannot leave. 2. A wife.
ballast 1. The coarse rock that supports the railroad ties. 2. Poor rock on a logging road. 3. Heavy articles on the bottom of a packsack. 4. Heavy bread in the cookhouse. (P.C.F.)
ballast car A dump car used to haul ballast on railroad logging shows.
ball hammer A small iron hammer carried on the harness hames, used for knocking out balls of ice and snow which formed under the shoes of horses while they were working. (N.F.) Also peen hammer.
ball-hooter One who rolls or slides logs down a hillside. (App.) (L.S.)
ballin' the jack 1. Going fast or highballin'. 2. Making a fast run with a logging train.

The **Baldwin** Locomotive Works was the most popular builder of **rod engines** for use in the woods. This 2-8-8-2 **Mallet** and a sister engine were the largest ever constructed for a logging company. They were turned out in 1929 and operated out of Longview, Washington, for the Weyerhaeuser Timber Company. *D.S. Richter photo*

ballix, bollix, bollicks, bollux 1. Any bad situation. 2. To ruin a good situation: "To ballix up the job."
balloon 1. A logger's packsack in the late 1880s. 2. A spark arrestor for a donkey or locomotive.
balloon it To pack up and leave camp. (P.N.F.)
balloon logging Logging or hauling lumber by balloon; preceded whirly-bird logging on the West Coast.
balloon stack The large round spark arrestor used on some locomotives and donkeys. See diamond stack.
balloon track, balloon wye A circular-shaped track for turning a locomotive or train.

Coos Bay Lumber Company's 10-spot, the first **Willamette** locomotive built, hauls a string of **ballast cars** built by Pacific Car & Foundry Company, near Powers, Oregon. *Willamette Iron and Steel Co. photo, courtesy of Jack Taylor*

Bridal Veil Lumbering Company operated in the Columbia Gorge, east of Portland, Oregon. "Peggy," their 3-foot gauge **Baldwin** 0-6-0T sported a large **balloon stack**. *Courtesy of C.T. Stoner*

balm A cottonwood tree. (P.C.F.)
balm of gilead log 1. The French-Canadian superstition of bad luck, based on the tradition that our Savior's cross was made from this kind of wood. If the lumberjacks found one of these logs (Balsam fir) used in the construction of their cabin, they would not stay in camp. (J. W. Morrison, "Lumberjack Rhetoric." American forests, No. 30 (December 1924), 722-24) 2. A small evergreen African and Asian tree with aromatic leaves.
baloney 1. The large wire rope used in logging. 2. A fat, heavily insulated electric cable used to run power out to electric-powered donkeys. See electric haulage.
balsam 1. Pitch from fir trees. 2. A name for several kinds of firs.
balsam myrrh Pitch from the balsam tree used for healing wounds. (N.F.) LL
banana belt 1. An easy show. 2. Abnormally good weather in the winter. (P.N.F.)
B and B gang The bridge and building crew on a big logging railroad. (P.C.F.)
bang juice Dynamite.
Bangor rule A scaler's rule. (N.W.)
Bangor tigers Experienced Maine loggers as they were called in the 1850s.
banick A type of bread used on log drives, made with flour, baking powder and water. (N.F.) See bannock. LL
bank 1. A reserve supply of logs held back to meet deficiencies in a daily delivery quota. 2. A landing to which logs are hauled. (N.F.) 3. To deck logs on a river bank which will be moved downstream on the spring freshet. 4. To leave money with the bartender so that all of it would not be spent at one sitting.
banking ground 1. The area beside a river on which logs were banked or decked, to be floated out later on the drive. 2. A rollway.
bank up to To pile up logs on a landing. Also roll up.
bannock A flat cake of breadstuff made in a frying pan. See banick.
Baptist cone A steel cone, invented by William Baptist of Green Bay, Wisconsin, which was placed over the front of a log in ground yarding to prevent hang-ups.
bar The strip of bark left between faces on a tree. (S.F.)
barber chair 1. A goose-pen tree that may collapse because of the rotting of the core. 2. A stump on which is left standing a slab that splintered off the tree as it fell. Also tombstone. 3. To make a tree into a barber chair or allow a tree to split from the stump in felling. The rear slab kicks back, and many loggers have been killed by being caught by a barber chair.
barber pole A tree with an extreme twist. (P.N.F.)
barb wire deal A tough situation to handle.
Bardon (hook) A type of choker hook. See ukelele choker, Wirkkala hook.
barefoot 1. Said of a logger without his corks. See slick-shod. 2.

Banking ground, where logs of the Mendocino Lumber Company were **banked** awaiting high water to carry them to the **booms**. This method of moving logs was common in the northern half of the country from the Atlantic to the Pacific, but Mendocino was the only company to employ the system in the Redwood area. They built a system of **splash dams** throughout the drainage of Big River, in Mendocino County, California, to provide the necessary head of water. *Photo courtesy of Nannie M. Escola*

Bardon hooks. *From Kloepfer Hardware Co. catalog, Glen Comstock collection*

Said of a truck driver who has to drive over dirt roads in wet weather. 3. Said of worn tires which slipped easily. (P.C.F.)
bare-root seedling A tree seedling grown in a nursery bed rather than in a container. When large enough for transplanting, the seedling is dug out of the nursery bed and the dirt removed from the roots before packaging.
barge boom A boom, the upstream end of which is attached to a barge in the stream; used on navigable streams to divert logs from one side of a stream to the other. (N.F.)
Barienger snubbing device A device used to slow the descent of logging sleds down iced roads. Used for distances of as much as 2,500 feet, and with rollers, any curvature under 90 degrees. The brake, which was snubbed to a stump at the top of the grade, had a heavy timber frame, mounted on steel-shod runners, which was faced on top with a 5/8-inch steel plate. Four or six friction bases with hard maple faces were fastened to the steel plate. A cast-iron grooved wheel with a smooth friction face on the under side was mounted on a vertical steel post above each hard maple face. A steel spring held the wheels above the maple blocks when the cable was running free. A control lever was provided for each two wheels by means of which they could be forced against the maple friction face and the speed retarded or the load stopped. Also called crazy wheel. (N.F.)
bark 1. The outside of anything. 2. The covering or bark on a log or tree. 3. The sharp exhaust of a steam engine. 4. To cut the bark off the underside of a log so it will skid more easily. 5. To scrape the skin off, specifically on the knuckles or shins.
bark beetle One of a group of small, cylindrical beetles of the family Scolytidae, the adult of which bores into and beneath the bark of various trees for the purpose of egg laying. This group includes many tree-killing species.
bark blazer A sharp ax or hatchet for blazing (marking) trees. (N.F.)
bark camp A camp roofed with bark. (Maine)

A **Barienger snubbing device.** *From "Logging," R.C. Bryant, Wiley & Sons, 1923, Philip Schnell collection*

A **ranking jumper**, or **bark dray**, hauls a load of hemlock bark to the railroad for the Lackawanna Lumber Company at Mina, in Potter County, Pennsylvania. *Courtesy of W. Casler*

bark dray A dray used to haul hemlock bark out of the woods. Bark was cut in four-foot lengths, and was used in the tanning industry. (L.S.) See ranking jumper. LL

bark-eater 1. A logger. 2. A sawmill worker. (P.C.F.&L.S.)

barked 1. Said of a man who has lost his hair. (P.C.F.) 2. Same as sniped.

barker 1. In the sawmill, the machine that peels bark. 2. A bark peeler in the woods, peeling bark off poles or pulpwood. 3. One who peels bark from redwood logs before milling. (Red.) 4. A bark-peeling tool or spud. 5. A logger who barked logs to make them ride better in the ox team days. 6. One who peels bark in gathering tan bark.

bark hack A chunk scalped out of the bark on a log, using an ax; an early method of marking logs before log brands were used. Also bark mark, side mark. See catchmarker.

barkie A pole sold with the bark on. (P.C.F.)

barking ax A small short-handled, broad ax for ringing bark before peeling with a spud or barking iron. (N.F.) LL

barking bar, barking iron A spud or tool for peeling bark. Also called idiot stick, spud.

bark is slipping Said when it is time to peel poles when the sap is up. When the sap is up, the bark slips off easily; otherwise it has to be shaved off.

bark ladder A platform or rack mounted on a wagon or sled, used for hauling tan bark. (N.F.) LL

bark mark See bark hack.

bark marker Before the log brands, the man who puts a specific mark on the bark of logs to show company ownership.

bark rack A frame used to hold bark on a sled. (N.F.)

bark slide A V-shaped trough used on steep hillsides to slide tan bark down to the roads. (N.F.)

Barkers with their **barking bars**, or **spuds** peel the bark from a redwood log in Humboldt County, California. *Courtesy of L. Carranco*

bark spud A tool for peeling bark off hemlock logs which were sold to tanneries and to peel logs destined for pulp mills. (N.F.)

barleycorn sprints Dysentery after drinking. (P.C.F.)

barn boss The man who was in charge of the animals at a logging camp. Also barn savage.

barn boy A barn boss's helper or a stable tender.

barn door gate In a logging dam sluiceway, a swinging door attached by hinges to the side of the sluice so that it can be swung across the opening to prevent the outflow of water.

barney 1. The power car on an inclined logging railroad. 2. A counterweight. (P.C.F.)

Barney Oldfield A railroad engineer who likes to go fast. Barney Oldfield was a race car driver of the 1920s.

Barnhart 1. An old steam log loader. 2. A brand name for a type of loader. These loaders were carried on the cars and were moved along from car to car as they loaded. (N.F.)

In areas where short hardwood and pine logs were harvested, the **Barnhart** loader was a popular item of equipment. Logs were yarded to trackside, rather than to a **landing**, and the logs were loaded as the train was moving along the track. The loader was mounted on small wheels, which ran along rails afixed to the cars. As each car was loaded, the loader moved ahead to the next car. This scene was on the operations of the Laquin Lumber Company, Laquin, Pennsylvania. *Courtesy of W. Casler*

barnyard Bergmans Rubber boots or slick-soled shoes as contrasted to the old loggers' favorite Bergman caulked shoes. (P.N.F.) Bergman was the name of a boot manufacturer in Portland, Oregon.

barrel 1. The part of a donkey drum on which the line is wound. 2. The swiveling center of a fairlead. 3. To go very fast.

bar room That part of a camp where the loggers slept. (N.F.)

basal area (BA) The cross-sectional area of a tree measured at breast height (4.5 feet above the ground), inclusive of bark and expressed

in square feet. Basal area per acre is a common expression of timber density or stocking.
base line The beginning line for a survey or timber cruise.
basket The hole in a socket into which melted zinc or babbitt is poured when placing a knob on a wire rope. See ferrule and nubbin.
basket hook A butt hook which used a Bardon-type choker ferrule.
bastard Anything not considered standard.
bastard fir Douglas fir, which is red fir when young and mature, but may be both red and yellow when old and mature (old growth).
bastard gauge 42-inch gauge railroad track.

Yeon & Pelton operated this 42-inch gauge **Climax** locomotive on their **bastard gauge** railroad above Rainier, Oregon. *Labbe collection*

bateau A flat-bottomed boat tapering toward the ends, used on river drives. It was a French boat used first in fur trading and built to stand heavy work. LL
bath house See wash house.
batten 1. A log less than 11 inches in diameter at the small end. (Maine) Also batts. 2. To patch anything.
battery 1. Two or more road engines placed at the beginning and at the end on a long haul from the woods. 2. A group of machines.
bayonet top A branch turning up sharply to make a new leader on a tree whose top was broken often resembles a bayonet attached to a rifle barrel. (P.C.F.)
beach camp A camp located at tidewater where supplies are unloaded for the inland camp of the operation. Men at the beach camp make up rafts for towing to the mills.
beach-comb To look for lost logs along the coast. (P.C.F.)
beads 1. Choker ferrules, the knobs on the end of choker wires. 2. A log chain used in loading logs. (N.F.)
bean burner A camp cook, usually a bad one. (P.C.F.)
bean can A lunch bucket. (P.C.F.)
beanery A cheap restaurant in the skidroad part of town.
bean hole A hole in the ground in which beans were baked, used as an oven in the early logging camps. Hardwood coals were raked into it and on top of the bean kettle, which had a tight lid; then it was covered with dirt. The beans were said to have tasted extra good. Also lumberman's trinity. LL
bean house The foreman's office at a logging camp. (Eastern Canada)
bean swagin A bean soup. (Maine)
bearberry bark Cascara bark which is harvested throughout the West Coast woods. See cascara and chittam.
bear cat 1. Anything that is very good. 2. A man who is a "character." 3. An old-time drag saw with two demountable wheels. (P.C.F.)
bear claw A flat hook used as a choker in the late 1800s.
beard The frayed-out whiskers made by loose, broken wires on a wire rope.
bearing tree A tree marked to identify a survey corner. See witness tree.
bear oil Shoe oil. (P.C.F.)
bear shit A mixture of asbestos and cylinder oil. See bucket of mud. (P.C.F.)
bear sign Berry jam. (P.C.F.)
bear story An exaggerated story. (P.C.F.)
bear trap 1. Any tricky situation. 2. An old-time gate used to back up water when splashing logs down a river. 3. Logs in a dangerous position that may roll on the bucker.
bear wallow A dirty or untidy bunk or bed.
beater A sledge or a maul.
beat himself out of a place at the table Said of a man who had fathered a large family. (P.C.F.)
beats In river driving, the rafting grounds were divided into a series of beats (areas on the shore), each made of anywhere from four to six joints (rafts of logs). (Maine)
beat the throttle with a stick Said of the attempt to get more steam out of an engine. (P.C.F.)
beaver 1. A clumsy axman. Also woodpecker. 2. A swamper. (Canada) 3. To chew up the undercut.
beaver oil 1. Grease from a beaver was said to be over 200 percent waterproof when used on caulk shoes. 2. A name for any oil.
beavertail bar A power-saw bucking bar, wider on the outer end than on the motor end.
Bechet hook 1. A hook used on a spool donkey that had a swivel neck and was tempered and designed to feed into the wood when a strain was put on it. 2. The hook was also used on the end of a yarding line to start logs down a chute. 3. A swamp hook on the end of a short piece of line used instead of a strap to fasten a yarding block to a stump. (P.C.F.)
bed a tree To level a path in which a tree is to fall so that it will not shatter.
bedding 1. A bed into which large trees are dropped to prevent breakage, made by leveling the ground surface by excavation or filling or both. See layout. 2. Construction of a raised mound on which seedlings are planted; a site preparation method used most extensively in the Southeastern United States.
bed down 1. Used as an order to settle down. 2. To make an overnight visit. 3. To move a bedroll into camp with the intention of staying. (P.C.F.)
bed ground 1. The area where the men who are off shift on a forest fire camp go to sleep, away from the main center of camp activity. 2. A very poor bunkhouse which is not much better than sleeping outside. (P.C.F.)
bedmaker Usually an old woodsman or mill worker who made the beds and kept the bunkhouses in order.
bed piece One of the skids placed under a pile of lumber. (N.F.)
bedroll 1. In the late 1800s and early 1900s, a roll of bedding was carried by the loggers from job to job. 2. A sleeping bag. (P.C.F.)
beetle 1. A small tractor used in light woods jobs. 2. A maul for driving wedges. (Maine) 3. Any one of a large number of insects which attack trees, specifically the western pine beetle, dendroctonus. The late Walter McCulloch quotes the following limerick from the Western Pine Country and says that, according to Colonel W. B. Greeley, it was originally in a Yale drinking song composed by Al Powers:

Oh see the dendroctonus beetle;
He dwells on the bark of the pine.
He lives on the pitch
The son of a bitch
And he's harder to kill than a lion.

before the sun sets At once. (Pennsylvania)
bell The socket which slides back and forth on the choker wire between the two nubbins. The nubbin is placed in the bell socket to make a circle around the log.
bell man See whistle punk. (S.F.)
bell ox A foreman. (L.S.)
belly 1. The slack in a wire rope. 2. The teeth side of a falling or crosscut saw. 3. The middle of anything. 4. The belly of a log which sags between the bunks.
bellyache 1. Usually transmission problems in a truck or tractor. 2. A complaint. 3. To complain whiningly or peevishly.

Benson rafts under construction in Wallace Slough, near Clatskanie, on the Oregon side of the Lower Columbia River. *Labbe collection*

belly guy See buckle guy.
belly pads Hotcakes.
belly robber 1. A cheap employer. 2. A poor cook.
belly-up 1. Said of a man who is finished, one who resembles a dead salmon floating belly up in the river. 2. To shove up to the table to eat. (P.C.F.)
belly wash 1. Poor coffee. 2. Thin soup. 3. Soft drinks. 4. Hot air or verbal baloney.
belly wrapper 1. A center chain on a truckload of logs. 2. A chain around the center of a load of logs to hold it firmly to the sleigh. Also gut chain. (N.F.) LL
belt man The man in charge of the belts used with motors in the sawmill.
benches Poor horses that were not able to pull a large load. (Maine)
bench mark 1. A reference point in a survey. 2. A starting point for anything.
bend the iron To throw a switch on the railroad.
Benson hook A type of choker hook used in the early days.
Benson raft The Robertson raft with innovations. Hugh Robertson, with W. E. Baines of San Francisco, built a cradle at Astoria, Oregon in the summer of 1894. The cradle was towed to Stella, Washington where a raft was built for A. B. Hammond in September. John A. Fastabend, who took over the Columbia River operation as a contractor, was responsible for most of the later improvements. A. B. Hammond (representing the Southern Pacific) formed the Oregon Rafting company in 1901. Hammond either bought out the Robertson patents or was licensed to use them. This company built rafts of piling for the San Francisco markets. Hammond shipped some 32 rafts to San Francisco before Simon Benson contracted to have the Robertson rafts built for him in 1906.
Simon Benson, who operated 15 camps and 75 miles of logging railroads along the Columbia River, built a large sawmill in San Diego in 1906 to serve the growing Southern California markets. In 1906 the so-called Benson raft successfully completed its journey to San Diego. Benson continued to ship some 116 rafts of logs to San Diego before World War II. See ocean-going raft and Robertson raft.
bent A single row of piles or timbers forming a unit of support for a trestle.
berm 1. The outside or downhill side of a ditch or trench. 2. The shoulder of a road.
berth 1. An area of 24 square miles of standing timber, used in Canada. Also limit. 2. In a camp, a large bed of boughs on which several men might sleep. (Maine)
better-steamer 1. A heavy iron bar illegally used by a lokie fireman to make a big hole in the spark arrestor screen in order to permit the engine to steam better and for sparks to escape better. 2. A locomotive which steams better, more steam with less fire.
between a rock and a hard place Said of one in a bad situation.
between the rock and the river Said of one in a bad situation. (P.N.F.)
between two days Night. (Pennsylvania)
bible pounder 1. A street corner evangelist wasting his time on the skidroad. 2. A religious fanatic making a nuisance of himself in camp.
big bear walks 1. Said when it is time to eat. 2. Said when it is time for something to happen. (P.N.F.)
bicycle A skyline carriage which runs on an overhead cable between the spar tree and the tail tree.
bicycle saw A portable circular power saw mounted in a light frame

It was not always necessary to drive piling in erecting a trestle. Here a **bent** has been prefabricated by fitting the piles to the **mud sill** and **cap** and then raising the completed unit into place with a **donkey**. The scene is on the Hammond Lumber Company operations at Oak Point, Washington, on the North Shore of the Columbia River. *E.W. Anderson collection*

on bicycle-type wheels. See Dow saw. (N.F.)
big auger Owner or high official in camp. (App.)
big blue butt 1. A log much larger at one end 2. A sinker. (P.C.F.)
big breeze 1. A helicopter used to seed cutover land. 2. A man who talks too much. (P.C.F.)
big casino The head man or foreman (from the card game). (P.C.F.)
big eddy 1. An old name for the Portland skidroad. (P.N.F.) 2. A pool in any river, e.g., the big eddy in the Penobscot or the Kennebec Rivers in Maine.

In 1899 the Bridal Veil Lumbering Company set up the first crude skyline used in Oregon, and probably in the Northwest, on Larch Mountain, a few miles east of Portland, Oregon. The **carriage**, or **bicycle**, was built by Bill Dyche; the log was held by a set of **crotch grabs**. *Courtesy of Oregon State University, School of Forestry*

bigger than a skinned horse Said of something very big. (P.C.F.)
bigger than Erickson's bar Anything too big to be measured by any other means. Erickson's bar in Portland's skidroad at Second and Burnside was known as the biggest bar in the world.
big hole 1. The lowest gear on a logging truck or donkey. 2. To work a steam engine at full power, with the Johnson bar in the last notch. Also dollar hole.
big-hole the air To make an emergency stop with the air, applying the brakes quickly by letting the air escape from the biggest hole in the valve.
bight 1. A sag or angle in a wire rope. 2. A loop in a rope.
bigness scale See full scale.
big shot 1. The boss or head of a logging outfit or lumber mill. 2. A heavy charge of dynamite.
big square The area of the stump is counted as a four-sided square around the stump when figuring the faller's pay when felling on the square or piece work on a square foot basis. The area is bigger than that of the actual circle of the stump inside such a square and gives the faller an advantage, thus referring to the "big square." Also falling on the square, government square.
big tyee The head man (from the Chinook Jargon). (P.N.F.)
big wheels Various types of two-wheeled log-haulers usually drawn by horses which were used in the Pine Country to bring logs from the woods to the landing or to the mill. Diameters ranged from 8 to 12 feet. Also high wheels, logging wheels and timber wheels.

The **big wheels** were found in the Pine Country throughout the West. This set is shown on the operations of the Hallack & Howard Lumber Company, near La Madera, New Mexico. *W.J. Duteau photo, courtesy of J. Hereford, Jr.*

billet 1. A short round section of log. (N.W.) 2. A split piece of wood for firewood.
bill raft Raft timber made to certain sizes. (App.)
billy goat 1. A small starting motor used on a big diesel engine. 2. A gas burner with coils used to get up enough steam to pump cold oil for starting a fire in an oil-burning donkey. See goat. (P.C.F.)
Biltmore stick A stick developed by the Biltmore forest school and used by cruisers in estimating height and diameter of trees. Also guessing stick.

bind 1. To get a saw stuck in felling and bucking by the sides of the cut. Wedges are used to eliminate the pinching. 2. To get a long log stuck in the sharp curve of a flume. 3. To get caught in a bad situation. 4. To be caught in the middle.
binder 1. A wire rope or chain around logs on a truck or car to hold the load in place. Also called a wrapper. 2. An iron handle or short stick used to get a purchase in tightening a binding chain or line. 3. A stick or short pole used for the same purpose as the iron handle.

Tightening the **binder** on the **belly wrapper**, before heading for the Simpson Timber Company plant in Humboldt County, California. *Peter Palmquist photo*

binders See runners.
binding chain 1. The chain or cable used to bind a load of logs or lumber on a truck. 2. A light chain to bind small pulpwood bundles for rafting.
binding logs Logs placed on the top of the chain binding a load, in order to take up the slack. (N.W.) LL
bindle A man's blanket roll.
bindle stiff 1. A traveling logger. 2. A camp inspector.
birch hook A small hook that resembles a hay fork and used for handling small timbers. Also bolt hook, pulp hook.
bird dog A small plane used in fire fighting to guide water bombers over target areas. (British Columbia)
bird cage 1. A wire rope that unlays, making the core too short and causing the rope to resemble a bird cage. 2. A screened shelter to keep meat cool. 3. A spark arrestor. 4. To unstrand or unlay a rope or wire rope.
birdseye tenderloin Ox meat showing the marks of the bullwhacker's goad stick. (P.C.F.)
birling The logger's game of log rolling on water. Two men wearing caulked shoes try to roll one another off a log. Also log roller and juggler.
biscuit The flat chunk of wood taken out in an undercut made with a power saw. (P.C.F.) See block, Humboldt.
biscuit shooter A waitress.
bitch 1. A skidding hook. 2. A bad deal. 3. To complain bitterly.
bitch chain 1. A short heavy chain which is used to tie down the butt end of a gin pole to a sled or car when loading logs. (N.F.) 2. A short, heavy chain connecting the mainline and the haulback of a yarding donkey, also serving as a point of attachment for the tackle fastened to the logs. See butt chain. (P.C.F.)

Hardwood logs of the Boswell Lumber Company head for the mill at Boswell, Pennsylvania behind a **Shay** locomotive. The logs are chained to the cars, with **binding logs** placed on top to hold the chains tight. *Courtesy of W. Casler*

bitched Said of something ruined or spoiled. (P.C.F.)
bitch hook A type of lock hook used in rigging.
bitch link A link which is wider at one end than the other. A chain can be pulled through an opening and when a link is turned sideways and pushed into the narrow slot, the chain is locked. In the ground lead days, the eyes of the chokers were sent back to the choker setters.
bitch strap Two eyes brought together with a strawline hook, used in place of a molle, to connect strawline sections, or the strawline to the haulback.
bite itself The tension of a pull that makes a turn of wire rope cross another turn on a donkey drum, causing the strands to wear. The wire rope is said to bite itself where the two strands cross.
bites on both sides Said of a double-bitted ax which cuts on both sides.
black and white A written agreement. (App.)
black cypress Cypress timber of heavy weight. (S.F.)
black gang 1. In the early days of redwood logging, the crew who burned the bark off the felled timber. (Red.) 2. Loggers salvaging timber that has been killed by fire and covered with charcoal.
blackjack A dangerous tool to work with.
black logs Burned timber.
black mike Stew. (P.N.F.)
black one A coarse measurement in forest surveys.
blanket 1. A measure of cold weather, as a two-blanket night, a four-blanket night, etc. (P.N.F.) 2. A tough hotcake.
blanket-busted Said of a man who has lost everything on a bet, including his bedroll. (P.C.F.)
blanket squad In the late 1800s and early 1900s, a new crew coming into camp with their blankets. (P.N.F.)
blaze 1. A marker for a boundary, made by chipping bark off a tree with an ax. 2. To cut a blaze.
bled timber Pine trees that have been turpentined. (S.F.)
blind conk, blind dog Hidden decay in a tree.
blind knot A knot overgrown with new wood.
blind pig A place that sold booze illegally. (P.C.F.)
blind shot Dynamite which fails to go off.
blister end The handle of a hand tool.

Early **ground-lead** logging for the Peninsula Logging Company, near Columbia City, Oregon.This log is being yarded out of the brush by means of a **bitch chain**, used to connect the **mainline**, **haulback**, and the **chokers**, equipped in this case with **bear claw** hooks. The log has been **sniped** to make it ride more easily. The haulback line can be seen receding to the right of the picture. *Labbe collection*

blob 1. The resinous exudation from the fir tree. (Maine) 2. A bad mistake. (L.S.)

block 1. A pulley. There are many varieties, named for the job, or the line which passes through them, such as bull, corner, fall, haulback, loading, mainline, pass line, skyline, snatch, squirrel, swing line, tail, Tommy Moore, transfer, traveling or yarding. 2. The flat chunk of wood taken out in an undercut. Also biscuit.

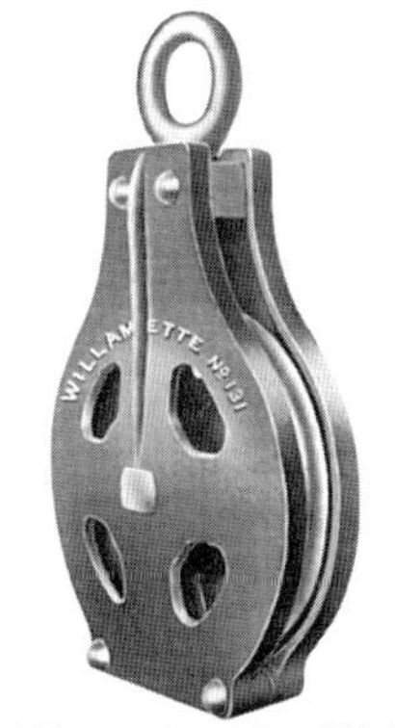

A typical logging **block**. *From Willamette Iron & Steel Works catalog,1920 , Glen Comstock collection*

block-and-a-half A setup using two blocks and one line. Commonly used in moving donkeys, it was rigged by anchoring a block to a stump, passing the mainline through the block and back through a second block attached to the donkey sled. The mainline was then anchored to another stump for the pull. (P.C.F.)

block and fall A block and the line leading from it to which the power is applied.

block and fall tackle Another name for the heel tackle used to tighten the skyline on a skidder show, or other tightline systems. Also tackle.

block and whip An arrangement of a cable and block, to secure added power for moving logs. The free end of the main cable, bearing a swamp hook, is passed through a block fastened to the log to be moved, and then attached to a stump. When the log has been pulled ahead as far as practicable, the cable and swamp hook are moved forward to another stump. Two-block and three-block holds differ in that two or three blocks respectively are attached to the object to be moved. (P.C.F.) See block hold.

block car A specially built railroad car to carry the smaller blocks located in a triangle which formed the main "block" used the Sessom's incline system. Loaded cars were coupled to the block car which was lowered downhill in the bight of the line. Also lowering car. See Sessom's incline system.

block cut 1. A cut made to take out the block which forms the undercut. 2. To cut a piece of timber in blocks or staggered settings. (P.N.F.)

block-cutting Said of a stand of timber which has been clear-cut, leaving alternate blocks uncut. See staggered settings.

blocker A worker who took care of the blocks at the turns in a road in the ground lead days when a donkey was used. He also built skidroads and chutes. (P.N.F.)

block heel A type of heel on a caulked shoe which prevents the spurs used in climbing from sliding back off the level.

block hold See block and whip.

blocking 1. Pieces of wood used to hold up some object. 2. In bucking, the chunks placed under a log.

block-out 1. To change the yarding blocks to move logs around a stump or rock. 2. To change the blocks in order to move anything to one side. 3. To take out surplus timber. 4. To set aside, such as to set aside a seed area or to lay out a patch of timber for logging.

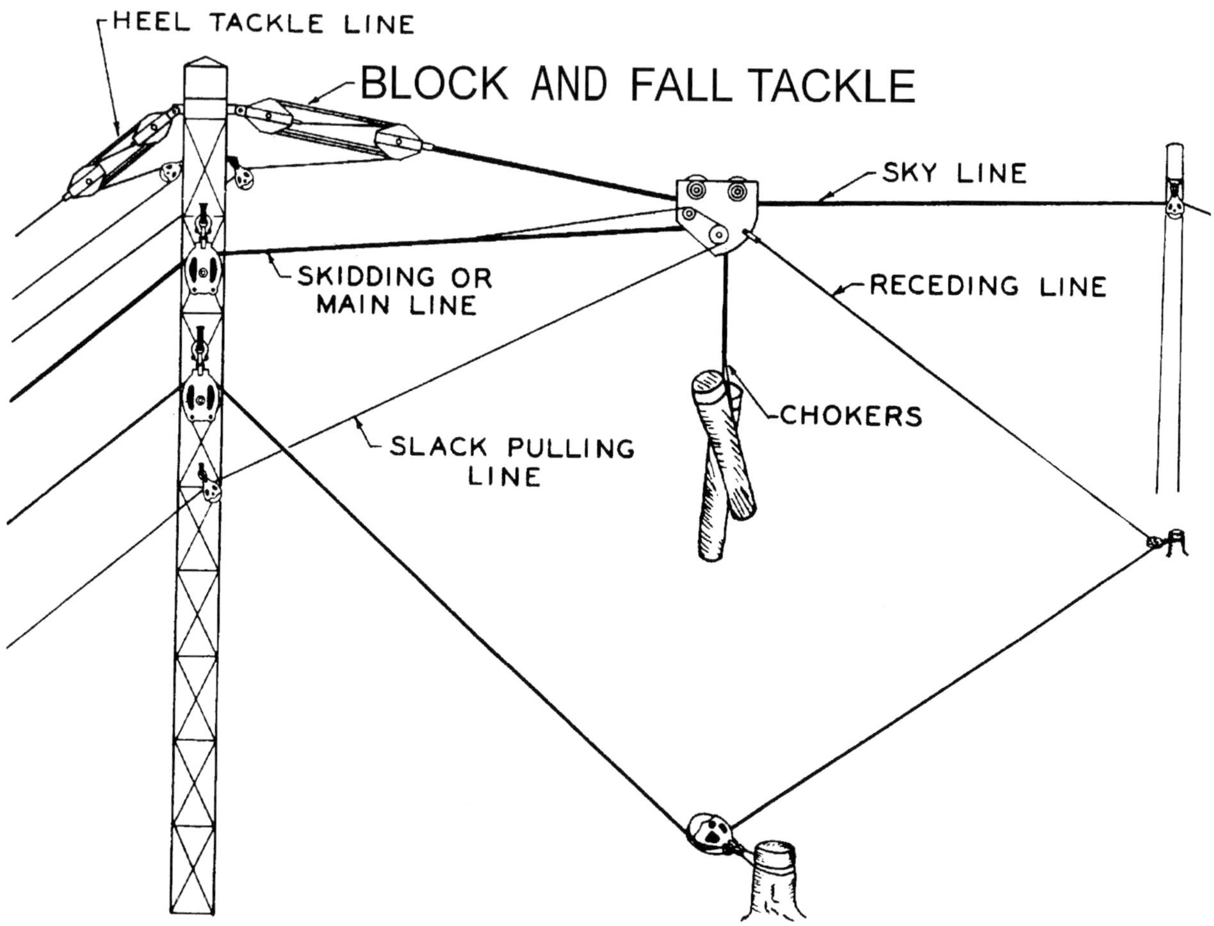

The **skidder** system showing **block and fall tackle**, used for tightening the skyline. *Labbe collection*

A **block car** at the head of the Ebey Logging Company **incline**, near Arlington, Washington. Built by the Willamette Iron & Steel Company, it posed for this company picture in June, 1921. This incline system, designed by H.W. Sessoms of Ebey Logging Company, and known as the **Sessoms system**, proved popular in terrain where the inclines were not too steep and a number of cars could be handled at a time. *Willamette Iron & Steel Co. photo, Labbe collection*

Coos Bay Lumber Company's 10-spot, the first **Willamette** geared locomotive, **blows a doughnut** as it starts a train of logs near Powers, Oregon. *Willamette Iron & Steel Co. photo, Labbe collection*

block tender See chaser.
blow A break in a dam, usually at the base because of the lack of proper toe piling.
blow a doughnut The spitting out of a perfect ring of smoke, which occasionally occurs when a throttle is quickly opened on a standing steam engine on a frosty morning. See doughnut.
blowdown 1. Either a tree or trees which have been blown down by the wind. 2. A specific area where the blowdown occurred. Also down timber, windfall, wind slash, windthrow. 3. A valve on a steam boiler used to blow down the boiler.
blow 'er down To remove trees or stumps with explosives.
blow for the push To blow the signal on the donkey engine, calling the foreman to investigate trouble.
blow his top To lose one's temper.
blowin' Said of one going to town to celebrate. (S.F.)
blow-ins Celebrations along the skidroads in town. (S.F.)

blow its top To blow the top out of a tree to be used as a spar by placing a circle of dynamite around the trunk at the point where it is to be cut off. A cable is wound around the tree below the point to keep the tree from tearing. (P.C.F.)
blow off steam To give vent on one's emotions.
blow-up A fire which suddenly blazes up, jumps control lines, and takes off across country.
blue butt 1. A large redwood butt, usually a sinker. 2. The term is also applied to other trees. (Red.)
blue it To spoil or ruin anything.
blue jacket Lice.
blue jay See road monkey.
blue nose, blue noser 1. A logger or millhand from Nova Scotia, so-called because of the cold climate. Generically all men from the Maritime Provinces (the Canadian provinces of Nova Scotia, New Brunswick and Prince Edward Island) could be identified as Blue Noses who came to work in the woods and in the sawmills on the West Coast. (P.C.F.) 2. A woodsman from Nova Scotia. (Maine)
blue one Said of anything big. (P.C.F.)
blue ox Babe, Paul Bunyan's faithful helper.
blue ribbon side The best working unit in camp, the highball side. (P.C.F.)
blue top A survey stake on a fill, the top to be flush with the finished grade. Usually colored with blue keel. (P.C.F.)
blue water rafting The method of towing log rafts in salt water.
blunt his ax To cut a man down to size.
board foot A measure of sawtimber and sawlogs: 12 inches wide, 1 inch thick, 12 inches long. One thousand board feet is the usual basis for buying and selling timber. In pulpwood the cord or cubic foot is more commonly used. Also board measure.
board holes Holes cut by fallers in the butt of a tree to set in a springboard. (P.N.F.) The term driver holes is used in the Redwood Country.
boarding car The mulligan, a cookshack on wheels which at one time carried hot meals out to the loggers at noon.
boarding with Aunt Polly Said of one who draws insurance. LL
board iron A metal dog or shoe used on the end of a springboard. (P.N.F.) Called a driver iron in the Redwood Country. Also spur.

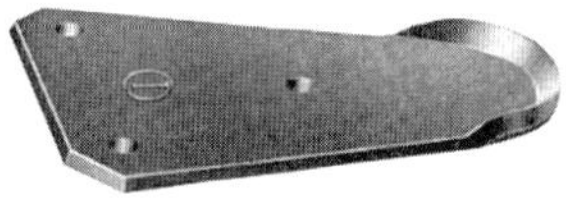

A **board iron**. *From Young Iron Works catalog, Glen Comstock collection*

board measure See board foot.
boards The boards which are put across the drivers to make a platform on which the fallers stand to fell the Redwoods. (Red.)
board up a tree To keep making higher springboard or driver holes in a tree until a good spot is found for the felling operation, above the swell butt, pitch, schoolmarm or other obstacles to sawing. Before the days of power saws, trees were sometimes boarded up 20 feet high, wasting much valuable wood in the stump. One of the most spectacular jobs of boarding up a tree in the Pacific Northwest was done by George Wasson, a long-time Weyerhaeuser employee, who boarded up to 186 feet with two springboards, and topped the tree at that point. Also rig up a tree.
boar's nest 1. An untidy bunkhouse. 2. Any untidy mess. Also rat's nest.
boat In ground-lead logging, another name for the pig which was dragged behind the log turns.
bob 1. A railroad truck which only has a bunk to haul long logs. 2. A disconnected truck. 3. To cut something off short. 4. To yard logs with a cat only, no arch.
bobber 1. A caboose. 2. A log that sinks on one end. See deadhead.
bob cat A cat being used to yard logs without an arch.
bob logs To transport logs on a bob or dray. (N.F.)
bob tail 1. A cat that yards logs without an arch. 2. A truck without a trailer.

Here in the Redwood Country of the Northern California Coast, **choppers** are preparing to start work after **boarding up the tree. Drivers** are set into notches cut in the stump, after which the **staging** is placed on the drivers. *G.J. Speier collection*

bogged down 1. To be stuck in mud or sand. 2. To be stalled.
bogie A dual axle truck rear end.
Bohemia boom The boom on a shovel loader, developed by Faye Stewart of the Bohemia Lumber Company, which has outriggers on each side against which a log can be heeled. This permits the loader to be close to a truck, and the truck can be loaded over the side.
bohunk 1. A foreign logger or mill worker from Central Europe, usually Austria or Bohemia. 2. At one time used disparagingly. See polack.
bohunk lingo Speech that is not clearly understood.
boiled shirt Town clothes.
boiler 1. A poor steam engine. 2. A poor cook.
boiler compound 1. A mixture used to neutralize water used in a boiler to prevent scaling. 2. Internal medicine. 3. A bad-tasting drink.
boiling dishwater A measure of cold. When a pan of boiling water is thrown out of a door and it freezes solid before it falls, it is very cold. (P.N.F.)
boil-up 1. To wash clothes. 2. To make tea. 3. To arrive in a hurry. (P.C.F.)
boil-up day Sunday in the woods when loggers and lumberjacks scrubbed their clothes, using half of a barrel as their tub and doing their boiling in a big lard can over a roaring fire. (N.F.) LL
bole A tree stem once it has grown to substantial thickness.
bolt 1. A chunk of wood which was split or cut off a larger piece or off a log. Thin slabs cut off the bolts were used to make shakes or shingles. 2. A short log, a four-foot length of pulpwood. (Canada)
bolt hook A steel hook used to handle pulpwood. Also called pulp hook and Birch hook. (N.F.) LL
bolt maker A man who made shake or shingle bolts.
bone crusher A chiropractor.

A **pond** holding 1,100 cords of shingle **bolts** somewhere in Washington State. *Darius Kinsey photo, Labbe collection*

bone yard A storage yard for machinery waiting to be scrapped.
bonus 1. More than one log yarded in with one choker. 2. A log that has no choker, but gets tangled in the rigging and comes in with the turn. (P.C.F.)
book (the) The book with all the answers that a woodsman wanted when he got into difficulty. (P.C.F.)
boom 1. The logs or timbers fastened together end to end and used to hold floating logs. Log pond is the more popular term now. 2. The complex of booms and cribwork piers arranged to contain logs and direct them to the place where they will be sorted. 3. A projecting arm of a log-loading machine which supports the log during loading. 4. A loose bag of logs in the water. 5. Cross member of a rigging of a vessel.
boomage 1. A booming area and all its works. 2. A charge made for catching in a boom and sorting by ownership the logs coming down a river from different camps. LL
boom auger A very large auger used to bore big holes in logs used as boom sticks to permit the passage of the boom chain toggle. Also rafting auger.

A **boom auger**. *From Mill & Mine Supply Co. catalog, 1915, Glen Comstock collection*

boom boat A short, powerful small boat which is used to move logs in a boom or pond, to assist in towing, etc. Also bulldozer boat, mule, sea mule.
boom boss A foreman on a boom or log pond. Also pond boss.
boom buoy See boom stay.
boom cat Specifically a man who is very agile on the logs.
boom chain A chain used in booming logs together for towing. It is equipped with a toggle so it can be slipped into or out of boom sticks. Also toggle with ring, toggle chain.

A **boom chain**. *From Mill & Mine Supply Co. catalog, 1915, Glen Comstock collection*

boom company An outfit that owns a headworks on a river and sorts out logs coming down river to various owners for a fee.
boom dog An iron wedge about 5-1/2 inches long connected by 4 to 8 (or more) links of chain. The dogs were driven into the ends of logs, chaining them together to make a floating log "fence" or boom. Also eye dogs, rafting dogs, raft shackles. (L.S.) LL

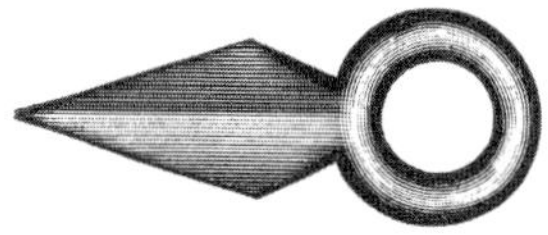

A **boom dog**. *From Kloepfer Hardware Co. catalog, Glen Comstock collection*

boomer 1. See drifter. 2. A man in charge of a boom. 3. The mountain beaver that forages on young coniferous trees and is a menace to forestry.
boomerang A type of steam donkey spark arrestor.

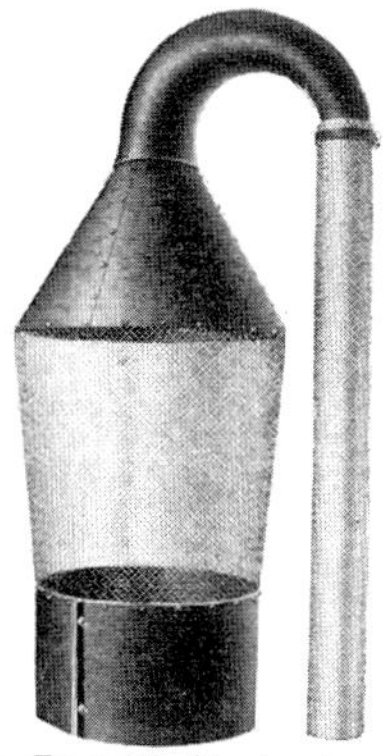

A **boomerang** spark arrestor. *From Washington Iron Works catalog, Glen Comstock collection*

boom house In river driving, the facilities or complex needed to accommodate the men, which included the dining room, the kitchen, storage room for equipment and areas available for maintenance work and administration. (N.F.)
booming grounds Areas at the mouths of the main streams and where the log drives ended. Here logs were sorted according to company log marks, pushed together into rafts with rope and pins and towed to the sawmills. The largest boom company in the Saginaw Valley was the Tittabawassee River Boom Company which owned over 12 miles of booms. (N.F.) (Harold M. Foehl and Irene M. Gargreaves, Logging the White Pine in the Saginaw Valley) Also rafting works.
boom jumper A motor boat with two keels or jump boom sticks designed to ride over logs, which was used in Maine.
boom lake Any lake where logs are stored in rafts or boom. (N.F.)
boom loader See boom.
boom man 1. A worker who poles floating logs to a sawmill. 2. A man who operates a boom loader. 3. A man in charge of floating logs.
boom of logs The logs enclosed within a boom awaiting reshipment or manufacture. LL
boom pin A wooden plug which is placed into the hole of a boom stick to keep the chain from working loose.
boom poke A man who works on a boom or raft. (P.C.F.)
boom rat One who works on a boom. LL
boom scale Footage of logs measured out of a boom. (N.F.)
boom stay A heavy weight used to anchor booms in deep water, its position indicated by a pole or float attached to it. Also boom buoy. (N.F.) LL
boom stick 1. A long log used to hold logs in booming. Holes are bored in both ends of a boom stick so that any number of them may be hooked together. 2. A narrow, wooden plank catwalk, held up just above the water level by piling. Sorters stood on boom sticks while sorting logs at the booming grounds. (N.F.)
boom stick car A skeleton car at least 60 feet between bunks to carry long boom sticks.
boom stick cutter The logger who trims and prepares boom sticks in the woods.
boom strap A short cable which holds a loading boom to a spar

Log **dump** of the Deep River Logging Company, at Deep River, Washington. The **boom boat** is waiting to handle the logs in the **booms.** Note the tiny **crummy** for the train crew, which is mounted on a single **disconnected truck**. *Courtesy of William Dippert*

tree. Also called heel strap.
boot caulks Spikes in a logger's boots. (App.)
boot jack A device made of a crotched limb or board to pull off leather boots. (N.F.) LL
Booth-Kelly cold shut A loop of wire used to make a temporary link; a kind of Molly Hogan. Booth-Kelly Lumber Company was a large operator in the Eugene, Oregon area.
bore In bucking, to cut into a log with a chain saw. The bottom half is cut first and then the top. (P.C.F.)
borers 1. The larvae, the worms of wood-destroying beetles, which are very destructive in fire-killed timber. 2. The teredoes or shipworms that riddle a raft of logs in salt water.
boss logger 1. The woods superintendent. 2. The best man in the crew.
boson's chair A sling seat which is used to carry the rigger up a spar tree. (P.C.F.)
bottle butt A butt which has a large swell in it. (P.C.F.) See churn butt.
bottle hook A hook wired to the neck of an oil bottle and used to hang the bottle to a tree or a stump. The bottle usually contains kerosene and lubricating oil and is used by fallers to free a saw from pitch when felling a pitchy tree. (P.C.F.)
bottom The lower tier or layer of logs in a joint, usually fastened together by boom poles and pins. (Eastern Canada)

A **bottle hook**. *From Kloepfer Hardware Co. catalog, Glen Comstock collection*

bottom loader See ground loader.
bottom sill See mud sill.
bourgeois The owner of a logging business. (Quebec)
Bouse block A small yarder block which is built with a stout yoke to which a heavy hook is attached by a link. It can be opened on one side to pay in a line. Also open side block, spring link yarder. See snatch block.
bow In manufacturing, a deviation flatwise from a straight line from end to end of a piece measured at the point of greatest distance from the straight line.

A **Bouse block.** *From Mine & Mill Supply Co. catalog, 1916, Glen Comstock collection*

bow and arrow An Indian. (P.N.F.)
bow-back A stooped man, usually because of an injury. (P.N.F.)
bow his back To refuse to do a job. (P.N.F.)
bowleg The wishbone frame of an arch, bowed up over the two tracks.
bowleg loader A loading jammer squatting over the tracks on hydraulic jacks or cribwork. (P.C.F.) See McGiffert.
bows The U-shaped pieces of white oak used to fasten timber to a lash pole. (App.) See lash pole.
bow saw 1. A hand saw which has a blade tightly stretched between the ends of a bow-shaped steel frame. 2. A chain saw running around a bow instead of a solid bar, and which is used chiefly in small pulpwood stands. Also bucksaw, fin saw, swede saw. (N.F.)
bow up To get mad, turn mean, refuse to do a job. (P.N.F.)
box 1. The water-carrying trough of a flume, supported by the trestle. 2. One section of the flume, usually 6 to 20 feet long. 3. A reservoir cut into the base of a pine tree in which crude turpentine is collected. (S.F.) 4. To make an undercut or to notch.
box canyon A dead end rock canyon with steep sides.
boxing ax See turpentine ax.
boxing glove An old-styled choker hook, called thus because of its shape.
boyhood diseases The popular story of a logger who, worried about his health, went to the doctor for a checkup. The doctor asked him what diseases he had had in his boyhood, and the logger replied, "The usual boyhood diseases: mumps, measles and the clap." (P.C.F.)
boy scout ax 1. A small, light ax. See surveyor's ax. 2. Any tool too small to do a job properly.
boy scouts 1. The high school or college students who plant trees or visit the logging operations. 2. The foresters.
bracket boom A stiff boom, three or four logs wide, the logs being fastened together by short boards placed crosswise and spiked.
bracket gate See needle gate.
bracket guy line A short guy from the spar tree to a bracket on a McLean-type loading boom. See nose guy, sail guy.
brail 1. A boom filled with logs ready to be towed. 2. A section of a log raft, six of which make an average tow. 3. To place logs end to end in a long flat raft or boom.

The brail system, which was introduced by W. J. Young of Clinton, Iowa, made possible a great saving of lumber, for the old system of boring and lashing damaged the logs and threw them into lower grades. Usually a deduction of 10 percent had been made for logs damaged by auger holes. By the new method of brailing, logs were saved from such blemishes. The logs were swung straight as they came into the pocket. They lay close together in tiers; when about 150,000 feet of logs had been packed side by side, the pocket was considered full. Then it was ready for the brailing crew, whose job it was to put the brail in shape to float down the river. (Agnes M. Larson, History of the White Pine Industry in Minnesota)

brainless wonder Usually the man who thinks he knows more than his coworkers. (P.C.F.)
brains 1. The boss, the super, the push or the head man. 2. A term of contempt for the conductor on log trains, used by the brakemen who think they know more than he. (P.C.F.)
brake, broke To work as a brakeman and used in the past tense: "I used to brake for Hammond's." "He braked for P.L." Broke is also used: "He broke for the N.W.P." (P.C.F.)
brake hickey A club for setting brakes on logging cars. Leverage is obtained by shoving the club through the spokes on the brake wheel. Also brake club and Jim Crow. Can be either wood or metal.

Retired **head brakeman** Vernon Goe here demonstrates how the **brake hickey** was used to brake the **disconnected trucks.** *Labbe photo*

brake sled A logging sleigh so constructed that when the pole team holds back, a heavy iron on the side of each runner of the forward sled is forced into the roadbed to hold the sleigh. (N.F.)
brakie A brakeman.
brand A log mark used to identify an operator's logs when several companies dump in one stream or area. Logs brands are registered.
branding ax, branding hammer A tool for marking ownership on a log. Also log stamper, marking ax, marking hammer, marking iron.

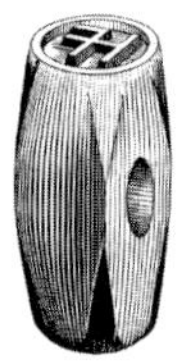

A **branding hammer.** *From Mill & Mine Supply Co. catalog, Glen Comstock collection*

branding man The man who marks the logs.
brashness A condition of wood characterized by low resistance to shock and by an abrupt failure across the grain without splintering.
brass 1. The mill owners or supervisors. 2. Impudence, marked by a contemptuous or cocky boldness or disregard of others.
brass check A capitalist newspaper or any newspaper not friendly to the I.W.W. (International Workers of the World). (L.S.) LL
brass collar The man who owned the trees that were being cut or the man for whom logs were being cut. (L.S.) LL
brass money Company tokens used instead of real money. LL
break 1. The edge of a draw or canyon. 2. The slope of a hill or mountain where it changes grade sharply. 3. A draw-bench used to hold a shingle slab while it is being worked with a draw-knife into a hand-rived shingle. (S.F.)
breakage The loss of merchantable logs due to breaks in falling on rocks, stumps or other logs. The heavy weight of a redwood tree itself will cause breaks.
break a ham string 1. To start logs in motion that have jammed during a river drive. (L.S.) 2. To do one's best.
break a jackpot In breaking a pileup, to grab hold of a stuck log, to reef on it, causing the logs to move about.
break a jam To start logs in motion which have jammed during a river drive. (N.F.) LL
break a landing To roll a pile of logs from a landing on a river bank

into the water. (N.F.) LL
break a rollway To push a deck of logs off a rollway into the river or bay for driving. (N.F.)
break a skidway To roll logs off a pile to be hauled away. (L.S.) LL
break cut A cut made in bucking to cut out a break in a log. (P.C.F.)
break down To stop a machine because of an accident. (P.C.F.)
break in a landing To roll logs from the landing into a stream. (R.M.F.)
breaking a center To dislodge key logs hung up on an island when they were being floated downstream. (N.F.) LL
breaking crew The men who had the dangerous job of going into the mass of logs above the jam piers and breaking the jam with cant dogs in river driving. (N.F.)
break-out 1. To get a stuck log out of a hangup. 2. To open up a road closed by a slide or snow. 3. To start up an operation after a shutdown. 4. To start a sled whose runners are frozen to the ground. (N.F.)
breaks The broken pieces of logs which are not long enough for most sawmills, but are relogged for pulp.
break the landings To roll the logs into the stream and to prod them into the current in the early 1800s. (Maine)
break-up A period of time in spring when melting snow creates soft soil conditions and high water in streams. Logging must usually be curtailed during this time.
breast 1. The cutting side of a crosscut saw. 2. The fore part of a jam. (N.F.) LL
breast line 1. A strap which is placed around the butt of a spar tree to hold it in place while it is being raised; this is done when the tree is too big to handle by tie-up lines alone. 2. A snub.
breast log A heavy log at the edge of the landing. See brow log.
breastwork log See sheer log.
breech-loader See bunks.
breed 1. A donkey that usually operates on a pole road between the yarder and the swing. See half-breed. 2. A logger of mixed blood, part white and part Indian.
briar 1. A falling or a bucking saw. 2. A small bucksaw.
bridge monkey A bridge builder. (S.F.)
bridle 1. A short line that connects the spreader bar to the fall blocks in crotch-line loading. See crotch-line loading. 2. A hookup using two chokers to get around a big log when one choker is not big enough for the job. 3. A rope wrapped around a log to act as a brake on a downhill haul in animal skidding. 4. See single coupler. 5. A chain wrapped around rear runners of a sleigh as a brake on steep icy roads. (Canada)
bridle chain A chain wrapped around a sled runner to serve as a brake when going downhill on iced roads. (N.F.) LL
bridle iron A flat plate with a big hole in the middle that is bolted to both the outside and the inside of a donkey sled runner. A cable is passed through the hole bored through the sled log to move and tie down the donkey. Also log plate, rope plate. See sled plate.

A **bridle iron**. *From Mallory Catalog No. 6, Glen Comstock collection*

bridle man The logger who followed logs down the skidroad and looked after the gear. (P.C.F.)
broad-arrow mark In colonial times, the mark of a broad-arrow that the king's surveyors put on the finest white pine trees in New England to be used only for the Royal Navy. See X-tree.
broad-arrow policy About 1695 the surveyors general of William and Mary and their deputies went dutifully about picking out the best and straightest trees and reserving them with the king's broad-arrow mark. This was in accordance with the new charter of 1691 issued to "The Province of Massachusetts Bay in New England," which set aside as property of the crown and for its use of the Royal Navy all trees more than 24 inches in diameter and 1 foot from the ground. Penalties were provided for violation of this provision, and the decree was soon spread throughout New England, Nova Scotia, New York and New Jersey. (Grant W. Sharpe, Clare W. Hendee and Shirley W. Allen, Introduction to Forestry)
broad ax A large-headed ax beveled only on one side like a chisel, for hewing masts, square timbers or railway ties.

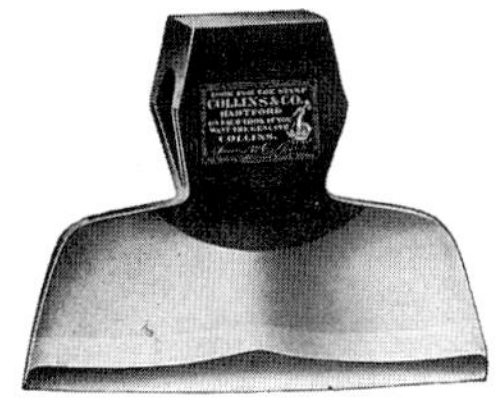

A **broad ax**. *From Kloepfer Hardware Co. catalog, Glen Comstock collection*

broadcast burn A controlled fire used to burn over a designated area within well-defined boundaries for the purpose of reducing fuel hazard, as a silvicultural treatment, or both.
broadleaf See hardwood.
brogan A lumberman's heavy moccasin. (Maine)
broke See brake.
brolo See two-hander.
broom boy Usually a boy who kept the skids clear of debris which could hinder the movement of the log turns in the 1800s.
broomer Said of a smashed log that has been banged against the rocks on a river drive, thus resembling a crude broom. (P.N.F.)
broom stick A long pole used to pocket logs in the log pond.
broom tail A horse. (P.C.F.)
brow Logs piled on a river bank to await the spring river drive. (Canada) See landing.
brow log A big log parallel to the road or track to protect trucks and cars during loading and unloading at landings and log dumps. Also breast log, draw skid.
Brownie A Brown-Lipe auxilliary truck transmission.
Browns Bay bonus A method of bonus payments for higher production used in some Washington camps at one time. The woods crew was paid at a flat rate for a set standard of production, with the bonus applied on a per thousand basis for all footage logged above the standard.
brow skid 1. A large log placed on top of a rollway, next to and parallel to the track; it prevented logs from rolling out of control. Also bumper. 2. The chief beam in a frame to which tackle for loading logs on cars is fastened. (P.C.F.) 3. To roll logs on cars off a rollway.
brush 1. The area in the woods other than the landing. 2. A logger working in the woods: "He's in the brush." 3. Another term for slash. 4. Any tree less than 3 feet in diameter (not worth logging).
brush ape A logger.
brush a road, brush out a road 1. To cover with brush the mud holes and swampy places in a logging road to make it solid. (N.F.) 2. To clear brush from a prospective road. LL
brush buster 1. A tractor. 2. A cat with a special plow to rip out brush before planting non-forested areas. 3. A man so scared by a bear that he runs blindly through the brush.
brush cat A logger. (P.N.F.)
brush-disposal To get rid of slash hazards on a logging operation in the Pine Country.
brushed Said of one who is crazy. (L.S.) LL
brusher 1. An operation where stave bolts are made. 2. One who cuts and piles limbs from felled trees. (S.F.)
brush hook A type of ax designed for cutting underbrush. It has a long bit, which is curved at the end to form a hook. Also hazel hook.
brush monkey 1. A logger. 2. A timber cruiser. 3. A brush piler in the California Pine Country.
brush-out 1. To clear off an area. 2. To clear a line for a survey

Early truck **landing** of the Crabtree Lumber Company, near Lyons, Oregon. Logs are being loaded with a **gin pole** and **end hooks**. The **brow log** prevents logs from dropping beside the truck or banging into it. In the foreground is the **back-around**, where trailers were turned before the days when they were loaded on the truck for the return trip. Logs are brought into the **rollway** tree-length and **bucked** at the **landing**. This arrangement was sometimes called a **bucking ladder**. *June Drake photo, Labbe collection*

party. 3. To clear a right-of-way for a road. Also bush-out.
brush pile 1. Small timber that is not worth logging. 2. A designated spot: "He's in the brush pile." 3. Piles of brush piled up to burn to make fire breaks in the Pine Country.
brush piler 1. A man with a strong back who does odd jobs around an operation, such as piling brush. 2. A bulldozer with heavy teeth on the blade to pick up brush, letting the dirt fall through.
brush rabbit See choker setter. (P.N.F.)
brush rake Teeth that are added to a bulldozer for brush piling.
brush snow fence A snow break to protect a logging road; used on wide marshes. (N.F.)
brush wren A wife in the woods. (P.N.F.)
brutting crew A crew that rolls logs down slopes too steep for teams. (App.)
bubble cuffer A log birler. (P.N.F.)
bucheron A French-Canadian lumberjack.
buck 1. To saw felled trees into shorter cuts. 2. To rebel. 3. To carry: "Buck the water." 4. To split ties. (S.F.) 5. See chore boy.
buck beaver 1. The boss of a crew that cuts trees and brush off a new road. (L.S.) 2. Assistant to the foreman, in charge of clearing and building roads. (Canada) LL
buckbrush Any type of brush that makes it difficult to cruise timber, run a line, or hunt for a corner post. Also snow brush.
bucked Said of logs cut up in the woods. See felled.
bucker A logger who saws felled trees into logs. (P.C.F.) Also sawyer. (Red.)
bucker's break A crack in a log caused by poor bucking. See alligator mouth.
bucker's coupling A log that has not been completely cut from the next log. Also Dutch coupling, Irish coupling, Russian coupling.

Bucking a big **yellow fir** log for the Big Creek Logging Company near Knappa, Oregon, on the Lower Columbia River. The **bull buck**, perched atop the log, is prepared to give a hand with the **wedges**, to keep the log from **pinching** the saw. *Weister photo, Labbe collection*

bucket of mud A mixture of asbestos and cylinder oil used in pouring hot lead or babbitt around the frayed end of a cable when a new cable is put in a yarder drum. Also bear shit.
buck fever A man who gets frightened at the thought of work; very

lazy.
buck for grade To cut a tree into various lengths to obtain the highest quantity of high grade lumber instead of cutting the tree into equal lengths regardless of grade. The value of a log may be raised by careful attention to grade.
bucking bar A power saw used by one man, with no handle or stinger on the end.
bucking board 1. A bulletin board on which a list of crews was posted. Superintendents pitted one crew against another to get more work out of them. (N.F.) 2. A spring board. (P.N.F.) 3. A record of a cutter's daily output. (N.F.) (Canada) LL
bucking chute 1. A chute along which logs were moved past a fixed bucking saw. 2. A short pole chute at a landing in which double-length logs are bucked before being loaded. (P.C.F.)
bucking ladder A series of short skids laid parallel at regular intervals, usually a rollway or landing. (N.W.)
bucking logs Cutting or sawing felled trees into standard lengths.
bucking prop A chunk or limb used to hold up a log so it will not bind the saw when bucking.
buckle guy A guy line fastened to the middle of a spar tree to reduce the strain in pulling a heavy load. Also belly guy, steadying guy.

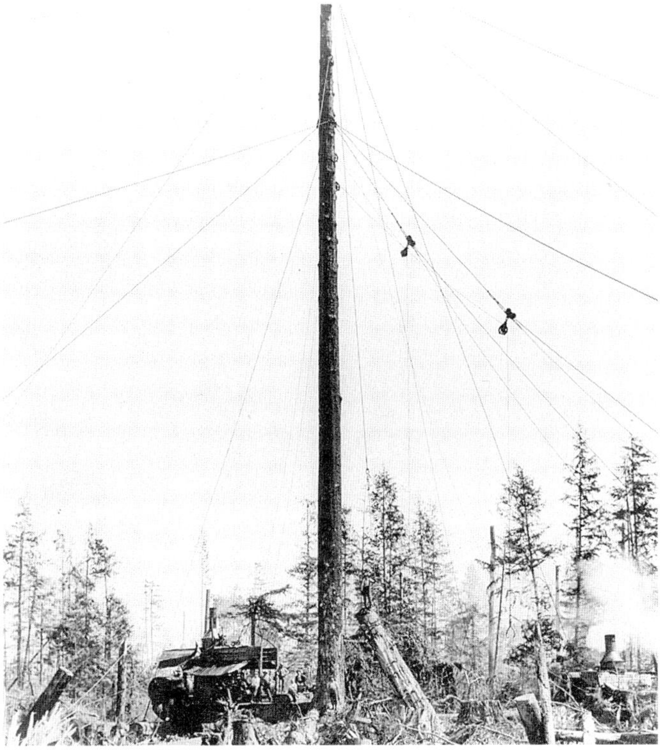

Greenwood Logging Company operated out of Aberdeen, Washington. Shown here at a **landing** is one of their **spar trees**, equipped with **buckle guys**. Greenwood used **duplex loading** to load logs. The **fall blocks** and **loading jacks** required for this system can be seen suspended from the buckle guys. *Philip Schnell collection*

bucksaw A thin-bladed, one-man saw for felling and bucking pulpwood. (Canada) Also pulp saw, swede saw, finn saw and bow saw.
buckskin 1. A peeled log. 2. A snag with no bark. 3. An old windfall or snag from which the bark has rotted away, but which still contains good wood.
buck teeth See rippers.
buckwheat To fell a tree so that it falls against another. (App.) LL
buckwheat bannocks Thick fritters made of buckwheat. (Maine)
buckwheater A green hand in the woods. (P.C.F.)
buckwheat eaters The generic term for the loggers from the Maritime Provinces of Nova Scotia, New Brunswick and Prince Edward Island; so-called because of their main diet staple. (Professor David C. Smith, Department of History, University of Maine at Orono.)
buckwheat show 1. A poor logging outfit made up of inexperienced loggers. 2. Easy logging. (P.C.F.)

buffaloed Said of one deceived, tricked or cheated.
buffer strip (zone) A corridor of vegetation located between a road and a treatment area or between two treatment areas to separate and/or screen the treatments from view, or to otherwise mitigate treatment impacts. Also green strip, leave strip.
bug 1. An electric switch used by the whistle punk to signal on donkeys, replacing the old-fashioned jerk-wire; also called signal grip, whistle bug. 2. A crude lantern made by cutting a hole in the side of a can and putting a candle through it. 3. An insect.

A **bug.** *From Mallory catalog, 1948, Philip Schnell collection*

buggin' To cut, peel or dispose of beetle-killed trees. (P.C.F.)
buggy 1. The carriage on a skyline logging system. 2. A small log car used in the pine country to handle 16-foot logs. 3. Said of the condition of beds in a bunkhouse. 4. Said of bug-infested timber.
buggy bell A choker hook. (P.C.F.)
buggy line The line that was used to haul in a skidder carriage before the slack-pulling system was invented. (P.C.F.)
bug kill A section of timber killed by insects.
bug picker An early three-wheeled tractor used in the pine country. See road locomotive.
bugs 1. Any insects or worms that kill trees or bore in the wood. 2. Troubles with a job or machine.
bug trees Trees infested with insects, usually bark beetles, in the Western Pine Country.
bulgine A locomotive. (P.C.F.)
bulkhead 1. A heavy wall in back of the cab to protect the driver of a logging truck. 2. A heavy fence at the end of a railroad log car used on an incline to prevent logs from sliding off when being lowered down the hill.

A **bulkhead** used on a **flat car** built for the Yosemite Valley Railway by Seattle Car & Foundry, for use on their **incline** near Merced, California. *Seattle Car & Foundry catalog, Labbe collection*

bull 1. The reliable ox that was used to skid logs in the 1800s. One old-timer said that at times these animals were extremely abused to make them pull the logs and that some loggers were even asked to leave because of their rough treatment of the animals. 2. Insignificant conversation. 3. A polite form for bullshit. 4. Big or strong, such as bull donkey, bull chain, bull edger, etc.
bull block 1. The lead block on the main line in high lead yarding. 2. In the ground lead days, a large block with a mouth wide enough to permit the butt rigging to pull through. Also butt chain block. (P.C.F.)
bull buck(er) The man in charge of fallers and buckers. (P.C.F.) (British Columbia) (L.S.) See saw boss.

bullbuckin' To be in charge of fallers and buckers: "I was bullbuckin' at Orick that summer." (Red.)
bull car A flat all-steel car for moving a donkey engine. See donkey mover, moving car.

A diesel **donkey** is loaded aboard a **bull car** and hauled out of the woods near Vernonia, Oregon, as Long-Bell Lumber Company closed out logging operations at Keasey, in August of 1957. *Labbe photo*

The John Vance Mill & Lumber Company, on Mad River, in Humboldt County, California was one of the early users of the **bull donkey**, or **Pine bull donkey** for roading logs. This one is unusual because of the wooden **lagging** used on the boiler. A little **gypsy** locomotive is in the background. *A.W. Ericson photo, Carranco collection*

bull chain 1. The big chain that hoists logs up the chute from the mill pond to the sawmill. 2. In the early days a heavy chain wrapped around a log to act as a brake when the log was skidded down a steep hill with oxen or horses. (P.C.F.) (App.) Called a jack chain in the Lake States.
bull choker A heavier choker that is used to move a very big log or to get out of a hang-up.
bull chokerman The unpopular rigging slinger who is paid a few cents an hour more than the rest of the crew to set the pace among the choker setters. (P.C.F.)
bull cook 1. The man who swept out bunkhouses and lighted heating stoves. So-called because in the early camps his first chore in the morning was to feed the oxen. 2. To do odd jobs: "He was bullcookin' at Camp 10." (Red.) Also lobby hog. (App.)
bullcook engine A locomotive assigned to work train service and odd jobs.
bull-cooking machine A cat rigged up to do odd jobs around the operation. (P.C.F.)
bulldog 1. A brand name for a railroad logging truck. 2. A horsefly. (P.C.F.) 3. A woodsman. (Maine)
bull donkey, also called **Pine bull donkey** Wire rope was introduced in the woods about 1884, when N. H. Pine of Eureka, California developed his bull donkey. The new donkey was equipped to haul logs over long distances of 1,000 to 2,000 feet. It was a portable machine with an upright boiler in back. The working parts of the donkey were two large drums. Two upright and two horizontal spools, which slid back and forth on a spiral-geared shaft to feed the line on and off the drums, were in front of each of these two drums. The main drum belt held over 4,000 feet of 3/4-inch steel line which hauled in the logs. The smaller drum held the 1/2-inch line which hauled the heavier line back to the starting point at the top of the skidway.

The early bull donkeys were comparatively small compared to the later bull donkeys which could carry up to 1-1/2 miles of 1-1/8-inch line and 2-1/2 miles of 3/4-inch haulback line. The two horizontal and the two upright spools were never too successful because the drums would get larger as the line was reeled in, causing the fairlead device to go out of timing. (P.C.F.) See Pine bull donkey.
bulldoze 1. To operate a bulldozer. 2. To knock the top off a rock with a mud shot. 3. To fool a man. 4. To charge through a job without thinking.
bulldozer A standard crawler-type tractor having a heavy steel blade mounted across the front at a right angle to the tracks. The blade can be raised or lowered by hydraulic or cable devices, and all pushing action is straight ahead. Also dozer. Ted P. Flynn, for many years with the Forest Service in Portland, Oregon, is given credit for making the first one in 1924. Outstanding improvements were made by Ed Stamm of Crown Zellerbach, who adapted the road-building bulldozer to logging use. George Drake, then of the Simpson Logging company, and Bill Dyche of the Algoma Lumber Company also modified bulldozers for woods work.
bulldozer groundman A man who helps on odd jobs, helping the cat skinner in road construction or similar work.
bulldozin' 1. To operate a bulldozer: "He's bulldozin' at G.P." 2. To put a big gob of mud over a dynamite charge on a smooth rock to direct the force of the blast downward.
bull gang 1. A rig-up crew. 2. A road building or repairing crew or any other crew that depends on strength rather than machines.

A **bulldozer**-equipped **cat** brings a **turn** of logs into the **landing** using an **arch**. The **spar tree** in the foreground uses a **boom** for loading which is suspended by a **boom strap**. *Pacific Logging Congress Loggers Handbook, 1948, Philip Schnell collection*

bull gang hooker The boss in charge of a rig-up crew.
bullhead A self-centered logger who will not work.
bullheadin' Said of the action of a saw (by the saw filers) when it leads in or out of a cut.
bull hook 1. An open choker hook used in the early days. 2. A skidding hook used on skylines. 3. A big hook once used on the butt rigging in some skyline systems.
bull it through To finish a job by strength and awkwardness.
bull line 1. The main line on a high lead show. 2. A heavy cable that connects the butt rigging to a winch on the back of a cat yarding with an arch. See arch line.
bull load A turn of logs ready for hauling with a road engine. (P.C.F.)
bull loading hook A large heavy loading tong used in the late 1800s and early 1900s.
bull logging The system of skidding logs with oxen.
bull moose 1. The push or head man on a logging show. 2. A term for anything large.
bull of the woods 1. The woods boss or the superintendent of the logging operations. 2. A walking boss or superintendent who had charge of several camps. (Canada) 3. A contest staged in logging shows at district fairs. Contestants, wearing boxing gloves and corks, plus regular logging gear, jump on a large log; the last man on the log is acclaimed "Bull of the Woods." 4. A big, tough man. 5. A top hand.
bull pen 1. A pen into which unsorted logs are dumped. 2. Any large room where men gather to talk. 3. Sleeping quarters. 4. The recreation room at camp.
bull pine Usually an isolated pine that has never been cut. (L.S.) LL
bull point A rock drill.
bull prick 1. A steel bar for punching a hole under stumps so that shot (dynamite or powder) may be inserted. 2. A spike that is used in splicing wire rope. Also marlin spike. 3. A compensating pin which allows for slack between the log trailer and truck. 4. A dead top sticking up in a tree. 5. A side sucker branch on a tree.
bull puncher The man who drove the ox teams and cared for the oxen in the 1800s. Also bull whacker, puncher, whacker.
bull road The main part of the skidroad that was just a dirt road.
bull rope A rope attached at each end of a log with a pig's foot hook to draw logs up onto the skidway pile. (Canada)
bulls The oxen.
bullseye fairleader A fairleader built to ensure a perfect lead at all times and swiveled to permit the line to run free at almost any angle.

A big diesel **donkey** of the Clemons Logging Company, near Grays Harbor, Washington. Note the **bullseye fairleaders**. *C. Kinsey photo, courtesy of S. Hauff*

bull skinner An ox driver.
bull team A team of oxen used for ground logging. See ox team.
bull teamster 1. A bull puncher. 2 A man who freighted with bulls.
bull whacker, bull whipper A bull puncher.
bull wheel The biggest gear in any gear train, such as on steam tractor wheels, on donkeys or in sawmills.
bully A foreman or boss of a logging camp. (N.F.) LL
bummer 1. A low-wheeler cart for carrying logs. (N.F.) (S.F.) 2. A self-loading cart that had a cross boom and a fairlead that was the

A **bull team** moves a **turn** of logs down a **skidroad** in Northwestern Oregon. The logs are fastened together with **dog hooks**. *Courtesy of H. Lonberg*

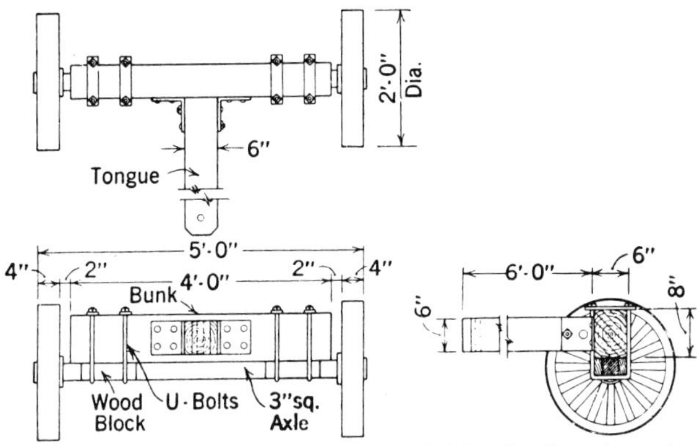

A two-wheel **bummer** used in skidding logs, particularly long lengths such as poles and pilings in the Northern and Eastern hardwood forests. *From "Logging," by Nelson C. Brown, Wiley, 1949. Philip Schnell collection*

forerunner of the arch. (S.F.)
bummery A nonsympathizer for the I.W.W. (L.S.)
bump 1. To take a job away from a man who has less seniority. 2. To replace a man in favor of another on any job.
bumper 1. A heavy spring device clamped on the mainline ahead of the carriage to act as a stop and prevent the carriage from ramming into the heel tackle or the spar tree. This has been replaced by a heavy rubber tire hung on the skyline. The tire is also used on some types of loading rigups. (P.N.F.) 2. Another name for a brow skid.
bumper shackle A heavy shackle to keep the carriage fended off the heel tackle on a skidder show in the days before old tires were common in the woods. (P.C.F.)
bumping knots Removing limbs from logs; limbing logs.
bunch 1. A turn of logs, one-load of logs ready to be pulled into the landing. 2. To assemble logs for yarding. 3. To encircle several logs with a chain to load them at once by steam or horsepower. (N.F.) 4. To put shingles into a bunch. (Maine)
buncher The man who skids logs to a gathering place. (P.C.F.)
bunching The process of rolling logs into compact piles for transportation with big wheels or other equipment.
bunching chain A log chain for skidding or dragging a log. (L.S.)
bunching ground In turpentine orcharding, a place in the woods where the cups are stacked before cup-hanging. (S.F.)
bunching hand line A light line used to pick up a number of chokers on a skidder show, getting them ready for the next trip of the carriage.
bunching team A team of horses used in the Pine Country to skid logs to a central point where they would be picked up by the big wheels.

bunch team A team used to bunch logs. LL
bunch the job To quit suddenly.
bunch-up 1. A locomotive engineer would slack the brakes in order to let the cars bunch up, giving slack in the couplers to get a roll on the locomotive. When the train was stretched tight, he had to move the entire tonnage from a dead stop. 2. To collect logs together at one point.
bundle To tie small logs together with straps for ease in handling, especially when booming. (N.F.)
bundle rafts Loose rafts made up of little logs put together in "asparagus" bundles.
bundle-toting Said of a man who was looking for work. In the 1800s and early 1900s, before the logging companies provided bedding, men carried their bedrolls from camp to camp.
bundling A method of transporting pulpwood down a steep slope. A chain is bound around about one half cord of wood that is then dragged down the slope to a central point. (L.S.) LL
bung The hole in the water tank where water is let out to sprinkle the icy roads. (N.F.) LL
bunghole The anus.
bung starter A light hammer. (P.C.F.)
bunion A burl or other swollen growth on a tree or log.
bunk 1. A bed. 2. A crosspiece of heavy metal or wood on which logs are placed on a car or truck. See Hercules. 3. Heavy timber crossbars forming the bed of a logging sleigh. 4. To place logs against the bunk when loading. 5. To coast downhill to start a sled.

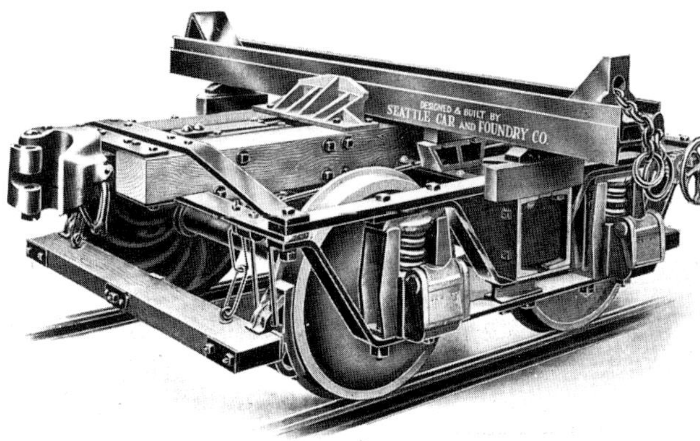

A "Snohomish" **disconnected truck**, manufactured by Seattle Car & Foundry Company. This model was equipped with an all-steel **Hercules bunk**. Visible at right are the **bunk chains**, used to tighten the **chocks** against the loaded logs. *From Seattle Car & Foundry catalog, Labbe collection*

bunk block A steel wedge that keeps the logs in place on a bunk. Also cheese block, chock, monkey block.
bunk-bound 1. Said of a truck that is overloaded on one side, causing the bunk to rub on the frame and making the truck difficult to steer. 2. Said of a logger or mill hand who will not get out of bed.
bunk chain A chain that runs through the bunk on a log truck to pull the bunk blocks together; this tightens the chocks against the bunk logs.
bunked-long Said of logs that are loaded with considerable overhang beyond the bunk on a logging truck.
bunked-short Said of logs that barely reach over the bunk on a logging truck.
bunk hook A hook that secures a stake that is attached to the end of the bunk to hold logs on cars. When knocked loose, it allows the stake, or cheese block, to fall free. LL
bunkhouse Living quarters that are built for the loggers or mill workers. Also sleeping shanty (L.S.), wickiup.
bunkhouse lawyer Usually a man who thinks he knows all the rules and works them to his advantage.
bunkhouse logger A logger who brags more than he works.
bunk load 1. A big log that makes one load. 2. A row of logs making a load, all touching the bunk.
bunk lock A pin that holds the bunk on a log truck straight across the frame while loading.
bunk log 1. A log next to the bunk. 2. The bottom layer in loading a truck or car.
bunko A man hired to steer lumberjacks into the whore houses and taverns. (L.S.)
bunk pin See bunk lock.
bunks 1. Supports on which units or lumber are placed. 2. Beds. Changes came to the lumber industry in the Lake States in the 1870s. The shanty gave way to a bunkhouse. Men no longer slept on the floor, on boughs or on a thin spreading of hay or straw. Bunks were built along the wall. Some camps had only a row of single bunks; some had one tier above the other; others had three tiers. The number of bunks depended upon the size of the crew. Bunks were of two kinds, the muzzle-loader and the "side-loader," called "breech-loader" by the lumberjacks. The muzzle-loader was built from the side wall out into the center of the room; the side-loader was built along the side and in an arrangement not unlike the berths in a pullman car. The bunks were built of cedar if it could be had, because that was insurance against vermin. (Agnes M. Larson, History of the White Pine Industry in Minnesota)
bunk spikes Sharp spikes set upright in the bunks of a logging sled to hold the logs in place. (N.F.) LL
bunting bar A battering ram to jar loose loaded sleighs or railroad car wheels that were frozen during very cold nights. (L.S.) LL
bunting pole A pole between the runners of a sleigh to hold them apart. (L.S.) LL
bureau A scaling bureau.
bureau scale The log scale that is set up by a bureau scaler; the accepted scale is standard for buyer and seller.
burl A swelling on a tree, usually near the base, caused from an injury to the tree or chemical or unnatural interference in the tree growth under the bark. Also spelled "birl." See knurl.

A grove of **redwood** trees, the **sequoia sempervirens** of the Northern California Coast. Note the **burls** on the tree at the right of the photo. *A.W. Ericson photo, Carranco collection*

burlap show A poor logging outfit; A haywire show.
burn 1. An area that has been burned by fire. 2. To make a line rub against an object, such as a tree or stump to change the direction of

pull. 3. To set a slash fire. 4. To siwash.
burn a stump To pass the mainline around a stump to move the front end of a donkey sled. (P.C.F.)
burn daylight To waste time and not get much work done. (P.C.F.)
burner 1. A welder. 2. A portable backfiring torch. 3. A person who deliberately sets fire to a forest.
Burnsider In the late 1800s, lower Burnside Street in Portland, Oregon was where all the employment agencies were located. Anyone sent out by an agency might be a Burnsider, one who might be a good man or a bum. (P.N.F.)
burnt piece A piece of woodland burned over after the wood has been cut off. (Maine)
burrs 1. The frayed wire on a cable. See jaggers. 2. Chunks or metal mashed off the striking surface of a wedge or sledge.
burton A tackle composed of two or more blocks that is used to increase the hauling power of the pulling line. The log is attached to a block in the bight of the running part. (P.C.F.)
bush The forest or back country. (Canada)
bush a road To mark the route of a logging road in the woods on ice by setting up bushes. (N.F.) LL
Bush butt hook An old-time hook used on butt rigging that had a heavy circle of iron like a doughnut that slid up and down the shank of the hook to lock in the eyes of chokers attached to it. Also doughnut hook.

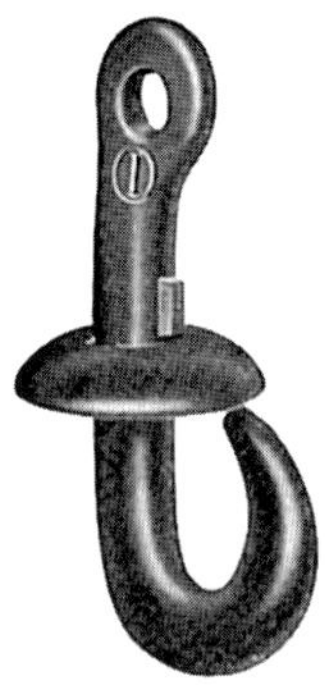

A **Bush butt hook**. *Young Iron Works catalog, Glen Comstock collection*

bushed 1. Tired. 2. Said of one lost in the woods.
bushel 1. A thousand feet of logs. 2. To fall or buck by piece work as opposed to day wage. 3. To accept a piece rate job or a small contract.
busheler 1. A contract worker, one who does piece work "by the bushel." Also piece worker. 2. A derogatory term for a logger from the Pine or Fir Country who cannot handle the Redwoods. (Red.)
busheler's enemy The scaler who measures what the busheler cuts, thus determining how much he gets paid. (P.C.F.)
bushel-hungry Said of the fallers or buckers who want to cut more timber each day. (P.C.F.)
busher See swampers.
bush-out See brush-out.
bush ranger A woodsman with a good eye for timber and special knowledge of the region. (N.W.)
bush monkey In the late 1800s and early 1900s, a man who piled tanbark. (P.C.F.)
bushwhacker A man who cannot be trusted.
bushwhackers Contracted men who cut and hack out hardwoods (alder and tan oak) competing with conifer regeneration. (Red.)
busier than a one-armed paper hanger with the crabs Said of a very busy worker. (Red.)
business end 1. The motor end of a power saw. 2. The engine end of a train. 3. The rear end of a mule or wasp. 4. The most important part of any machine.
bust a gut 1. To snap a belly guy on a spar tree. 2. To break the middle wrapper in a load of logs. 3. To make a great effort.
busted his eggbasket Said of one who had sustained an injury in the woods. (P.N.F.)
busy as the devil in a gale of wind Said of a very busy man. (Maine)

butt 1. The base of a tree. 2. The large end of a log.
butt chain 1. At one time a short chain used in place of chokers. 2. See bitch chain. 3. In butt rigging, the assembly of a swivel, link and a Bardon butt hook, onto which the chokers are fastened. See butt rigging.
butt chain block In the ground lead days, the short heavy block up to 20 inches in diameter, with a wide open throat that permitted the butt chain to pull through. See bull block, Tommy Moore.
butt chain block tender A logger whose job it was to unhook the chokers, pass the butt rigging trough the throat of the butt chain block, then couple on the chokers again on the donkey side of the block permitting logs to be dragged right up to the landing.
butt chain head block, butt chain lead block See butt chain block.
butt cut 1. The bottom log in a felled tree. 2. The cut that separated the bottom log from the top logs. 3. In gathering tanbark, the section of bark taken from the butt of a tree before felling it for further peeling. (N.F.) 4. To cut off a sidewinder or blowdown from the stump or butt end.
butte Usually an isolated hill or small mountain with steep sides.
butterfly hook A choker hook used on skidders. The butterfly, that could be opened or closed to fasten in the choker, was operated by a push button on a spring.
butt hook The heavy hook on the butt rigging to which the chokers are attached. The early rigging butt hook was shaped like a fish hook without the barb, and the chokers often fell out. Many changes were made as loggers experimented with a positive lock hook. At present the butt hook is built like a socket into which the ferrule or knob of the choker is seated. Among butt hooks currently used are the Bardon-Wirkkala, and McGovern. The first butt hook was used on the trace, connecting this part of a horse's harness by a small chain to the single tree. The butt hook was unhooked and the chains left on the single tree when unharnessing at night.
butting timber To round off the log ends to make them skid more easily. (L.S.) See snipe. LL
butt lash The proclivity of a tree, when felled, to move forward, sideways or backward off the stump, and very dangerous to the fallers. (L.S.) LL
butt line A name for the tag line or short line connecting the skidding tongs to the butt rigging. See tag line.
butt log The bottom log in a tree.
butt-off 1. To cut a defective portion from a log, usually from the butt. See long-butt. 2. To square the end of a log. (N.F.) LL
button An iron clamp placed on a skidding line as a stop for the swivel of the slack-pulling line. The position of the button determined how much slack could be pulled.
butt plate A heavy steel plate used in making up the butt rigging in some systems of logging. The mainline, haulback and butt hook are all tied to the butt plate. Also diamond. See butt rigging.
buttress A wall or abutment built along a stream to prevent the logs in a drive from cutting the bank or jamming. (N.F.) LL
butt rigging 1. In a cable logging operation, the equipment (swivels, clevises, etc.) that connects the mainline and haulback and from which chokers are hung. 2. Short lines between the chokers and the mainline.

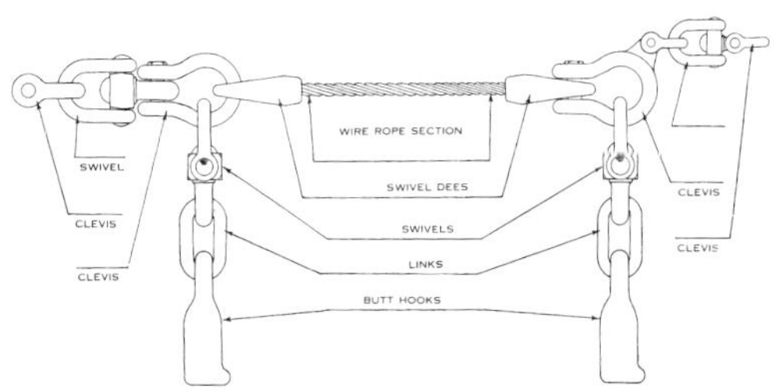

Butt rigging. *Young Iron Works catalog, Glen Comstock collection*

butt rot The decay in the base of a tree that is cut off. See stump rot.
butt team In the 1800s the pair of oxen or horses next to the logs when skidding. See wheel team. Old-timers in the Pacific Northwest preferred bull for the butt team and oxen for the rest because they

were quicker on their feet and kept out of the way of logs better. Also pull team, snub yoke.

butt the skid 1. In animal skidding, to miss the cutout groove in the center of a skid and to bump either side of skid, thus tearing up the skidroad. 2. To stupidly upset a job. (P.C.F.)

buy 'er 1. To leave or throw away equipment or logs. 2. To give up. 3. To be killed. (P.C.F.)

Buzzard hook A self-locking choker hook used on skidders.

by the book To work at a job by rules instead of using common sense.

by the inch 1. An expression used to describe work by piece rates as in busheling. 2. A term to describe the way overtime pay is earned. Also by the bushel, by the piece, by the mile. (P.C.F.)

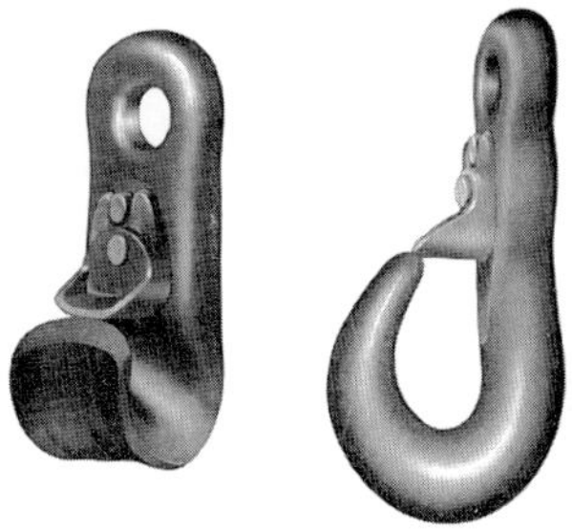

Two types of **Buzzard hooks**. *Young Iron Works catalog, Glen Comstock collection*

C

C.I.F. A marine shipping term, "cost, insurance and freight," meaning that the seller delivers the goods to the carrier and agrees to pay all charges to bill-of-lading destination. On delivery, to buyer, of bill-of-lading with insurance policy attached, the seller's responsibility ceases.

C.I.F.E. Cost, insurance, freight, exchange.

C.I.O. A lumber industry union. See I.W.A.

cabbage stack A big round spark arrestor on the smokestack of a locomotive.

cabin car A caboose.

Medco's 4-spot **Willamette** poses near Butte Falls, Oregon, towards the end of her career. She sports a **cabbage stack** spark arrestor manufactured by **Gerlinger**. *Lamar Fergusen photo*

cabin fever 1. Said of a man who lives away from other people too long and begins to talk to himself. 2. Said of a crew who work out in the brush for a long period of time and begin to argue among themselves.
cable cutter A tool for cutting wire rope.

A **cable cutter**. *From Mallory catalog, 1948, Philip Schnell collection*

cable locomotive One of many types of cable locomotives. See Fouts' grip wheel and walking Dudley.

A **cable locomotive** of the Tillamook Spruce Company, near Tillamook, Oregon. The locomotive was built by the Electric Steel Foundry Company, of Portland, Oregon, following a design originated by F.C. Walters at Veneta, Oregon. *Courtesy of Miller-Freeman Publications*

cable logging Transporting logs from where they are felled to a landing, using steel cables so the logs are partly or wholly lifted off the ground.
cable road Any cable-operated road, not necessarily on an incline. The walking Dudleys were used on level ground.
cable saver A donkey built by the Washington Iron Works. A heavy idler wheel riding on the drum was supposed to keep the cable perfectly spooled, saving wear.
cable tram An endless cable that brought logs in from great distances, over hills and valleys. Used especially in Idaho and Vancouver Island. Also aerial tram.
cable way Any system of logging using wire rope.
cableway machine An early name for a skidder. (P.N.F.)
cableway yarding system See skidder system.
caboose 1. On a mainline railroad, a car for the train crew to ride in. 2. On logging roads, it was used to haul freight, equipment or stores. See crummy and mulligan. 3. A bunkhouse. (P.N.F.) 4. The back end of anything. 5. Before the 1870s, the open fire in a shanty. (N.F.) 6. After the 1870s, a big stove in which cordwood was burned. (L.S.)
caboose hop A train consisting of engine and caboose only.
cache A storehouse for logging camp supplies. (N.F.) (Canada)
cackleberries and grunts Eggs and bacon. (L.S.)
cackler A white-collar worker. (L.S.) LL
calaboose A jail. (L.S.) LL

A Washington Iron Works **instantaneous change donkey** equipped with the **cable saver** device. The lines to the rear drums pass over the small sheaves, while the bar above holds the lines steady during sudden speed changes. *Washington Iron Works photo, Labbe collection*

A **cable tram** in operation. *Willamette Iron & Steel Co. catalog, 1920, Glen Comstock collection*

calico A woman: "She's a good piece of calico." (Maine)
California sling shot A name given to crotch line loading. In this logging system two poles are used with the rigging in the middle, which resembles a sling shot. (Red.)
California white pine 1. White fir in the Redwood Country. 2. Also western white pine, sugar pine or western yellow pine.
caliper rule A scale rule for measuring logs. (L.S.) LL
calk 1. To drive moss or oakum between logs in a building. (N.F.) 2. To tighten the seams of a steam boiler by upsetting the metal around the rivets with a special chisel.
calking iron A wooden or iron bar used for driving moss or oakum into the cracks between the logs of a building to keep out the cold. (N.F.) Also caulking bar. LL
calks, caulks Shoes with short spikes in the soles to give safe footing in the woods and log pond. The popular term is corks. Also caulk shoes or caulk boots. The term "caulk" is also used for the spike.
camboose, cambuse 1. The open fire bed or fireplace directly be-

neath the smoke hole in the roof of the bunkhouse cook shanty of the 19th century pine camps. (N.F.) (Eastern Canada) 2. The bunkhouse. 3. The railroad car for the train crew to ride in. LL
camel back See rerailing frog.
camera A term of contempt for the transit by those who are not engineers.
camp The buildings and site of the logging community located in the woods, or near the point of operation.
camp car 1. A building on railroad trucks. These cars were often grouped together to make up a camp, which in turn could be moved as the operation progressed. (P.C.F.) 2. A flat car equipped with seats and used to haul loggers back and forth between camp and the logging operation.

Medco **camp cars** sit on a siding near Butte Falls, Oregon in June of 1957. *Ivan Ergish photo*

camp dog The helper who cared for the bunks and sleeping quarters in a camp. (L.S.) LL
camp foreman One who has charge of a logging camp and the logging operations conducted from that camp.
camp helper See bull cook.
camp hunter The hunter who kept the camp supplied with meat. (L.S.) LL
camp inspector An itinerant logger who moved continuously from job to job, never staying long in any one camp.
camp lawyer The man who is always arguing in camp.
camp louse 1. A bull cook. 2. A chore boy. 3. A handyman. (P.C.F.)
camp push The woods boss or resident superintendent.
camp robber The Canadian Jay that follows the loggers, always looking for food. Also lumberjack bird, whiskey jack.
camp run A basis for selling all the logs produced by a camp, without sorting them for grade.
camp site The location of a logging camp. (L.S.) LL
camp strawberries Beans. (S.F.)
camp watchman The man who guarded the logging camp during the summer months when there was no logging. (N.F.) LL
Canada To use a peavey or cant hook to get the logs moving in a straight line when loading on a sleigh or a railroad car. (L.S.)
Canada greys Very heavy woolen or felt stockings. German socks. (L.S.) LL
Canada peaker A system of arranging logs on a load in the form of a triangle. The load was built by nesting logs in the saddles of the other logs. (L.S.) LL
Canada side The opposite of the face side. (L.S.) See gin-pole side.
Canadian eye A crude rolled eye splice, made in a hurry. (P.N.F.)
Canadian peaker In the Klamath Country (along the Klamath River in Northern California) the top log in a load. (Red.)
canal wrench A hand shovel. Also armstrong plow.
canary 1. A long rod hooked at one end that is used in reaching under a log to pull the eye of the choker through. 2. A long rod used in the Pine Country to pull a chain or cable under a load of logs.
candy Anything especially fancy or well rigged. (P.C.F.)
candy kid The boss's favorite man. (P.N.F.)
candy run A short or easy haul. (P.C.F.)
candy side The side or operating unit that has the best timber or best logging conditions. The other side is the haywire side.
candy wagon 1. A station wagon or bus that transports men to and from work in the woods. Crummy is the more popular term. 2. A camp commissary. (P.C.F.)
canned cow Tinned milk. (P.C.F.)
cannon A log that ends up at an angle to the load while loading logs by crosshaul. (P.N.F.)
cannon a log To send up a log that swings crosswise instead of parallel to the load. Also gun a log.
cannoned 1. Said of a log being sent up the skids to a rollway. 2. A load out of control, pointing skyward. (L.S.) LL
can opener artist A very poor cook. (P.C.F.)
canopy 1. The cab over a cat. 2. The rock guard over the cab of a dump truck. 3. The continuous cover of branches and foliage formed collectively by the crowns of adjacent trees and other woody growth.
cant 1. A large section or a large piece of wood cut from a log in the sawmill. 2. To turn a log. 3. To make something lean to one side.
can't cut'er, can't hack'er, can't make'er Said of a man who cannot handle a job.
cant dog A short peavey.
canter A cant hook man, usually one who rolled logs on the landings.
cant hook A tool similar to a peavey, but having a toe ring and lip at the end instead of a spike and used to turn logs and cants. Also gilleary, mooley, swing dog.
cant hook man A logger or lumberjack who works on a rollway.
canuck, canucker 1. A logger or mill worker from Canada. (P.C.F.) 2. A French-Canadian. (Maine) See blue nose.
canyon carriage A skidder carriage with a specially rigged slack puller to lower the chokers into a deep canyon. (P.C.F.)
cap 1. A cone of sheet iron or steel, with a hole in the end through which a chain passes, which is fitted over the end of a log before snaking it, to prevent catching on stumps, roots or other obstacles in steam skidding. (S.F.) 2. The timber placed across a row of pilings to form a bent.
capering clothes Clothes that were worn to town. (P.N.F.)
capstan donkey The early vertical spool donkey. See Dolbeer.
capstan spool A spool on a donkey that was used to pull a line. The rope or wire rope was wound several times around the capstan and then coiled on the ground as it came in.
captain Jim Crow The boss of the tie men. (S.F.)
captain of the goad A teamster in the early 1800s. (Maine)
car camp A railroad logging camp, see railroad camp.

The logging camp of Bloedel, Stewart Welch, at Franklin River on Vancouver Island in British Columbia was a **car camp**. *Courtesy of Campbell River Museum*

carbide Bad coffee. (P.C.F.)
card man A union man. LL

This **car-mounted unit**, built in September, 1925 by Willamette Iron & Steel Works for Schafer Brothers Logging Company of Montesano, Washington, consisted of a **high lead skidder** and a **duplex loader**, mounted onto a **unit car** built by Pacific Car & Foundry Company. *Willamette Iron & Steel Company photo, courtesy of Jack Taylor*

card room The place with gaming tables where woodsmen or mill men gather while in town. LL
car frog See rerailing frog.
cargo To prepare a load for pack horse shipment. (P.N.F.)
car knocker An inspector and repairman of railroad cars. Also knocker. (P.C.F.)
car line A line with a winch or donkey drum that was used to pull railroad cars into position for loading. Also car spotter, snubbing line, spotting line.
car-mounted unit A yarding and loading machine mounted on a railroad car instead of on a sled. See unit.
car replacer See rerailing frog.
carriage A two-wheeled device that rides back and forth on the skyline in yarding and loading. Various blocks and lines hang beneath the carriage. In operation it is raised to carry the logs above the ground most of the time. There are various sizes and shapes, but usually the V-shaped ones are used on slacklines and the L-shaped ones on tight line systems. Skidder carriages may go over 1,500 pounds. Also trolley.

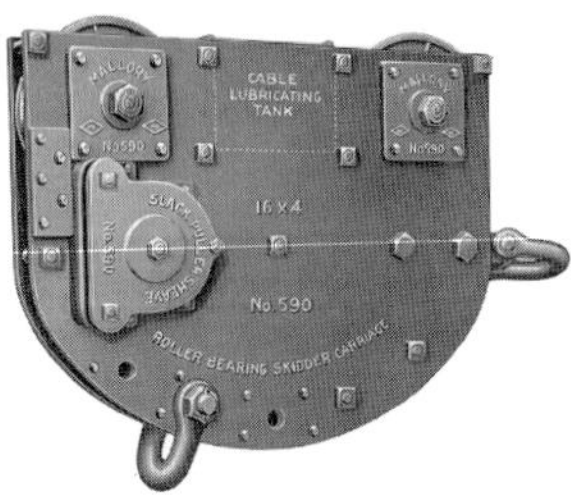

A roller bearing skidder **carriage**, equipped with double **slack puller** attachment. *Mallory Catalog No. 12, 1925, Glen Comstock collection*

carriage line A skyline.
carrier A carriage on a skyline.
carry blankets To move from one camp to another. (P.N.F.)
carrying line A skyline. (P.C.F.)
carry the balloon To search for employment. (L.S.) LL
car spotter See car line.
car stake 1. A vertical pole on a flat car to hold logs or lumber. 2. The arms at the ends of the bunks in place of cheese blocks, which can be dropped for unloading. Also drop stake.

Cascade Lumber Company, operating out of Casland, Washington, used **car stakes** to keep the logs on their flat cars. *Jack Whitnall photo, courtesy of Ralph Craig*

carters Teamsters on the tote or supply roads. (N.F.) LL
carting grab A crotch grab used on the first log of a log train. (N.W.)

A **Carting grab**. *From "Logging and Lumbering," C.A. Schenck, Biltmore Forest School, Glen Comstock collection*

carting team The team used on the tote or supply roads. (N.F.)
car tinkerer A repairman. (P.N.F.)
car toad Same as car knocker. (P.N.F.)
car whacker Same as car knocker. (P.N.F.)
cascara The cascara buckthorn of the Pacific Coast whose dried bark is used as a mild laxative. Also bearberry bark. See chittum.
Casey Jones 1. A locomotive engineer. 2. A gas speeder or rail motor car. 3. A speeder operator. 4. A trade name for a speeder.
cashed in his chips Said of a man who died.
cast-off hook A hook used on a tramway that could release logs quickly.
cat 1. A powerful tractor that hauls logs out of the woods. The word is from Caterpillar, but is now applied to all tractors. A cat with a single front wheel for steering, and with rear wheels based on tracks, was used in log-hauling in 1912. The loggers were instrumental in developing the cat for yarding logs. 2. An easy woman. 3. A cougar or bobcat.

A Holt tractor, an early version of the well-known Caterpillar, or **cat**, here handles a train of loaded **log wagons** on steel wheels. The scene is on the operations of M.G. Ashley, at Cle Elum, Washington, in 1912. *From R.A. Slaughter, courtesy of M. Koch*

catamaran A small raft carrying windlass and grapple and used to recover sunken logs. Also gun boar. (N.F.) See monitor and pontoon. LL
cat bummer A log carrier mounted on endless treads and pulled behind a cat. The head ends of the logs rested on the bummer and the hind ends dragged on the ground.
cat car See cat-moving car.
catch basin A small reservoir along the line of a flume where logs can be stored temporarily when there is a surplus of logs. Also used to feed in more water in dry spells.
catch boom A boom fastened across a stream to catch and hold floating logs. (N.F.) Also trap boom. LL
catcher's ax A very narrow ax with a five-foot handle; used to cut side marks into logs before floating them downriver. (N.F.) LL
catchmarker The man in a log boom who swung his long ax to reproduce the bark mark in a log where it could be seen more readily as it moved swiftly down the river and to indicate that it had been received in the boom. Also bark hack. (N.F.) LL

A 10-ton Holt Caterpillar tractor drawing logs loaded on a **cat bummer** in Idaho. *From "Logging," R.C. Bryant, Wiley & Sons, 1923, Philip Schnell collection*

catchment facilities All the facilities of a boom company, such as wing booms, sorting booms, storage booms, etc. (N.F.) LL
cat doctor A tractor mechanic.
catface A fire scar on a tree. (P.C.F.)
cat-heads Biscuits. (West Virginia)
cathedral woods 1. A beautiful stand of pine trees with a soft carpet of pine needles. (L.S.) 2. A redwood grove. (Red.)
cat hook A hook used on the arch line when yarding logs. It has a fishtailed end to prevent the chokers from falling out when coming back to the woods empty, or from slipping off when the line is slacked. Also arch hook, tractor hook.

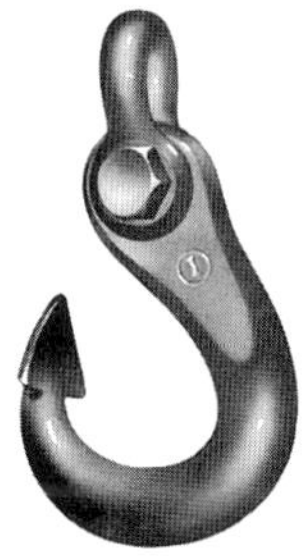

A **cat hook**. *From Young Iron Works catalog, 1952, Philip Schnell collection*

cat hooker A head man of a crew of choker setters on a cat show.
cat man 1. A logger who knows how to handle and repair a cat. 2. A term of praise.
cat-moving car Usually a car with a wooden deck for handling cats and trucks. Also cat car.
cat house A whore house.
cathouse A disturbance or a fight. (Maine)
catpiece A small stick in which holes are made at regular intervals, placed on the top of uprights set firmly in floating booms. The uprights are fitted to enter the holes in the catpiece so as to narrow or widen the spaces between the booms at the entrance to a sluiceway or sorting jack. The catpiece is held by the uprights high enough above water to allow logs to float freely under it. (N.F.)
cat road 1. The wide road that the tractor makes in yarding logs. 2. A road prepared for cat logging.
cat's ass 1. A knot in a line. 2. The tangle when a choker cinches up tight. 3. A closed choker hook that cannot be unhooked. 4. A round scar on a tree trunk that has almost completely healed.
cat show 1. Timber that can be logged by tractors. 2. A logging operation in which tractors are used instead of a cable system.
cat skinner A driver of a tractor that hauls logs from the woods, the modern version of a mule skinner. May be called a dozerman if moving dirt. Also pussy pusher, skinner.
cat's paw 1. A poor splice on a wire rope. 2. A knot used on a steel

cable. 3. A line put through an eye and joined to itself for a temporary quick tie.
cat track The tread on which the tractor rides.
cat trail A path cleared around a fire or slash area by a tractor or bulldozer.
catty 1. Said of a logger or mill worker who is fast and light on his feet. 2. A donkey that is quick and easy to handle. 3. A good steamer.
catty-corner See antigodlin.
cattyman An expert river driver. (N.F.)
catty on his caulks Said of a good river pig on a river drive. (L.S.)
cattywampus Said of anything out of line or off the track. (P.C.F.)
caulks See calks.
caulking bar See calking iron.
caught in a bind Said of a man caught in a difficult situation.
cedar The Western red cedar is among the largest and finest of the cedars grown anywhere in the world, its heartwood famed for resistance to decay. Its wood is of exceptionally light weight, one of the lightest of all commercial species, of soft texture and close, even straight grain. The wood is highly regarded for exterior and interior uses. Cedar logs are shipped to Japan to be used for interior purposes.
cedar ax A light single-bitted ax.
cedar fever An itch that sometimes affects men working with Western red cedar logs or poles. Men also came down with cedar asthma that was supposedly from the dust from cedar lumber and shingle mills. (P.C.F.)
cedar pole camp A logging operation that cuts mostly cedar poles. (L.S.) LL
cedar products log A Western red cedar log otherwise falling into the classification of a cull log with the exception that it contains a shell or core of sound fiber suitable for splitting into merchantable products.
cedar savage A logger who works in a cedar pole camp, or for a shingle bolt outfit. (L.S.) LL
cellulose The organic matter making up the cell wall of the wood fiber.
center Logs jammed on an obstruction during a river drive. LL
center jam A log jam on an island in the center of the stream. Also stream jam. LL
center roller A line roller with three sheaves that was used on an incline.
centipede A term used to describe the use of two trucks under long poles or piling, rather than a truck and trailer. (P.N.F.)

A system devised by truckers for handling long loads of **piling** was sometimes called a **centipede**. However, it was soon discouraged, since it proved dangerous on the highway. *From "The Timberman" collection*

chain 1. The flexible saw chain used on a power saw. 2. A unit measurement of 66 feet used in land surveys—five chains equal a tally, the standard measure of distance in logging. 3. To measure the distance on a survey.
chain block See Tommy Moore block.
chain boom Any boom in which individual boom sticks are held together by boom chains. (L.S.) LL
chainer A dogger in the horse-skidding days who helped get the turns ready for skidding; comparable to a choker setter except that a chain was used in horse-skidding.
chaining Dragging bundles of pulpwood wrapped with a chain crossways down a skid trail. Used in winter logging on steep slopes. (N.W.)
chain plate See bridle iron.
chain saw Usually a power saw run by gas. The teeth are on a flexible chain which runs around a long oval bar. A portable gasoline chain saw was demonstrated at Eureka, California as early as 1905.
chain swifter A light chain used to tie a log raft together.
chain tender 1. A term used in the late 1800s for the head rigger or hook tender. (P.C.F.) 2. The man who assists in skidding or loading operations. (N.F.)
chair Part of the tree trunk left on a stump as a result of the tree break between the stump and butt log. See barber chair.
chaland A barge for hauling men, horses and supplies. (Quebec)
chambermaid In river driving, an older man who would get the men out of bed in the mornings and who did odd jobs around the boom house. (N.W.) See bull cook.
chance A logging operation or a logging show.
change ends 1. To swing a log around when loading so that it will fit better on a truck or car. 2. To reverse a wire rope so that the heavy wear becomes uniform. 3. To move a donkey to the other side of the spar tree. 4. To take a fall out of a man. 5. In the days of ground lead yarding, to move a donkey to the other side of the spar tree to yard a fresh area.
change road To move the rigging to a new unlogged area where the timber is felled and bucked.
change their gait To do a job in a different way. (L.S.) LL
changing line A line used in setting up rigging for a new road in skidder logging. See transfer line.
chantier A shanty or bunkhouse. (Quebec)
chase 1. To unhook or hook logs at the landing. 2. To ask for food or drink in the cookhouse: "Chase the canned cow." (P.C.F.)
chaser 1. A logger who loads logs. 2. A logger who unhooks the chokers from the logs at the landing. 3. A logger in the days of ground lead yarding who chased or followed the logs from the woods to the landing to see that the logs did not hang up or to replace chokers that were always falling out of the old open choker hooks. (P.C.F.) (R.M.F.) Also block tender, hook chaser, pigman.
chase the cow To pass the milk. (S.F.)
cheater 1. A scaler. 2. A timekeeper. 3. A bar or length of pipe used to elongate the handle of a jack or other machine. The cheater helps improve leverage by lengthening the movement of the arm.
cheat stick, cheating stick A scaling stick used by a log scaler in determining the contents of a log; usually had a metal tab at one end to hook over the end of a log and a scale rule or inch scale imprinted on it. Also gyp stick, robber's cane.
cheat stick rigger A scaler.
check A crack in a log or tree or finished lumber.
checker 1. A tallyman. 2. The man who sorted the logs according to log marks in river driving. (N.F.)
checkline In river driving, a heavy cable strung across a river or to a pier to support a checker who sorted the logs. (N.W.)
checks The small lengthwise separations in the wood which are caused by too rapid evaporation of moisture from the surface; checks usually occur across the rings of annual growth.
check scaler A man who rescales logs to detect errors on the part of the scaler. (N.F.) LL
cheechako A greenhorn. (P.N.F.)
cheek plates The steel plates that enclose the sheave or wheel of a logging block (pulley).
cheeky as a man on the town Said of an overbearing man. (Maine)
cheese block 1. Usually a triangular wooden block that is placed under a loaded log or loaded lumber to prevent moving or rocking. 2. Steel blocks used on log bunks.

cherry picker 1. A loading machine for picking up logs that are lost along a railroad. Occasionally used in prelogging and relogging. 2. A mobile crane.

Cherry Valley Logging Company, a Weyerhaeuser subsidiary out of Monroe, Washington, built this **cherry picker** to load logs along their right-of-way. In this view, the machine rides on a **bull car**. *D. Kinsey photo, Labbe collection*

chew the fat (hog) Said of a person who is talking.
chickadee A road monkey whose job was to clean the horse manure and debris from the iced roads. (N.F.) (Canada) LL
chicken coup A truck weigh station.
chiken shit outfit A poor show, no place for a good logger to work.
chicot A dead, dry, standing tree trunk. (Canada) See rampike.
China boy In the early days of logging on the West Coast, the Chinese were employed in the woods as swampers and fire tenders. About 1875 they were brought to Humboldt County, California to construct railroads. On February 6, 1885, a Eureka councilman was killed by a stray bullet fired from the gun of one of two quarreling Chinese men. All of the Chinese were forced to leave the county by April 1886, and it was not until after World War II that Chinese families again began to move into the area.

There were many Chinese loggers at one time in the British Columbia camps: one Chinese man would contract to put out logs, and did all the bargaining for the whole crew; they had their own quarters; did their own cooking; worked on their own piece of ground; and did not mix with the other loggers.
China house The combined cookshack and bunkhouse used by a crew of Chinese loggers. (British Columbia)
chin chopper A tree that splits in falling. (L.S.) LL
chin jaw Small talk; social conversation. (Red.)
chink 1. Moss or mud. Also stog. 2. To fill the crevices between logs with soft clay, oakum or moss in a logging camp. (N.F.) LL
chinook 1. An early jargon or trade language of the Pacific Northwest. 2. A warm winter wind that can turn ten-below-zero weather into rain overnight. 3. A logging donkey made by the Puget Sound Iron and Steel Works of Tacoma, Washington. (P.N.F.)
chin-whiskered Said of a poor logging setup: "A chin-whiskered outfit." (P.N.F.)
chin-whiskered jobber See gyppo.
chip 1. The narrow portion of sapwood that is removed from above the box of a tree that is being held for crude turpentine. 2. The refuse strained from scrap. (S.F.) 3. To cut away a thin section of sapwood above the box or cup in order to stimulate the flow of crude turpentine. (S.F.)
chipper grain A type of saw chain having the cutting faces at right angles to the direction of travel. (P.C.F.) Also Oregon chain.
chipping their teeth Said of a group of woodsmen or mill men doing a great deal of talking. (P.C.F.)
chippy A woman of easy virtue; not a professional.
chittam, chittem, chittern 1. The bark of the cascara tree. 2. The cascara tree. At one time the bark was collected on Sundays by old

Puget Sound Iron and Steel Co. of Tacoma, Washington built the **Chinook compound geared yarder**. *From Mallory Catalog No. 6, Glen Comstock collection*

loggers, but now the professionals have taken over. The dried bark is used as a mild laxative by druggists. In the Bible (Isaiah 41:19), the shittah tree was a tree belonging to some species of acacia, from which shittim wood was obtained.
choc A low-grade beer originally made by the Choctaws. (S.F.)
chock 1. A tapered block to prevent loaded logs or lumber from rolling off a car or truck. 2. Any type of block to prevent logs or lumber from rolling or falling. See cheese block.
chock block A cast-iron angle bracket used to bolt a donkey engine's frame securely to a sled runner.

A **chock block**. *From Mallory Catalog No. 12, 1925, Glen Comstock collection*

choke To put a line around a log and to draw it taut.
choke-bore pants Pants worn by beginners in the woods (tight at bottom) and laughed at by the loggers. (P.N.F.)
choke-bored Said of a thin man or long-legged skinny man. (P.N.F.)
chokem Cheese. (P.N.F.)
choker 1. A short cable with a ferrule on each end. One end is put around a log and then socketed into the choker hook that runs on the choker cable; the other end is socketed into the butt hook. When the choker is pulled, the cable around the log cinches tight, choking the log as it is pulled to the landing. 2. A necktie.
choker gun 1. A tool for pulling the end of the choker under a log. 2. A shovel used for the same purpose.
choker hole A hole dug under a log so that the choker may be passed under and around the log, getting it ready to yard. At one time some choker holes were made with dynamite, now outlawed.
choker holes Doughnuts. (P.C.F.)
choker hook Not a "hook" but a socket that slides up and down a short cable, called the choker. The "hook" is slotted to take the ferrule of the choker that can be stuck into the socket. The following hooks have been popular: Ashdown, Bardon, boxing glove, butterfly, Dennison, jail latch, Peters and Wirkkala. Also buggy bell.
chokerman A logger who sets chokers. (P.C.F.) (R.M.F.) (British Columbia)
chokerman's brute A term to describe an unknown sickness. (P.N.F.)
choker setter The logger who puts the choker around the logs to be yarded. (P.C.F.) Also brush rabbit, tongsman, riggin' man, riggin' slinger.

The **Bardon** or **Wirkkala hooks** are the standard types of **choker hooks** used by the present day logger. The **nubbin** at one end of the choker is slipped into the **bell**, which slides freely on the line. A nubbin on the other end slips into a similar bell on the choker hook attached to the **cat**. *From "The Timberman" collection*

choking Binding logs together with a cable when hoisting them to load a sleigh or railroad car. (N.F.) LL
choking limit The steepness of a grade on an incline. Beyond this point it is necessary to choke or tie logs on cars to prevent them from sliding off, generally about 35%.
chop 1. To fell trees. (Red.) 2. To cut trees, logs or limbs.
chopper 1. A faller. (Red.) (App.) 2. A logger. (N.W.) 3. The man who felled trees and cut them into logs. (N.F.)

Choppers at work on the **undercut** on a large redwood tree for the John Vance Mill & Lumber Company near Arcata, California, in the 1880s. They are standing on **staging** supported by their **driving boards**, better known among Northwest loggers as **springboards**. *A.W. Ericson photo, Labbe collection*

chopping A tract of woodland where the trees have been felled. (Maine)
chopping board A springboard in the Pacific Northwest Fir Country. See springboard.
chopping boss A bullbuck or foreman of fallers and buckers. (Red.)
chopping ground knots Hitting the ground with an ax instead of a tree while limbing a fallen tree. (P.C.F.)
chopping platform Staging used instead of a springboard to provide a working place when felling big timber. Rigging is the popular term in the Redwood Country.

chore boy The person who cleans up sleeping quarters and tables in a logging camp, cuts firewood and builds fires. Pronounced and spelled "show boy" in Quebec. See bull cook and flunky.
chuck 1. Food. 2. A shortened form for "saltchuck," the ocean, from the Chinook jargon. (P.C.F.) 3. To clear the ground of obstructions which cannot be removed by hand. (P.C.F.)
chuck boat The boat with the food to feed the river drivers. (L.S.) Also wanigan.
chuck bucker The logger, in advance of felling, bucks up merchantable windfalls and other down timber that may interfere with yarding. (P.C.F.)
chuckhouse A cookhouse or cook shack.
chuke A knit cap worn by cruisers in the early 1800s in the Lake States. LL
chunk 1. A piece of a log that was sawed or broken off. 2. To clear the ground of obstructions that cannot be removed by hand.
chunk bucker 1. A good bucker who tries to salvage short logs out of broken logs. 2. A bucker working on a road-clearing crew.
chunk line A line used in chunking out a road or clearing land. (P.C.F.)
chunk-out 1. To remove trash, stumps and logs from any area. 2. To clear a route.
chunk-out donkey A light donkey used in road building and land clearing. (P.C.F.)
churn butt A swollen butt of a log or tree. Also bottle butt.
chute 1. A V-shaped trough built of poles for log transportation out of rough country, used widely in the days of horse logging and in the early days of steam logging. In steam logging, the logs were dogged to one another and pulled by a donkey. The famous Ankeny chute on the Lower Columbia in the 1880s was faced with iron for more speed and wear. The Pokegama chute southwest of Klamath Falls dropped logs 2,650 feet to the Klamath River. See log slide. 2. Any dry slide to send logs down a steep hill. 3. The rapids in a river. 4. To pull logs along a chute, or to let logs move by gravity when steep enough.

Along the rugged coast of Northern California there are few harbors and the lumbermen loaded out their products from the rocky shore. The tiny schooners sailed right up to the rocks in the **dog holes** where the lumber was loaded from the **chutes** directly onto the decks of the ships. *Courtesy of Nannie M. Escola*

chute boar See riggin' sled.
chute grease 1. Cheap oil or grease that was used to make the chutes slippery on the light grades where logs might stall. 2. Poor butter.
chute greaser Usually the skid greasers who became chute greasers when loggers turned to chuting their logs. (P.C.F.)
chute tender A man who took care of the chutes.
cigar raft An ocean-going raft of logs, such as the Robertson raft, that was shaped like a cigar. The first cigar raft on the Pacific Coast was launched at Noyo, Mendocino County, California in 1895. See

The **Robertson raft** was the first successful **ocean-going raft** to be developed. Often called **cigar rafts**, they transported many millions of feet of logs and piling between points in Oregon and Washington and California ports. This rafting ground at Stella, on the Washington shore of the Columbia River was the origin of most of the piling rafts, while a similar arrangement across the river served to build the log rafts of the Benson Timber Company, the company that made the greatest use of the Robertson rafts. The rafts were assembled in the **cradle**, as shown, which was hinged in the middle and could be dropped away from the completed raft. *Gifford & Prentiss photo, Labbe collection*

Benson raft, Robertson raft.
cinch line A light line used in raising the heel tackle and to tighten the skyline on a skidder show. See block and fall tackle.
cinch rope See cinch line.
circle hook A large circular-shaped hook on the end of a skidding chain that loosened easily in skidding logs. (L.S.) LL
claim A piece of timber that belongs to someone.
clam A machine used in loading and unloading logs. (L.S.) LL
clam digger 1. A poor workman or an inexperienced logger. 2. A bucket on dirt-digging machinery. (P.C.F.)
clam gun 1. A shovel. (P.C.F.) 2. A power shovel. (L.S.)
clamp In fluming lumber, a short iron bar with recurved ends that is used to bind several boards together into a bundle. (California)
clamp ring A metal ring that is clamped over a choker on a Peters choker hook to prevent the choker from pulling through the eye of the hook should the choker slip off the hook.
clap Gonorrhea.
claw A latch on a skyline to hold the trolley in position at the end of the line.
claw bar A heavy steel bar with a claw turned up at the bottom. Used in railroad work and in pulling track spikes that have been driven into a stump.

A **claw bar**. *From Mill & Mine Supply Co. catalog, 1916, Glen Comstock collection*

claw bark To grab onto something and to hang on for dear life. (P.C.F.)
clean drive A term to describe all the logs washed down the river in one freshet, or on one splash. See splash.
cleaner teeth The rakers on a crosscut saw, as opposed to the cutting teeth.
clean the clock, clean the gauge To apply the emergency air brakes on a locomotive. This draws all the air out of the line and causes the indicator needle on the gauge (clock) to drop to zero and the "clock" is empty or cleaned. Also wipe the gauge, wipe the clock.
clear To cut trees in order to clear an area for a road or landing.
clearcutting The harvesting of the entire standing crop of trees. In practice some unsaleable material may be left standing.
clear-cut 1. An area in which all the timber has been removed. 2. To cut all timber in an area.

West Oregon Lumber Company at one time logged out of Clatskanie, Oregon, to supply their mill at Linnton, near Portland. In 1910, they took delivery of their No. 1, a 30-ton **Climax** locomotive. Their No. 3, also built in 1910, was nearly identical. In this view, No. 1 appears to be in the process of **wooding up**. Note the **bucked** fuel logs in the foreground. *Courtesy of Walt Grande*

clearer The foreman of a chunk-out crew. (P.N.F.)
clearing An opening in the forest.
clear year Any year when river drives came down with no log jams.
clears Logs of the best quality.
clevis See shackle.
Climax A company name for a geared locomotive, which was favored by many loggers. The cylinders were parallel to the frame of the engine but sloped at an angle of about 30 degrees. The drive was through a propeller shaft to geared axles. Other geared engines used in the woods were the Heisler and Shay.
climax forest A plant community dominated by trees that represents the culminating stage of natural succession for its locality and environment.
climb 1. To work as a climber. 2. To work as a climber and high rigger. 3. The action of a floating log riding the outside of a flume on a curve. (P.C.F.)
climber 1. A logger who tops trees so that a rigger may rig them for yarding. 2. A logger who both tops and rigs a tree. 3. On an incline, a big drum donkey mounted on a car used to climb up a hill, pulling itself up by a cable. (P.C.F.)
climber hole A temporary hole cut in the base of a redwood tree to put in the drivers in order to get up to where the driver holes are cut in the tree. The drivers are put into the driver holes and the staging boards are placed on the drivers. The faller stands on the staging boards to fell the tree. (Red.)
climbing irons Irons with sharp spurs that the high climbers strap to their legs at the ankles and below the knees to climb up the trees for rigging or topping. Also irons, spurs.
climbing rope A rope with a steel core that is looped around a tree and anchored to the belt of the high climber while he is climbing and working at the top of a tree; the steel core reduces the danger of a climber cutting the rope accidentally.

A **climber** starts up a tree to be prepared for a **spar tree** for Big Creek Logging Company, near Knappa, Oregon. His **climbing irons** are dug into the tree for support, while his saw and ax are suspended by ropes below him. As he climbs and the tree gets smaller, he will take in on his rope to compensate. *Weister photo, Labbe collection*

Climbing irons. *From Mallory Catalog No. 48, Philip Schnell collection*

clip A cable clamp.
clogged flues Said of a logger when he has sexual urges and has been away from the red-light district of the town skidroad for some time. (P.C.F.)
clone A group of genetically identical organisms derived asexually by somatic reproduction from a single individual. May be natural, such as an aspen clone, or may be created in the laboratory from plants or animals.
closed area Part of the forest area that is closed to the public because of the fire season.
closed To shut down a logging operation for lack of timber, too much wind or too little humidity.
closed season The dangerous period from spring to late fall when permits are needed to set fires and when fire prevention measures are enforced.
close-grained Said of wood with narrow and inconspicuous annual rings. The term is sometimes used to designate wood having small and closely spaced pores, but in this sense the term fine textured is more often used. Wood with more than six rings per inch.
close-off 1. The last tier of logs in a boom. 2. To finish up a boom by stretching a tail stick across the open end of the last section, and pounding it against the ends of the stuck-out logs until they become fitted into place.
close to the blanket Said of a shady deal.
closure An order that closes a forest during high fire danger.
Clyde Equipment manufactured by the Clyde Iron Works of Duluth, Minnesota. Clyde manufactured a wide range of unusual logging equipment, including the McGiffert loader, decker, four-line skidder, and a track layer.

A **Clyde four-line skidder** at work for the Crookston Lumber Company, near Bemidji, Minnesota. Built by the Clyde Iron Works, of Duluth, Minnesota, these skidders were common throughout the pine producing regions of the United States. *Courtesy of Omar Long*

Clyde skidder (or jammer) A steam skidder used in early railroad logging; built by the Clyde Iron Works of Duluth, Minnesota and widely used in the Pine Country.
coal off To cut a forest clean for charcoal wood. (N.F.) LL
coal oil Kerosene. See saw oil.
coarse grain Widely spaced rings due to rapid growth.
coasting Said of logs moving rapidly down a steep pitch in a flume.
cobbled up Said of anything poorly put together. (P.N.F.)
codfisher A logger from a camp near the ocean or from a float camp. (P.N.F.)
coffee mill See corkscrew.
coffee pot 1. A small steam locomotive. 2. A small steam donkey.
coffin nails Cigarettes.
cog-wheel drive In 1883, John Dolbeer of the Dolbeer & Carson Lumber Company in Eureka, California invented a geared locomotive on which the drive gear could be shifted into neutral and the power applied to a front mounted drum for loading. Forward and reverse were obtained by reversing the engines. These locomotives, which were used successfully in the Redwood Country, would run up to the landing and throw the drivers into neutral, and the gypsy head on the pull-in winch worked like the donkey engine. (Red.) Also donkey locomotive. See gypsy locomotive.
cold camp Said of camping out overnight in the woods without blankets because of an accident. (P.C.F.)
cold deck 1. A pile of logs in storage as a reserve supply in the woods or at the mill. Also log deck. 2. To stack logs in a pile. 3. To fire or reprimand a person: "They cold-decked him." 4. To knock someone unconscious. (P.C.F.)

Cold decking logs beside the **mill pond** of the Oregon-American Lumber Company, at Vernonia, Oregon. A **skyline** is being used, with the **loading tongs** hanging from a **spreader bar**, which in turn is suspended by a **fall block**. *Harold M. Brown photo, Labbe collection*

cold-decker A gas-driven donkey that was used mostly to build cold decks. (P.N.F.)
cold-shut 1. A link for joining pieces of chain. 2. A tough fried cake or doughnut. (L.S.)

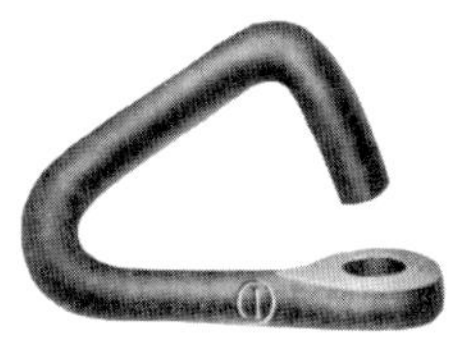

A **cold-shut**. *From Young Iron Works catalog, Glen Comstock collection*

cold trail To build a fire trail along the dead edge of a fire just to be sure. (P.C.F.)
collapse Irregular shrinkage in wood caused by collapse of wood cells as free water is drawn out of the cell cavities.
Columbia An early donkey made by Columbia Engineering Works, Portland, Oregon. (P.C.F.)
Columbian pine Douglas fir. (P.N.F.)
Columbia River market The log-marketing area served by the Columbia River and its tributaries as far upstream as logs could be towed economically to the mills down river. Independent loggers could find a buyer on the river. Today, logs are contracted before they hit the water, and the Columbia River market has almost disappeared. (P.N.F.)

Columbia River Rule Variations of the Spaulding Log Rule which was used in Washington and Oregon. The Scribner Decimal C is now the official rule of scaling in these two states. (P.N.F.)
Columbia River snipe A specific kind of bevel hewed on the end of a log for easier skidding. (P.N.F.)
Columbia River spruce Sitka spruce. (P.N.F.)
Columbia River yarder A long-haul Tacoma donkey; manufactured by the Puget Sound Iron & Steel Works. (P.N.F.)

A **Columbia River yarder**. *From Mallory Catalog No. 6, Glen Comstock collection*

comber A man who cruises the beaches in a small power boat looking for stray logs from broken booms. (P.N.F.)
combination engine An engine that is used both for yarding and roading.
combination spool A wide sheave used both in yarding and roading to guide the lines that are used.
comealong A levered device for applying a heavy pull by hand to a line by one man.
come-along carry-all 1. A grappling hook used by two men for dragging ties or timbers. (L.S.) 2. A yoke. (P.C.F.)
come-back line See haulback.
come-back road 1. Another road that empty trucks use to return to the woods when the outgoing road is too steep. 2. A return road used by cats yarding on very steep ground. (P.C.F.)
commercial thinning A thinning where all or part of the felled trees are removed for useful products and the value of these products is greater than the cost of the thinning operation.
commissary A general store for supplying lumbermen. (App.) (S.F.) See van.
commissary jimmie A clerk or store manager. (S.F.) (L.S.) LL
compaction soil Increased soil density resulting from the packing effect of machines moving over the soil or other causes. Compaction disturbs the soil structure and can cause decreased tree growth, increased water runoff and soil erosion.
company man A man who works for the interest of a company in order to get ahead.
company money Script printed by the company and generally accepted in the company town except by the U.S. post office. LL
company notch The position of the Johnson bar on a locomotive or donkey where the greatest efficiency was achieved.
company town A community owned by a lumber company.
company warden A fire warden on a logging operation, as opposed to a state or federal warden.
competition The struggle between trees or between trees and other vegetations for the limited nutrients, light and water present on a site. Competition can cause reduced tree growth. Severe competition in very dense stands may cause stand stagnation.
compound-geared yarder A yarder with gears between the drums that were "compounded," to increase pulling power. The donkey had power to haul in logs, and could return the butt rigging to the woods quickly. See Humboldt yarder.
compression wood Abnormal wood that may form on the lower side of branches and of leaning trunks of softwood trees. It shrinks excessively lengthwise as compared with normal wood.
condition powder See snoose.
conifers The botanical group of trees having needles, or scale-like leaves and cones. They are usually evergreen.
conk A wood-destroying fungus in a tree. See elephant ears.
conk inspector 1. Usually the bull buck who decides whether questionable trees should be cut. 2. The man who inspects logs for rot before they are loaded out.
conk knots Rotten knots that tend to spread conk inside a tree. Conk knots tell the logger or cruiser that the tree is defective before it is felled.
conky 1. Said of a tree that has conk. 2. Said of any poor situation.
connected engine A rod locomotive, not geared. Also direct-connected.
connected truck A railroad logging car consisting of two trucks connected by one heavy center beam or girder. See Seattle car and skeleton car.
connecting rope A wire rope that connects the heel tackle on the main cable to the anchor shackle on a skidder skyline.
conservation The protection, improvement and wise use of natural resources according to principles that will assure their highest economic and/or social utility.
contour road A road that is on a horizontal line, following the contour of the ground.
contract hauler An independent truck owner, working for a logging operator, who hauls logs from the woods to the dump; the truck owner may have drivers working for him.
contract logging Usually a small operator doing all or part of the logging for a large company. Also gyppo logging.
contractor A man who has a contract to log all or any part of a logging job.
contremaitre 1. A camp foreman. 2. A walking boss who had charge of several camps. (Quebec)
controlled burning The planned application of fire to natural fuels, including timber slash, with the intent to confine the fire to a predetermined area.
convertible An extra drum on a donkey that was used for a high lead or a skyline. (P.C.F.)
cook camp The building used as kitchen and dining room in a logging camp. LL
cookee A cook's helper and a dishwasher in a logging camp. LL
cookhouse The building where the food is prepared and which also includes the dining room. LL
cookhouse cost The usual loss caused by serving good meals below cost in order to keep the crew happy.
cookie pusher A hasher, same as a biscuit shooter. (P.C.F.)
cook shack 1. The kitchen. (West Virginia) 2. The company store. (Montana) 3. The cookhouse. Also hash house.
cook shanty See cook camp. LL
cooning One who plays smart tricks (smart as a raccoon). (P.N.F.)
coon log A log with a big hole. (P.N.F.)
coon shit A combination of decayed wood, dust and bugs that usually collects in a hollow within a tree and that causes a saw to bind. Water is used to soak the mixture and to allow the saw to draw. (P.C.F.)
coonskinner 1. The person who cuts logs by the thousand board feet. (N.W.) 2. A subcontractor who cuts timber. (West Virginia) See busheler.
coon tree A hollow tree. (P.C.F.)
Coos Bay logger A logger who wore his pants inside his boots. (P.N.F.)
Coos Bay splice Any crude splice. (P.N.F.)
Copenhagen Snuff, a very strong chewing tobacco. See snoose.
copper hammer A soft hammer used by machinists.

cord A unit of measurement for stacked wood. A standard cord equals a pile of stacked wood containing 128 cubic feet within outside surfaces. Standard dimensions are 4'x4'x8'.
corduroy, corduroy road 1. A road made by placing small logs across the way of travel. See puncheon. 2. A gravel road that has been waved by use. 3. To build a corduroy road.
cordwood splitter A steam-operated heavy wedge that resembles a huge ax and used for breaking up chunks of fuel wood for donkeys and lokies. Also wood splitter. (P.C.F.)
core The center of wire rope that may be hemp or wire.
cork pine The fine old growth Idaho white pine. The term is from the lumberjacks of the Lake States who meant top quality eastern white pine. (N.F.) (P.C.F.)
corks The popular term for sharp spikes in the soles of loggers' shoes to prevent slipping. Also calks, caulks.
corkscrew 1. A geared locomotive, usually a Shay. (P.C.F.) See screwdriver, sidewinder, stemwinder. 2. See wind.
cork shoes Leather boots fitted with corks.
cornbinder 1. Any piece of woods equipment made by International Harvester. 2. A railroad logging truck equipped with cornerbind chains to hold logs on the truck.
corned Said of a drunk person. (Maine)
corner 1. The place where the haulback line is turned sharply. 2. A section corner which is the basic point in a land survey. 3. A line formed by the intersection of any two surfaces of a piece of lumber.
corner binds Four stout chains, used on logging sleds to bind the two outside logs of the lower tier to the bunks, thus giving a firm bottom to the load. (N.F.) (Canada) LL
corner block A block at the back of the haulback line. Also tail block.

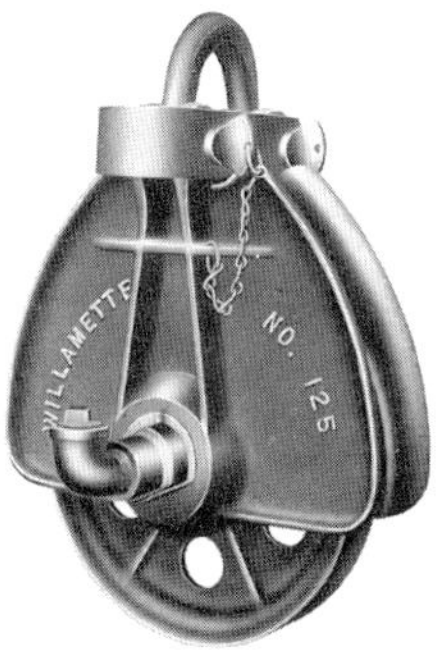

A **corner block**. *Willamette Iron & Steel Co., catalog, 1920, Glen Comstock collection*

corner man The man who notches the logs so that they will fit closely and make a square corner in building a camp or a barn of logs. (N.F.) LL
corner post See corner.
corner up To cut up to the corners of the undercut before the center of the tree is cut, in felling timber. (P.C.F.)
corridor skidding A logging procedure using skyline skidders in which narrow clearcuts are made through a stand. Cables are strung in the clearcut corridors to transport logs from the woods to the landing. Between corridors only a portion of the trees in the stand are removed, and these harvested trees are laterally skidded to the transport cables in the corridor.
corset Irons placed around a crack in a spar tree and tightly cinched with wire rope. (P.C.F.)
Cosmopolis flush A four-card flush in poker; a busted flush. The term originated in the lumber town of Cosmopolis, Washington. (P.N.F.)
cotton picker 1. A mill worker or logger from the Southern states. 2. A dark complectioned logger. (P.C.F.)
cougar den A bunkhouse.
cougar juice See coal oil.
cougar milk Any kind of home brew liquor.
couldn't fire a cookstove Said of a poor fireman. (P.C.F.)
couldn't run a sewing machine Said of a poor donkey puncher. (P.N.F.)
counterbalance system A logging incline railroad laid up the face of a mountain on a very steep grade. It might be either double tracked or with a passing track in the middle so that loaded cars coming down would counterbalance empties coming up. This is opposed to the single track incline where a lowering engine controlled the entire movement of both loads and empties. (P.C.F.)

The Porter-Carsten **incline** at Estacada, Oregon used the **counterbalance system**. The weight of loaded cars descending would hoist empties back up the hill. Travel was controlled by a large **gypsy lowering engine** at the top of the hill. *Labbe collection*

counter-jumper A clerk. LL
country An area, such as the Redwood Country.
count ties 1. To quit. 2. To be fired (in order to leave camp one had to walk down the railroad). 3. To walk to or from camp. (P.C.F.)
coupled onto a turn too big to handle Said of a logger who married an overbearing woman or one much bigger than himself. (P.N.F.)
coupling dog A short chain with a dog on each end to fasten logs end-to-end in the early days of skidding with animals. See single coupler.
coupling grab 1. See coupling dog. 2. A short tong on each end of a small chain and used also for hooking a couple of logs together.
coupling-up man Usually the logger who uncoupled dogged logs coming in to a landing down a fore-and-aft road. (P.C.F.)
cover 1. The type of timber or other growth on an area. 2. A hiding place. 3. To play a dirty trick on a choker setter. (P.N.F.)
cow Milk. (West Virginia) LL
cowcumbers Pickles. (P.N.F.)
cowhorn cedar A northern white cedar of low quality characterized by its curved and flared butt.

cow's mouth A notch axed into a tree to fell it. (Canada)
coyote hole A horizontal hole dug for blasting. (P.N.F.)
crab A small raft bearing a windlass and anchor, and used to move log rafts upstream or across a lake. (N.F.) (S.F.) LL
crabs Body lice.
crack-a-jack A light Willamette yarder with a large bull wheel driving two narrow drums side by side on one shaft, instead of one ahead of the other as on a standard yarder. (P.N.F.)

A **crack-a-jack yarder**. *From Willamette Iron & Steel Co. catalog, 1920, Glen Comstock collection*

crack the throttle To open the throttle slightly.
cradle 1. The wooden framework or large metal brackets to hold small chunks, poles, pieces of pulpwood, being bundled for transport. 2. The frame in which an ocean-going log raft is built. See Benson raft, Robertson raft. 3. A frame in which shingle bolts are towed.
cradle knolls 1. Small knolls which require grading in the construction of logging roads. (N.F.) 2. Small knolls that must be avoided in pointing a tree for felling. (P.C.F.)
crane 1. A machine for lifting heavy loads or machinery. If rigged to lift logs along a railroad, it is called a cherry picker. 2. A horizontal beam that travels in a building or over a mill or factory yard to lift timbers or stack and unstack units of lumber.
crane hooker The head man on a crane loading crew.
crane talleyman A man who tells the crane operator where the lumber is in the yard and keeps ticket files up-to-date. A ticket is attached to each unit of lumber. (P.C.F.)
crank jack A jack operated by turning a crank.
crawdad A lift loader mounted on a cat that looks like a crawfish waving a pincher in the air. (P.C.F.) See grapple.
crawler 1. A tractor that moves on endless tracks. 2. An arch that also operates on endless treads. 3. A power shovel on tracks.
crawlers The endless treads of cats, arches or loaders.
crazy as a shithouse mouse Said of a very eccentric person.
crazy chain The short chain used to hold up the spout of a water tank that is not in use. (N.F.)
crazy wheel See Barienger snubbing device.
cream 1. The best of anything. 2. To high grade a stand, taking only the best logs. (P.C.F.)
creamy show A good logging show. (P.C.F.)
crease 1. When building rafts, a notch cut into logs to lay a swifter,

The **crank jack**, or **screw jack** was the early logger's primary power tool. Here a log is being rolled from **disconnected trucks** with the aid of the jack, while a second man applies a **sampson** at his end. *From Steven R. Gatke collection*

a cross timber that joins the boom sticks on either side of a raft. 2. A ring cut around the upper part of a tree, around which a belt containing dynamite is placed to blow off the top of the tree, thus saving cutting.
creep The tendency of machines to move forward or backward under constant heavy use.
crew car See crummy.
crew truck A truck with seats and sometimes a stove to carry men to and from work in the woods. (P.C.F.) See crummy and mulligan.
crib 1. A square structure of small timbers that is used to hold pieces of wood, poles or bolts being towed from camp to a mill. 2. A big timbered form in which ocean-going rafts are built. 3. The pier or bridge made of cribbed logs and usually filled with rock. 4. A raft of logs. (N.F.) 5. A whore house. 6. A frame, or heavy crate, designed to hold lumber tightly in place. A crib was built on a slanting platform or on rollers resting on the shore of some stream where it could be readily slid into the water. It was 32 feet in length and 16 feet in width. (L.S.)
cribbed-up Said of anything that is braced, supported or shored-up with timbers.
cribber A horse that gnaws his feeding crib, or any other object made of wood. (N.F.) LL
crib bridge A bridge built by criss-crossing logs instead of using piling.

A **crib bridge** on the Columbia & Nehalem Valley Railroad of the Peninsula Lumber Company near Columbia City, Oregon. Such bridges were common in the days when logs and labor were cheap. Builders were careful not to use any metal in the construction of these bridges, because when the railroads were removed, the bridges were taken up and hauled to the mill to be sawn into lumber. *Courtesy of Mrs. Burns*

crib dam A dam built with cribs of logs, filled with stones, and planked on the upstream face. LL
crib logs To surround floating logs with a boom and draw them by a windlass on a raft for towing with a steam boat or tug. (N.F.) LL
cribwork The various structures built of criss-crossed logs, such as trestles or retaining walls.
crimp 1. A kink in a wire rope. 2. An employment agent. (P.C.F.) 3. To squeeze the end of a blasting cap on a fuse.
crimpy Said of cold weather. (P.N.F.)
cripple 1. A leaning tree. 2. An unsafe fastening. 3. Any broken or damaged tool. (P.C.F.)
critter 1. A machine. 2. A no-good man. 3. A domestic and harmless animal. (P.C.F.)
crock A bottle of whiskey. (P.C.F.)
crooked elbow A disease enjoyed by thirsty loggers hitting town after a long dry spell. (P.C.F.)
crooked steel A cant hook. (P.N.F.)
crop In turpentine orcharding, a working unit comprising the number of cups or boxes worked by a chipper. A crop may consist of any number of cups or boxes, ranging from 6,000 to 10,500. (S.F.)
cross as a badger Said of one who is very ill-tempered. (N.F.)
cross chains Chains that crossed in the form of an "X" and connected the front and rear sleds of a logging sled. Also lead chains. (N.F.) LL
crosscut, crosscut saw 1. The long saw used in falling and bucking by hand. A twelve-foot crosscut was used to buck the huge redwoods. (P.C.F.) Also harp, misery harp, misery whip, swede fiddle and wood harp. 2. To work as a bucker: "He's crosscuttin'." (P.C.F.)

Two types of **crosscut saws**. Above: **bucking saw**; below: **falling saw**. *From Kloepfer Hardware Co. catalog, Glen Comstock collection*

crosscutter A logger who bucks logs. See bucker.
cross-ditch A shallow channel laid diagonally across the surface of a road so as to lead water off the road and prevent soil erosion. Also water-bar.
cross girt A cross timber used in building a donkey sled.
crosshaul 1. A special road built for a team to work on to pull the cable or chain used while loading logs with a jammer or gin pole. 2. The cleared space in which a team moves in crosshauling. 3. An imaginary article used in logging that a greenhorn is often sent to the boss to get one. (Maine) 4. To parbuckle, or to load a log by passing a line over and under, then pulling on it to roll the log up a set of skids. 5. To pull at right angles to the main haul. 6. To work at cross purposes. 7. To get a payload on a return trip. 8. To load cars or sleds with logs by horsepower and crotch or loading chains.
cross link a chain To tighten up a swifter by pulling the boom chain through the hole, then turning a link of the chain crossways to plug the chain in the hole.
crossover 1. A railroad crossing. 2. The specific spot where the wire rope changes from the bottom layer on a drum to begin a new layer on top. 3. A railroad switch that allows a train to cross over from one track to another.
crossplank road A logging road made by laying short heavy planks on small timbers crossways of a road, on pole or timber stringers.
cross-rib A skid laid in the skidroad.
cross skids The skids or small poles placed across the road to keep the logs from digging into the dirt in skidding with oxen or horses. A shallow groove was cut in the middle of the skids to guide the logs.
crotch 1. A fork in a tree. 2. Any type of fork. 3. A single sled used in dragging logs. (N.F.) See go-devil. 4. To cut notches on opposite sides of a log near the end where the dogs are fastened.

crotch chain A tackle for loading logs on sleds, cars or skidways by crosshauling. (S.F.)
crotch dogs, In the bull team days, two dogs connected by a short link to attach the skidding line to the logs. Similar to crotch grabs.
crotch grabs Two grab hooks connected by short chains to one ring, thus forming a "V". (West Virginia) Also yarding grabs. (P.C.F.) LL

Crotch grabs. *From Mill and Mine Supply Co. catalog, 1916, Glen Comstock collection*

crotch line loading A power system of loading logs in which two tongs are hung from a spreader bar that is controlled by wire ropes that are attached to fall blocks on the loading and haulback lines. The spreader bar is swung back and forth between two spar trees and can be raised or lowered. Sometimes only one tree was used, in which case one block was fastened to a guy line. (P.C.F.)

An unidentified scene showing the use of **crotch line loading**. In this case, **end hooks** are used instead of **loading tongs**. Note how the left **fall block** is hung from a **carriage** riding on the **skyline**. Also note the **fairleads** on the **donkey** at lower right, used to direct the loading and **car lines**. *Mallory Catalog No. 12, 1925, Glen Comstock collection*

crotch tongue Two pieces of wood, in the form of a "V", joining the front and rear sleds of a logging sleigh. (N.F.) LL
crow bait 1. A poor logging team. 2. A worn out horse. (P.C.F.)
crowd 1. Said of a top log which forces a lower log outward, off the truck or car. See shit out. 2. To force the dipper stick of a power shovel out to the last notch of the rack on the boom.
crow-foot 1. A mark occasionally used as a log brand or cruiser's

mark. 2. A log that contains crow-foot or figured grain that resembles the tracks of a crow's foot, specifically in Western hemlock. 3. Cracks extending out from the center of a log.
crown 1. The living branches of a tree. 2. Said of a fire that jumps from the ground up into the branches of a tree.
crown class A designation of trees in a forest with crowns of similar development and occupying similar positions in the crown cover.
crown fire A fire which runs through the tops of living trees or brush.
crown grant In British Columbia, the ownership of the surface rights to land; used often to distinguish between actual ownership and a license to cut timber.
crown sheet A flapjack full of bubbles that resembles a crown sheet of a steam engine (the sheet in the firebox which holds the flues). (P.N.F.)
crown timber Timber owned by the province in British Columbia, Canada.
crown wires The outer wires on a strand of wire rope.
crow's nest 1. A thick growth of twigs or limbs on a tree. (P.C.F.) 2. A falling tree that lodges against another. (L.S.)
crude turpentine The resinous substance that exudes from the wounds of certain pine trees, and from which turpentine and rosin are secured by distillation. Also dip, gum, pitch, soft gum. (S.F.)
cruise 1. A timber inventory. 2. To estimate the quantity and quality of standing timber in a given area.
cruiser 1. A person who makes a timber cruise. Also estimator, landlooker, valuer. 2. A logger in town looking for trouble. (P.C.F.) 3. Lightweight logger's shoes.
cruiser coat A short wool coat, shower repellant, with many pockets and a double back, that is very popular in the Pacific Northwest woods.
cruiser's ax A lightweight double-bitted ax. See Hudson's Bay ax.
cruiser's marks The personal trade marks used by cruisers to mark up witness trees (trees scribed and recorded in a survey to mark the location of a section corner) that they have found. Sometimes the cruisers established the corners.
cruiser vest A canvas vest with a double back and many pockets.
crumb boss 1. The bedmaker in a lumber camp. 2. A kind of bull cook. 3. A conductor on a log train. (P.N.F.)
crumb chaser A cook's helper.
crumbs Lice. LL
crummy 1. A railroad car, station wagon or bus that transports men to and from work in the woods. Originally a transportation car on a railroad, so-called because of its usual condition. (P.C.F.) Also candy wagon, crew car, dog house, mulligan. 2. A shirt that does not belong to the person wearing it. (P.N.F.) 3. The man who runs the bunkhouse. (S.F.)

Crummies of the Oregon-American Lumber Company await their cargoes at Camp McDonald in the Coast Range of Northwest Oregon. These cars were designed to carry the men into the woods in the morning and back to camp in the evening. The racks on the sides and beneath the floor were for carrying tools, such as saws and axes. *Labbe photo*

cubic scale An estimate of the cubic foot volume of wood fiber in a tree, log or product.
culch Refuse. (Maine)
cull 1. A tree or log of merchantable size rendered unmerchantable because of poor form, an excess of limbs, rot or other defect. 2. A poor piece of lumber. 3. A worthless piece of machinery or logging equipment. 4. To reject a log. 5. To reject any unwanted material.
culler The man who graded and sorted lumber and who picked out the culls or rejects. (N.F.) (S.F.) LL
cull or peck A grade of cypress comprising pieces below No. 2 common and also those cut from the "pecky" part of a log. (S.F.)
culls Hash in the cookhouse. (P.N.F.)
cultus Said of something no good or bad; from the Chinook jargon of the Pacific Northwest and used widely. (P.C.F.)
cultus lumber Lumber of poor quality. (P.C.F.)
cumshaw A Chinook jargon word meaning a handout.
cunt hair A unit of measurement used facetiously in the woods and in the lumber mills. Also frog hair. See red one, pink one.
cunt splice 1. A makeshift splice for temporary use; formed by separating the strands of a cable into two braids which were wrapped with similar braids on the second piece of cable. 2. A wild eye or temporary loop in a cable.
cup 1. A curve in a piece across the grain or width of a piece in manufacturing. 2. A galvanized sheet-iron or clay receptacle placed on a tree in which crude turpentine is caught. (S.F.).
cup hook, cup point An end hook with a center spike set in a cup, used in loading smaller or medium sized logs; required in later years by the Accident Commission. Also grip hook, Martin grip hook, saucer hook, spiked bell hook. See end hooks.

A cup hook. *From Young Iron Works catalog, Glen Comstock collection*

cup'n saucer The fifth wheel on a logging truck. (P.F.C.)
cupola 1. The lookout hatch on a caboose or speeder. 2. The ventilator over a cookhouse or engine house.
cupping ax In turpentine orcharding, a large broadax with a concave cutting edge, used in cutting an incision at the base of the face into which the apron is to be inserted. (S.F.)
current slash The slash made in the current year in a logging operation, after the fall slash report is made up. Current slash is not burned nor released by the state forester. The logger is held responsible if a fire starts. (P.F.C.)
cut 1. In the woods, the volume of logs produced each day, month, year, etc. Also output. 2. A truck road or railroad excavation through a rise of ground. 3. A string of cars in a switching position. 4. The wear on the sheave of a logging block caused by the wire rope running through it.
cut a box To cut a box on a pine tree that is to be bled for crude turpentine. (S.F.)
cut a forked stick To open the throttle wide. The expression originated from the old trick of placing a forked stick under the throttle of a steam donkey. (P.N.F.)
cut a log To slide one end of a log forward or backward so that the log will roll in the desired direction. (L.S.)
cut'er To be able or not able to handle a job; "He can cut'er;" "He can't cut'er."
cut-off An artificial channel by which the course of a stream was straightened to facilitate log driving. (N.F.)
cutover A logged over area of land, usually clearcut.
cutover stand A stand from which part or all of the merchantable timber has been harvested.

cutter 1. A logger or lumberjack who cuts down trees. (N.F.) 2. A busheler or gyppo (S.F.) 3. A three-cornered instrument used in sharpening hacks and pullers (S.F.)
cutter teeth The cutting teeth on a crosscut saw.
cut the mustard To be able or not be able to handle a job; "He can cut the mustard;" "He can't cut the mustard."
cut the throttle To coast, to take it easy.
cutting bar The bar that carries the cutting chain on a power saw.
cutting crew The men who fell timber on a logging show.
cutting end The cutting end of a chain saw.
cutting line 1. The boundary of the felling areas that the bull buck marks off for the cutting crews. 2. The boundary of a logging area where it leaves off against the marginal timber.

Unidentified **cutters**, probably in the Lake States at the turn of the century. *From Steven R. Gatke collection*

D

D In the late 1800s, a D-shaped iron used in the butt rigging connections. (P.C.F.)
D.B.H. The diameter breast high: four and a half feet above the ground.
D.I.B. The log diameter inside the bark.
D.O.B. The log diameter outside the bark.
dabster A versatile person. (Maine)
dangle To travel or move quickly. (P.N.F.)
dap 1. A notch cut in a timber. 2. To cut a notch in a timber so that another timber will fit into it.
dashboards The farmers who work part-time in the woods, so-called because of the bib overalls which they wear. (P.N.F.)
daub See mud.
Davenport A standard rod engine built both with saddle tanks and

The **Davenport** locomotive, while of high quality, never proved popular with the loggers. Here the 3-spot of Columbia Timber Company is working the woods back of Goble, Oregon, on the Lower Columbia. *Ira Withrow collection*

tender and used on mainline logging roads. This fine-quality locomotive, which was built by Davenport, at Davenport, Iowa, never became popular in the woods.
Davis raft An ocean-going log raft which was first built by Gilbert G. Davis at Port Renfrew, British Columbia, and who later moved to Portland, Oregon. In 1923 Davis and the Multnomah Lumber & Box Company formed the Davis Ocean Towing & Rafting Company. Although of good design, the Davis raft did not have the capacity of the Robertson raft. See ocean-going raft.

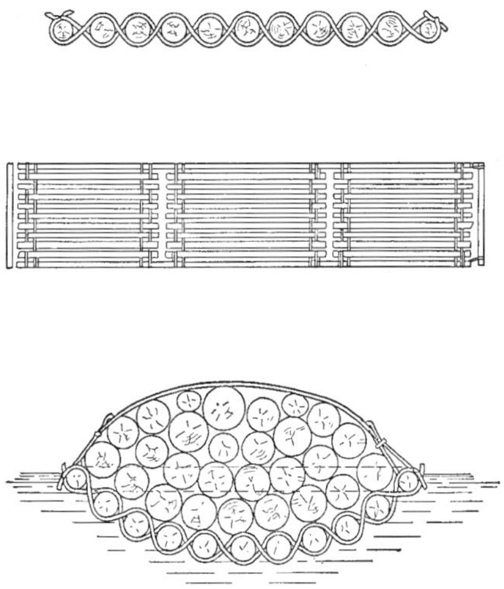

Construction details of the **Davis raft**. *From "Logging," R. Bryant, Wiley & Son, 1923, Philip Schnell collection*

day faller A man who fells timber by the day for a daily wage, not working on a piece-rate basis.
day in the hay 1. An easy day's work. 2. A rainy day. 3. A day off.
day late (a) An expression that was used around the employment offices when a job was filled or no work was available. (P.C.F.)
daylight in the swamp 1. Timber that is being cut down lets light into the forest. A popular expression in the Fir Country of Oregon and Washington and in the Lake States. 2. Time to get up.
daylight robber A log scaler. (L.S.)
day man A laborer who received a designated daily remuneration. (N.F.) LL
day work See day faller.
deacon's seat 1. Usually the bench in front of a row of bunks in the 19th century shanties. 2. The seats in the middle of a room around a big stove.
dead and down Standing trees and trees on the ground either dead or living. (L.S.)
dead bunk In early truck logging with single axle trailers, a bunk placed behind the swinging bunk, used to keep the reach from breaking when the brakes were applied. Also false bunk.

Labbe Logging Company truck on the bay front at Newport, Oregon, in 1935. In the days of single axle trailers the **dead bunk**, or **false bunk**, behind the swinging bunk was necessary to keep the **reach** from breaking when the brakes were applied. The trailer could move back and forth along the reach to accomodate logs of various lengths. *Labbe photo*

deadener A heavy log or timber with spikes set in the butt end that was fastened in a log slide so that logs passing under it came in contact with the spikes and would retard their speed. LL
deadening An area where the timber has been killed by fire, flooding, insects or disease.
deader'n hay Said of a man lacking energy. (Maine)
deadhead 1. A sinker, a log under water except for one end, which barely floats. Also bobber. 2. To be transferred, as "The railroad crew was deadheaded to Camp 20 to pick up the train." 3. Said of a non-revenue trip, such as a returning empty truck or taking a locomotive from one point to another without a train.
dead horse A useless job. (L.S.) LL
deadman 1. A short log buried securely in the ground and used as an anchor for guy lines or to anchor a cable. 2. A fallen tree on the shore, or a timber to which the hawser of a boom is attached. (N.F.)
deadwater A spot in a river where there is no current. Also stillwater.
death warrant A hospital ticket. (L.S.) LL
decay Disintegration of wood due to the action of wood-destroying fungi. Also dode, dote, rot.
deciduous Said of perennial plants that annually lose all of their leaves, usually in autumn.
deck 1. Any flat surface, as the deck of a flatcar. 2. An abbreviation for cold deck. 3. To store logs by piling them.
decker 1. In the 1800s and early 1900s, a man who built piles of logs for loading or for rolling into a river or bay. 2. The man who rolls logs upon a skidway or log deck. 3. A type of railroad log loader built by the Clyde Iron Works of Duluth, Minnesota. Similar to the McGiffert loader, except that rail ramps were used to bring log cars up into the "tunnel." See McGiffert.

The **decker** was manufactured by the Clyde Iron Works of Duluth, Minnesota. This one is shown loading short logs somewhere in Mississippi. *From Clyde Iron Works catalog, 1912, Larry Nyce collection*

decking chain See loading chain.
decking hook A light peavey used by a top loader. (App.)
deck-up To pile logs upon a skidway.
deduct 1. The amount deducted from a log scale because of a defect. 2. Deduction from a paycheck.
deer foot 1. An iron claw that holds the chock chain in the opposite end of a log truck bunk. 2. The end of a single-bit ax handle.
deer foot hook A double pointed loading hook.
defect The split, conk, decay or other injury that decreases the amount of wood that may be obtained from a log.
defective Said of timber that is partly rotten.
defect percent Volume of defect expressed as a percentage of the gross volume.
deflecting block A small block used to deflect the heel block line away from other lines around the spar tree in skidder logging.

deflection 1. The amount of sag in a wire rope below a straight line. 2. The sag in the ground which permits deflection.
defoliators 1. Insects that destroy foliage. 2. Chemicals which cause plants to drop their leaves.
degrade Any defect that lowers the grade of a log or piece.
dehorn 1. Any kind of drinking liquor. (P.N.F.) 2. To knock limbs off logs. 3. To water down bad alcohol. (P.C.F.) 4. To saw off the end of a log to be stolen by removing the brand. (Kentucky)
dehorners Alcoholics. (P.N.F.)
delimber A machine that removes the limbs from felled trees.
Dennison lock hook In the late 1800s and early 1900s, a line hook that could be locked to keep in the cable.
density rule An authorized and approved set of specifications of the Southern Pine Association under which southern yellow-pine timbers are graded.
dentist In the early days, the sawfiler of crosscut saws who fixed the teeth. (P.C.F.)
den up To quit active work because of stormy weather or during the winter.
depending sheave On a skidder carriage, a sheave through which runs the skidding line.
depot The headquarters or supply house for a logging operation. (N.F.)
depot camp 1. A logging camp comprised of several buildings that are to be used for more than one year. (Eastern Canada) 2. A central storage supply location.
derail 1. Usually a 90-day grace period used by the railroad companies to determine the qualifications and background of recently hired employees. (P.C.F.) 2. A device or switch designed to derail a runaway item of equipment before it can cause an accident.
derailer Any obstacle on the railroad.
derrick The loading boom on rigs that loaded logs on railroad cars.
detached truck See disconnected truck.
deuce The number two locomotive. Also two-spot.
devil dust False smoke in the forest that causes a fire alarm. (P.C.F.)
devil's club 1. Specifically a stout spiny plant found in the damp West Coast woods. 2. Any spiny object, such as a frayed piece of wire rope. (P.C.F.)
dhobie tongs Skidding tongs used with bummers. (S.F.)
diameter breast high See D.B.H.
diameter limit The minimum diameter of trees to be cut, as in a timber sale contract.
diameter tape A tape measure specially graduated so that the diameter may be read directly from the tape when it is placed around a tree stem or log.
diamond 1. A diamond-shaped steel plate from which rigging was hung in some loading systems. (P.C.F.) 2. See butt plate.
diamond hitch 1. One of the most popular and important hitches used to load packhorses. 2. A disparaging term to describe the mess which a green man makes in tying up a bedroll.
diamond lead The area yarded from a point beyond right angles (square lead) to the donkey and making a sharp angle back toward the donkey.
diamond line A short section of wire rope used in a crotch line loading system.
diamond stack A big diamond-shaped spark arrestor on the stack of a locomotive. See balloon stack.
dibble A spade-like tool used to prepare planting holes for seedlings.
didapper A log pond worker. (S.F.)
diesel loggers Huge diesel trucks that carry logs only on private logging roads from the woods to the log dump.
dinging A loader riding with one foot in the loading tong, hanging on to the loading line, in violation of the safety code. (P.N.F.)
dingle, dingle room 1. A term for the roofed-over space between the kitchen and sleeping quarters of an old-styled logging camp, commonly used as a storeroom. See alley. (N.F.) 2. The shed-like structure for storing food supplies. LL
dingmaul A mythical animal in a lumber camp. (Maine)
dinkey 1. A small locomotive. 2. A small cat.
dinner hole The location where the noon meal was served out in the woods. (L.S.) LL
dinner horn In the early camps, a long tin horn to announce that dinner was ready. Later the gut hammer beaten on an iron triangle was used. Also gabriel.
dinner triangle The heavy iron rod bent into an unclosed triangle that was rung to call the lumberjacks and loggers to eat. (N.F.) (P.C.F.) See gut hammer.
dip See crude turpentine.
dip barrel In turpentine orcharding, a tight-coopered barrel with a detachable head, in which crude gum from dip buckets is placed for transportation to the still. These barrels hold about 550 pounds gross weight. (S.F.)
dip bucket In turpentine orcharding, a wooden vessel holding about five gallons, in which dip is placed when taken from cups or boxes. (S.F.)
dip iron See dipper.
dipper 1. The bucket on the steam or gas shovel. Also dip iron. 2. A man who shovels in much food. (P.C.F.) 3. One who collects crude turpentine from boxes or cups. (S.F.) 4. The tool or paddle with which the workman removes the crude turpentine from the box or cup. (S.F.)
dipper lip The tooth edge of the bucket on a power shovel.
dipper stick The movable arm that holds the bucket on a power shovel.

A steam **shovel** building **grade** somewhere near Astoria, Oregon, shows off its **dipper stick**. *Brown photo, from Steven R. Gatke collection*

direct-connected 1. Said of a yarder with both the main and haulback drums driven directly from the crankshaft without any intermediate gears. Also direct-geared yarder. 2. Said of a steam locomotive with pistons directly connected by rods to the drivers. (P.C.F.)
direct-geared A rod locomotive, as opposed to the real geared engines.
direct-geared yarder See direct-connected.
direct seeding Spreading seed over the forest seedbed by hand or machine; a practice used to assist or supplant natural seed fall and achieve regeneration.
direct-yard To yard with one donkey directly to the landing. (P.C.F.)
dirt daubers Grade builders (in building a railroad route) in railroad work.
dirt-hider A road grader. (P.C.F.) LL
dirtproof Said of anything that is very good or very tight. (P.C.F.)
dirt work The work of grading a right-of-way.
dirty blue dick A euphemism for swearing. (P.C.F.)
dirty sidewinder A person who is not to be trusted.
disconnected truck A four-wheel railroad truck carrying sharp-faced bunks. Logs laid across the two trucks made up a car to be

coupled into a train and the weight of logs spanned between the truck was sufficient to keep the train together. Also detached truck.

A pair of **disconnected trucks** at the Crown-Zellerbach Corporation camp near Cathlamet, Washington. The trucks are connected with a **rooster**, thereby forming a car that was not in danger of pulling apart. A temporary deck across the bunks could convert it to a flat car for special service, if desired. Note the swinging **bunks** and the **cheese blocks** to retain the logs. Each set was so designed that the footboards were at either end of the car and the brakewheels always on the same side of the train. *Labbe photo*

dish-face Anything concave or which has a surface hollow in it.
district 1. Pacific Northwest logging camps were at one time classified by districts in which there were some differences in pay scales. The main districts were Columbia River, Grays Harbor, Puget Sound and Oregon Coast. 2. A region over which local unions have jurisdiction. 3. A forest region, such as a ranger district or fire protection district. (P.N.F.)
divide A ridge between two drainages.
D link A link shaped like the letter "D" in the end of the trace chain which hooks to the whiffletree. (Maine)
dock 1. A raised deck of any type. 2. A loading platform. Also ramp. 3. To shorten or to take something away as "to dock his paycheck."
doctor balsam The balsam pitch that lumberjacks often used as a medicine for colds and sore throats or as a salve on cuts or cracks. (L.S.) LL
dode, dote See decay.
dog 1. A spike with an eye through which a line is passed to tie logs together. 2. A stop, as the pawl of a ratchet wheel. 3. A sharp edged metal plate on the tree end of a springboard to give a bite into the wood and to provide a safe hold for the board. 4. A short, heavy piece of steel, bent and pointed at one end with an eye or ring at the other. It is used for many purposes in logging, and is sometimes so shaped that a blow directly against the line of draft will loosen it. 5. A poor piece of equipment.
dog bed A mackinaw or three-quarter length jacket. (West Virginia)
dog boat See pig.
dog chain A chain used to hold the logs together in the log pond.
dog driver The man who drove dogs into logs so they could be hauled back into deep water.
dog'er 1. To put chocks under the wheels of a locomotive. 2. An order to stop.
dogger 1. The man who drove in the dogs for the dog tripper to knock out. 2. The man who drove dogs into logs to make up a turn ready to be hauled down the skidway by animals. 3. The man who pried dogs out of logs when they came to the landing in the animal-skidding days. (S.F.) (P.C.F.)
dogging jaws See logging dog.
doggin' it Said of a man not working at a job.
dog-haired Said of a dense stand of saplings or second growth, usually too young to log. (P.C.F.)
doghole 1. A contemptuous maritime term for the dangerous shallow bays or harbors, specifically along the Mendocino Coast in Northern California. From the 1850s to the early 1900s, small lumber schooners came along the coast to take on cargo dropped to the ships by means of wooden chutes or wire rope rigging in "a hole barely large enough for a dog to turn around in." 2. In river driving, a break or gap in a boomstick. (N.F.)

A **doghole** on California's Redwood Coast. Slings of redwood lumber are lowered down to the decks of Caspar III, a steel-hulled steam schooner belonging to the Caspar Lumber Company, of Caspar, California. *National Maritime Museum*

dog hooks 1. A short coupling chain with hooks, used to tie logs together when skidding with animals. 2. The strong hook on the end of a dog warp. (N.F.) 3. A hook on the end of a haul-up chain of a size to permit its being hooked into a link of the chain when it is looped around a log or other object. (P.C.F.)
dog house 1. A caboose or crew car on a logging train. See crummy. (P.N.F.) 2. A small enclosure on the tender of many logging locomotives where the head brakeman could sit and keep an eye on the train.

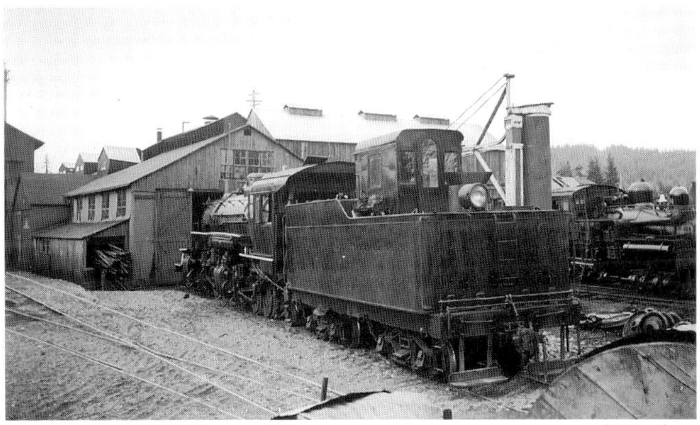

A **dog house** on the tender of Rayonier, Inc., locomotive 120, a 2-6-6-2 **Mallet** from **Baldwin**. *Labbe photo*

dog iron 1. An iron rod with a right angle turned at each end and sharpened. The pointed ends were driven into one edge of a block being sawed off a log and then into the log itself, to steady the block when sawing. 2. Any one of the various kinds of dogs.
dog knocker A light hammer or mallet used to force dogs out of logs.
dog knot 1. A tightened knot in a line or cable. 2. Any unusual bulge.
dog leg A bend in a cable caused by a heavy strain on that part of the cable when kinked.
dog leg crane A log-loading rig that has a bent or gooseneck boom.
dog line Any line that is used to hold fast to dogs in booming logs.
dog robber A camp cook who fed the scraps to the loggers, leaving nothing to feed the dog. (P.C.F.)
dog room A lounging room in a logging camp. (N.W.)
dogs See skidding tongs.
dog tripper The man who knocked dogs out of logs on the landing, on the river, or in the log pond.
dog warp A rope with a strong hook on the end, that is used in

breaking dangerous jams on falls and rapids and in moving logs from other difficult positions. Also hand dog. (N.F.) LL

dog wedge An iron wedge with a ring in the butt, that is driven into the end of a log and a chain hitched in the ring for skidding the log by horsepower. Also used in gathering up logs on a drive by running a rope through the rings and pulling a number of logs at a time through marshes or partially submerged meadows to the channel. (N.F.) LL

doing The condition of the roads for sleighing. (Maine)

Dolbeer The original donkey, the Dolbeer Steam Logging Donkey, was invented by John Dolbeer of the Dolbeer and Carson Lumber Company of Eureka, California and first tried out in the Salmon Creek operations in July of 1881. The Dolbeer Steam Logging Donkey was patented on April 18, 1882. The donkey had an upright boiler with a horizontal single cylinder with a gypsy head on a horizontal shaft. Manila rope was first used to haul in logs, and the line was taken to the woods by a line horse. The first Dolbeers were used only to yard 150 to 200 feet, to bring logs to the edge of the skidroad where the bulls could reach them easily. Donkeys were used in the Pacific Northwest as early as 1888. The rig was also known as a dolberry and dolby.

Bridal Veil Lumbering Company, just east of Portland, Oregon, was quick to accept the donkey engine when it appeared. Shown here is one of the original **Dolbeer** donkeys, which made use of a horizontal shaft with **gypsy drums** on either end. Note the **lead block** that kept the line in line with the drum, while the **spool tender** made sure the line was coiled properly beside the donkey. Just behind the log is the **line horse**, waiting to take the **butt rigging** back into the woods for the next **turn**. *D.L. Stearns collection*

The second version of the **Dolbeer** donkey proved very popular in the Redwood Country where it was developed, as well as in the woods of the Northwest. The vertical **capstan** would lead very well in any direction, and the **fairlead** could swing in an arc around a frame at the base of the drum. In this view taken in Humboldt County, California, in the late 1880s, the loggers are still using manila rope for the mainline. *A.W. Erickson photo, Carranco collection*

Dolbeer locomotive The Dolbeer locomotive was the most popular of the gypsy locomotives. The shaft driving the gypsy drums, or niggerheads, was geared directly to the engines, while a sliding pinion gear on the axle could be shifted into neutral while the gypsy was in use. Invented by John Dolbeer of the Dolbeer & Carson Lumber Company. See gypsy locomotive.

dollar a day The pay rate for ordinary labor in camps in the 1890s. (L.S.) LL

dollar hole The compound low gear on a truck. See big hole.

dollar short (a) Said of a logger who did not have the fee to pay for finding out which camp needed men. In the late 1800s and early 1900s, a logger who was looking for work was forced to depend on the employment sharks to take him out to the camps. (P.C.F.)

doll buggies Two-wheeled carts. (S.F.)

dolly 1. A two-wheeled log cart. 2. In the lumber yard a small four-wheeled cart on which units of lumber are placed to be moved to a new location. 3. A two-wheeled truck for hauling timber.

dolphin Several piles driven in a bunch, wrapped together with wire rope, to form booming grounds for logs, or to make docks. (P.C.F.)

donald See squaw hitch.

donkey 1. Originally a stationary steam engine. When Dolbeer conceived his machine, it bore little resemblance to the common hoisting engine. Instead he turned to the shipboard steam winch, from which came the name "donkey." There were many varieties of these power plants that were run by steam, gas, diesel or electricity, and which had drums to hold wire rope. They were used to haul logs from the woods, to load logs at landings, to move equipment or to rig trees. In the late 1800s and early 1900s, a donkey was also used to lower cars down inclines. See bull donkey. 2. The word "donkey" was also applied to a small turntable used unsuccessfully in place of a fairlead. (P.C.F.)

A typical logging operation somewhere in the Pacific Northwest. Here a **donkey** is **ground lead yarding** logs to a **landing**, where they will be loaded onto railroad cars. *Philip Schnell collection*

donkey block A small yarding block.

donkey bucker In the early days of the donkeys, the man who bucked long logs that only the donkeys could handle, being too long for horses or oxen to handle. (P.C.F.)

donkey camp In early logging, a logging camp where logs were yarded by a donkey in the Pine Country.

donkey crew Usually a crew that consisted of the engineer, fireman, two choker setters and head slinger. (P.C.F.)

donkey doctor A man who makes mechanical repairs on a donkey or other logging equipment.

donkey driver A donkey engineer.

donkey float A raft on which a donkey is mounted for logging along the shore of a lake or tidewater. Also float. (P.C.F.)

donkey jammer 1. In the Pine Country, a donkey-powered loading rig that could be used any place. 2. A donkey engineer.

donkey locomotive See cog-wheel drive.

donkey logging Yarding on the ground with a donkey engine as contrasted with animal logging or other power-logging methods.

A **donkey float** anchored in the Coweeman River, near Kelso, Washington, where it was busy logging into the river. *Mr. Park collection*

donkey mover A heavy steel flat car used to transport big donkeys on a logging railroad. (P.C.F.) See bull car and moving car.
donkey pounder A donkey engineer.
donkey puncher The popular term for a donkey engineer, one of the key men in the old steam camps. Also steam choker.
donkey setting 1. An area to be cleared for a donkey and a landing for logs. 2. An area that has been yarded by a donkey.
donkey show 1. A logging show yarded by a donkey. 2. An area requiring the use of a donkey for logging.
donkey skinner The man who runs a small engine in the woods. (L.S.) LL
donkey sled The heavy frame on which the donkey sits. Two heavy logs are usually used and big cross timbers are bolted to the two heavy logs. Also sled. (P.C.F.)
donkey snubber A donkey engine used to lower loaded cars down an incline, partly braking against air compressed in the cylinders. (P.C.F.)
donkey tender 1. A fireman. 2. The man in charge of a donkey crew.
donkey wood Wood fuel, usually cull logs, for the steam donkey that was usually supplied by a logger working by piece work. (P.C.F.)
Donovan unloader A rotating boom that had four horizontal poles that swung around and knocked logs off railroad cars. T. H. Donovan of Raymond, Washington was the inventor and first used by the Willapa Lumber Company. Also rapid unloader. (P.N.F.)

Donkey sleds under construction somewhere in the Coast Range of Oregon. *Martin E. Hansen collection*

A **Donovan unloader** at the dump of Rayonier, Inc., at Sekiu, Washington, at the Northern tip of the Olympic Peninsula, as it appeared in September, 1958. This type of **jill-poke** unloader was fast and efficient. As the train moved along the track, the unloader turned and the spokes pushed the logs from the cars. *Labbe photo*

dose of hike (a) An order meaning to leave the job. (P.N.F.)
dote, doty See decay.
double-bitted ax An ax with two cutting edges. These axes used in the Redwood Country were the longest in the United States because they had to reach into the huge redwoods to complete the undercut.
double block 1. A traveling block with two sheaves, to reduce wear on the skyline carriage. 2. A single block with two sheaves.
double couplers, double crotch grabs Two log grabs joined by a short chain or cable, used in pulling heavy logs and in fastening logs together. (P.C.F.) (App.) Called double crotch grabs in the Northern Forest. See fourpaws.

In the woods of the Henry Colvin operation at Marshland, Oregon, on the Lower Columbia, loggers have **sniped** the logs for **ground lead yarding**. In the background a logger is chopping a **ride** on a large log. The line is attached to the log in the foreground with **double couplers** or **crotch grabs**. *M. Morgan collection*

double cut To cut a new path through the wood instead of following the old cut when sawing.
double deck fairleader A fairlead or cable guide through which two lines can pass. Often used with tandem loading drums.

A **double deck fairleader**. This example uses two **bulls eye fairleaders**. *Willamette Iron & Steel Works catalog, 1920, Glen Comstock collection*

double drum 1. An early two-drum steam donkey. 2. A cat with two drums for holding wire rope, used as a yarder or loader. The late Ted Flynn of the U.S. Forest Service is given credit for using the first one in 1923.
double-haul To handle logs with two donkeys, one taking the logs from the other.
double-head To couple one locomotive onto another for more power.
double header A place from which it is possible to haul a full load of logs to the landing, and where partial loads are topped out or finished to the full hauling capacity of teams. (N.W.) LL
double length log A long log in pine country, 32 feet or longer.
double line 1. In overhead yarding, the doubling of the haul-in line through the fall block. The butt rigging is so heavy that it has to be spotted almost over the logs to be choked. 2. An incline system using a cable dead-ended on the top of the hill, reeved through a block on a car, and back to the lowering engine. (P.C.F.)

A Willamette-Hyster **double drum** winch mounted to a Cat 30 tractor. *Philip Schnell collection*

double line skyline A slackline system using a light skyline that could be raised and lowered readily, in order to get logs out of deep canyons. The skyline ran from a big drum out through a large block on the tail tree (or a stump) and back to a second large drum on the donkey. This double heavy duty cable logging system was so big and heavy that it was limited to railroad shows. Also called double slackline, or dudley skyline system. (P.C.F.)

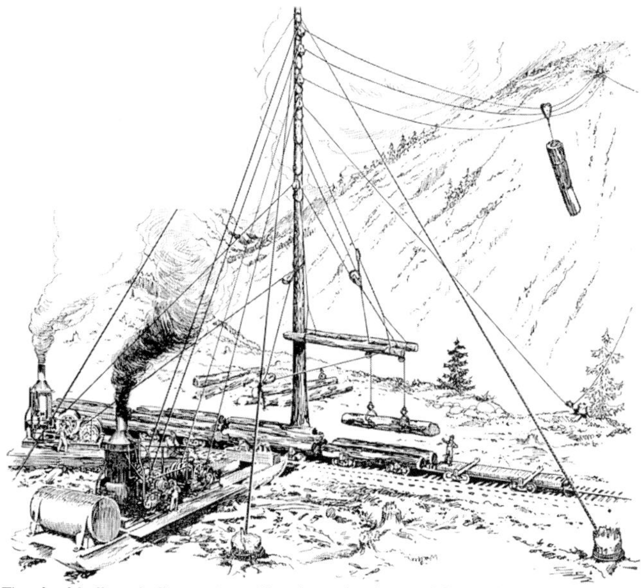

The **double line skyline** system. Also shown is a **hayrack boom** loading system. *From Willamette Iron & Steel Company catalog, 1925, Glen Comstock collection*

double main cable A skidder system using two mainlines, one laid in advance along the next road to be yarded. (P.C.F.)
double rack A sled body designed to carry two parallel tiers of pulpwood or other short bolts. (N.W.) LL
double slackline See double line skyline.
double standing line An overhead cable system used on very steep grades or long hauls. It had two parallel main lines close together and two counterbalanced trolleys. The loaded one was lowered by a donkey that pulled the empty up the hill, saving the return time on the trolley. (P.C.F.)
double the hill 1. To double the power to get a load up a hill, such as a cat pulling a truck up a steep pitch on a logging road. 2. On the

A **double rack** sled hauling 4-foot spruce pulpwood in the Central Eastern forest region. *From "Logging Principles and Practices," Brown, Wiley & Sons, 1934, Philip Schnell collection*

railroad, cutting the train in two and taking up one half at a time. 3. To cut a truck load in half and to take half the load.

double tong loader See McLean boom loader.

doubletree A term borrowed from the farm to mean a crosspiece of wood to which singletrees were coupled when pulling with two horses. Once an important part of logging equipment, it is becoming popular again as horse teams are now being used for the light, young timber that is being logged.

double trip Running a half raft a certain distance, tying up and walking upstream to the other half, running it to or below the first to where the stream is large enough to carry the two half rafts coupled end to end as a full river raft. (App.)

double-yard To yard logs to a yarder tree, then to either swing or road them to the landing. (P.C.F.)

doughdabs Small flat cakes. (Maine)

doughnut 1. A rubber tire used as a bumper on a skyline system to prevent the trolley from bumping into the spar tree. 2. A heavy weight which was used at times in the butt rigging to pull the chokers down to the ground on a skidder show that had no slack-pulling line. 3. A smoke ring from a locomotive. See blow a doughnut.

doughnut hook A hook with a locking ring which was used in the different types of butt hooks and choker hooks. See Bush butt hook.

Douglas fir One of the strongest woods, pound for pound, that has ever been tested. It enjoys a reputation throughout the world as an all-purpose wood of great dependability, and is supreme for both heavy and light construction. Originally sold as Oregon pine, the Northwest lumbermen were afraid that a new species would not compete with the popular pine lumber in the eastern markets. Also bastard fir, B.C. fir, Douglas spruce, nootka pine, Oregon fir, Puget Sound fir, Washington fir, Oregon pine. See red fir, yellow fir.

Douglas spruce See Douglas fir.

down-easter Anyone born northeast of the Susquehanna River. (App.)

down-going car A load being lowered down the hill on a logging railroad incline. (P.C.F.)

down-hill clevis A brake on a logging sled, consisting of a clevis encircling the runner, to the bottom of which a heavy square piece of iron is welded. (N.F.)

downhill drag A good or easy show. In cat logging, a downhill drag is preferred to take advantage of gravity; an uphill haul is preferred for high lead logging. An uphill drag was preferred for donkey logging to keep the logs from running behind stumps. (P.C.F.)

downhill haul 1. An easy job. 2. A smooth-running show. 3. Said of

Ground lead yarding at Camp 4 of the Sunset Timber Company, near Raymond, Washington. Mr. Dunason stands beside the **doughnut hook**, which he invented. *Nelson photo, Vera Jogtich collection*

one who has an advantage or the upper hand of some kind. 4. Said of a yarding operation where the logs are moved downhill.
downhill North Bend system See South Bend system.
down the hill! 1. A faller's warning when a tree is about to fall. 2. To leave camp.
down timber 1. Timber that has been felled. 2. Trees that have fallen in a windstorm. See blowdown.
down time Time lost due to shut down for weather, operating troubles or anything else that stops production. (P.C.F.)
Dow saw A make of a gas chain saw mounted on a buggy with rubber tires, and once used in the Pine Country of Oregon and California. See bicycle saw. (P.C.F.)

Demonstrating the **Dow saw** for the Diamond Match Company officials at the operations near Stirling City, California, about 1935. This was a **bicycle saw**, made necessary because of the bulk of the gasoline motor. Until lightweight motors were developed, power saws were too bulky to be practical, except in the pine woods where there was little underbrush to hinder their operation. *From "The Timberman" collection*

Doyle log rule A formula log rule used to predict lumber yield from logs. The Doyle log rule tends to underestimate yield from small logs and overestimate large logs.
do you speak the two-spoken? A greeting of one Nova Scotian to another, referring to the two languages, French and English, spoken in the homeland. (Eastern Canada)
dozer An abbreviation for bulldozer.
draft chain A light chain used to bunch together some small logs to make up a turn for horse skidding.
drag 1. A train of loaded cars. 2. A brake. 3. A road leveller, usually a wooden surface plow. 4. An older term for a turn of logs. 5. The raker or cleaner tooth on a crosscut saw. 6. A stone boat.
drag cart A small truck with two low wheels used in skidding logs. (N.F.) Also bummer. LL

Loggers of the Warren Lumber Company, near Irvine, Pennsylvania, at work in the white pine that once blanketed the state. They are making use of the **drag cart**, or **dolly**, in yarding the logs. *W. Casler collection*

drag chain A butt chain used in horse skidding.
drag chute A chute in which logs are pulled by horses. Often called a pole road in the West and a log slide in the East.
drag day The day of the month on which a man can draw wages in advance of payday. LL
drag'er To quit. LL
drag freight To pull out, quit camp. (P.C.F.)
dragging the rear Cleaning up the river of logs in river driving. (N.F.)
drag-in To skid a log. (N.F.) Also dray-in. LL
drag road A road over which skidding teams return to the woods after having delivered their load at the landing. Also dray-road. (R.M.F.)
drag saw A crosscut saw powered by a small gas engine. In October 1875, John Dolbeer of the Dolbeer & Carson Lumber Company of Eureka, California invented a portable saw to be used for both felling trees and for bucking logs. Although the trial run was successful, there is no later evidence to indicate that the machine became part of the equipment of the industry.

Most of the early efforts were related to the drag saw or pitman-type machine with gasoline, steam and air power. In 1905, near Henry Melde's green house in the vicinity of Sequoia Park in Eureka, California occurred the first experiment with a gasoline drag saw. The saw was driven by a two-cylinder, water-cooled marine-type engine set at 90 degrees from its normal position. A tank cooling system was suspended on the tree above the engine. Drag saw felling and bucking were established in the redwoods by 1937, but the popular chain saw, which was more efficient, took over in 1949. In the Pacific Northwest, the drag saw was most commonly used for wood-cutting, such as cordwood, donkey fuel or wood for the cookhouse. See Vaughn, Wade.

An early version of the **drag saw**. Later refinements reduced the bulk of the engine, but the principle remained virtually unchanged. *Vera Jogtich collection*

drag saw man 1. A wood buck who kept the camp in firewood. 2. A logger who handled the drag saw. (P.C.F.)
drag sled A two-runner sled used to haul logs out of the woods. (N.F.) LL
draw 1. A gully. 2. An advance on pay. 3. To draw money on the next paycheck. 4. To pull a falling tree away from its normal fall by using wedges as needed.
drawbar The connection between the source of power and the load to be moved. Couplers on a log car are attached to the drawbar.
drawbar hook A fishtail hook, fanned out to hold a number of chokers on the back end of a cat. (P.C.F.) See gooseneck.

A **drawbar hook**. *Young Iron Works catalog, 1952, Philip Schnell collection*

draw day A day in which money could be drawn against the next pay check. Also dray day.
drawed in his horns 1. To back out of a deal. 2. To back down from a stand for any reason. (P.C.F.)
draw head The frame through which a drawbar is attached to a railroad car.
drawing a tree 1. Notching a tree to make it fall in a desired direction. (N.F.) 2. Forcing a tree to fall where the faller wants it to fall, and not where gravity would normally take it. (P.C.F.)
draw-skid See brow log.
draw your time To get your pay because you quit or were fired.
dray 1. A small sled that was used in logging with animals. The front end of logs were placed on a bunk on the dray to keep them off the ground when being yarded to the landing. 2. Two runners with a bunk in the center to haul logs out of the woods. (N.F.) Other names are bob, crotch, drag-sled, lizard, scoot, skidding-sled, sloop and yarding sled.
dray day See draw day.
dray-dog To seize the rear end of a ranking jumper with a peavey and to turn it around. (N.F.)
dray-in To skid a log. (N.F.) Also drag-in. LL
dray-road A long skid road to drag logs from the woods to the skidway. (N.F.) Also drag-road, gutter road, runway. LL
Dr. C.C. (compound cathartic) A camp doctor. (S.F.)
dress a tree To rig a tree for logging. (P.C.F.)
Drew rule A log rule that was once the legal standard in the State of Washington.
drift 1. Logs and flood trash piled up on a river bar. 2. In turpentine orcharding, any convenient working unit of a crop. Formerly it comprised 2,100 cups or boxes, five drifts constituting a crop. (S.F.) 3. To shut off engine power and coast. 4. To leave camp.
drift drive Unmarked logs sold after the drive is in; in the Northeast auctioned for the benefit of the boom or for all. (N.W.)
drifter A logger or lumberjack who wanders from camp to camp. Also boomer, roustabout, short staker. LL
drift pin 1. A pointed iron rod that is driven into the timber like a large spike to anchor timbers together. 2. The pin that couples the arch to the cat.
drift smoke A light smoke that drifts, making the area of a slow-burning fire difficult to find.
drive 1. The system of floating logs on a river from the forest to the mill or shipping point. 2. That part of logging that consists of floating logs or timber. 3. A row of trees partly cut and then driven down by a heavy tree felled against the end of the row. (Maine) 4. To float logs down a river on the high water of fall rains or on the spring freshet. River driving was widely used in all parts of the United States as long as timber was within cheap hauling distance of a drivable stream.
drive-out See turnout.
driver 1. A logger who drives his logs down the river to market. 2. A donkey puncher on a yarder. (P.N.F.)

Drivers at work using their **pike poles**, location unknown. *Steven R. Gatke collection*

driver boards See staging boards.
driver holes Holes cut by fallers in the butt of a tree to put in the drivers.
driver iron See board iron.
drivers 1. The wide iron bars that are inserted into the driver holes to support the staging boards on which the faller stands to fell a tree in the Redwood Country. 2. See driver, river driver. 3. The powered wheels on a locomotive.
driving boots Pegged boots for riding logs on drives. (L.S.)
driving camps Station and buildings along a stream to take care of an outfit in lieu of arks. (N.F.)
driving crew In river driving, the crew in the vanguard who were spread along the streams for ten or fifteen miles, urging the logs along, keeping them out of shoal water, steering away from behind inlets and keeping them in the mainstream. (N.F.)
driving head, driving pitch A term used for high water that is suitable for driving logs down a river.
driving hooks A lightweight peavey, i.e., with a stock about two inches in diameter at the first iron clips. (App.)
drop a tree To fell a tree.
drop landing A landing where the logs were decked by hand, such as on a side hill. (L.S.)
drop line A long lay line spliced on the main line to give greater flexibility to that part of the line that passes through a fall block, as on the downhill North Bend system. (P.C.F.)
drop log 1. A log at the yarder tree on which the incoming logs were dropped so that the hook-on men could more easily attach chokers for swinging to the main landing. (P.C.F.) 2. A log used on top of a dam gate to raise the head of the water. (L.S.) LL
dropping off In river driving, said of men who cut the joints loose from the shore logs, tying them together and attaching them in long swings to buoys in the middle of the river. (N.W.) See joints and swings.
drop seat The union suit type of underwear.
drop stake A patented car stake fixed to the bunk on a logging car, rigged to be tripped from the opposite side of the car as a safety measure. See car stake.
dross Foreign matter that is strained out of crude turpentine and resin at the mill. (S.F.)
drum 1. A reel or power unit on a logging machine on which a cable is wrapped. 2. To yard logs with a drum on a cat.
drum line See arch line.
drug logs To yard logs a short distance to a road, using the winch or drum on a cat.
drum man 1. A donkey puncher. 2. A spool tender, a much lower job than running a donkey. (P.C.F.)
drum narrower A heavy clamp fastened on the barrel of a drum to keep the wire rope on only part of the drum instead of having it wind across its entire width.
drum tender The man who guided the manila rope or wire on or off a spool. (P.C.F.) See spool tender.
drunkard The loggers' passenger train returning to camp from a night or weekend in town. (P.N.F.) See Sunday night train and whiskey special.
drunker than a thousand dollars Takes about three days to get this drunk.
dry ass A sack filled with straw to make a seat cushion when sitting on wet or cold metal equipment.
dry belt 1. The country east of the Cascades (P.N.F.) 2. The pine belt in general.
dry-kill Trees killed by flooding, often found in areas flooded by beaver dams. (N.F.) LL
dry-kye The driftwood and dead trees in the water along the edge of a pond or lake. (N.F.) There are various spellings. LL
dry land lokie A steam tractor used to haul log carts or lumber carts. (P.N.F.) Also road locomotive.
dry log chute A chute, trough or pole road on which logs move by gravity alone, no water or power.
dry-pick As applied to a log jam, to remove logs singly while the water is cut off. (N.F.) See dry-roll. LL

dry-rear Logs of a river drive left stranded on the sand of the stream that must be dragged or rolled back into the water. LL
dry-roll To roll stranded logs into the water behind the drive. Also dry-pick. (N.F.) LL
dry room Usually a separate room where loggers could dry their wet clothes.
dry rot A term applied to many types of decay, especially to that which permits the wood to be easily crushed to a dry powder.
dry shack Usually a shanty for drying clothes.
dry slide See slide.
dry-sloop To drag logs on the bare ground when the slope is so steep that it would be dangerous to drag on snow. (N.F.) LL
dry top A dead top on a tree.
dude 1. A new man in the woods. 2. A man working in the woods who is dressed in store clothes. (P.C.F.)
dude ranch A camp full of greenhorns. (P.C.F.)
dudler, dudley 1. The power car on an incline railroad. A cable was attached to the top and bottom of an incline and wound several turns on a big drum on the donkey or power car. The engine on the car turned the drum and moved the power car and a string of logs behind, skidding on the ties. See walking dudley. 2. The counterweight that was used sometimes on a counterbalance or incline. (P.C.F.)
dudley skyline system See double line skyline.
duff Forest litter and other organic debris in various stages of decomposition, on top of the mineral soil, typical of conifer forests. In the Redwood Country, where the rate of decomposition is slow, litter accumulation exceeds decay.
duffel 1. The personal belongings of a logger. 2. Any loose material. (P.C.F.)
duffel bag A canvas sack used to carry clothing and personal belongings. Also dunnage bag, packsack, turkey, tussock.
duffel shuffler A bunkhouse nuisance who takes up more room than he needs. (P.C.F.)
duke A tool for pulling a choker under a log. (P.C.F.)
duke's mixture Anything mixed up, any confused situation; the term taken from an old-time pipe tobacco. (P.N.F.)
dumber than Adam's off ox A popular expression used by the older people. Oxen were driven from the left side (the near ox and supposedly the smart one), and the off ox was on the right. See off ox. (Red.)
dummy 1. A disconnected truck. 2. A steam locomotive with a housing, or coach body, covering the boiler. Also steam dummy. (P.C.F.)

This Baldwin-built steam **dummy** locomotive handled log trains for the Cathlamet Logging Company, near Cathlamet, Washington. In its heyday it headed suburban trains of the City & West Portland Park Motor Company out of Portland, Oregon *J.F. Ford photo, Frank Potter collection*

dummy couplers Steel couplers that were fitted to the ends of squared timbers to make a skeleton car out of a pair of disconnected trucks. (P.C.F.) Also rooster.
dummy line 1. A dead end spur road or track. 2. A short spur, commonly called a "tail track," leading off a landing where the empties are stored and from where they run down to the mainline by gravity and are spotted for landing. 3. A light cable or strawline. (P.C.F.)
dummy log See squirrel.
dummy tree A standing tree that is used for leverage in raising a spar tree (raised tree) into position. The dummy tree is then cut as a merchantable tree. See raise a tree. (P.C.F.)
dump 1. A log dump where trucks or cars are unloaded. 2. A gravel truck. 3. A railroad ballast car. 4. A portable mill lumber yard. (New Jersey) 5. To roll logs from a flat car into the mill pond.
dumper A man who unloads logs at the dump.
dump hook A levered chain grab-hook attached to the evener to which a team is hitched in loading logs. By tripping the lever it re-

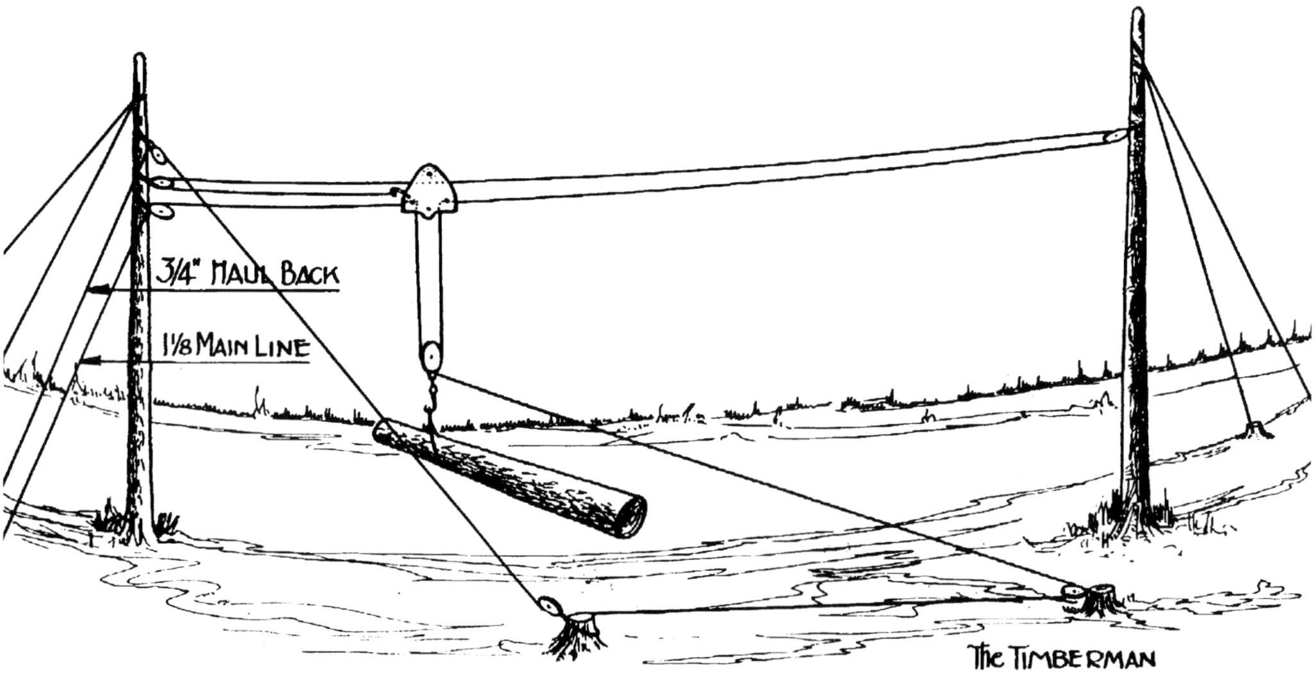

The **Dunham skyline system**. *"The Timberman," November, 1924, Glen Comstock collection*

leases the hook from the logging chain without stopping the team. (N.F.) LL

dump logs To roll logs over a bluff, or from a logging car, or a sled into the water. (N.F.) LL

dump the pan To throw everything away. Taken from the usual practice of dumping the ash pan after cleaning the fire on a steam engine.

dunderjunk In the early 1800s, bread and pork cut into pieces and mixed with molasses, all baked in a kettle in the ground. (Maine)

dunghister A farmer. (L.S.) LL

dung out To clean out a bunkhouse or camp building or to dispose of trash; taken from the early days when stables were part of every logging camp or mill yard. (P.C.F.)

dung snuffer A teamster. (L.S.) LL

Dunham skyline system A slackline system powered by a two- or three-drum donkey. The mainline acted both as a skyline and haul-in line, doubling through blocks on both the head and tail trees. (P.N.F.)

dunk 1. To reach down through the hole in a boomstick to get the toggle chain and to fasten it to the next boomstick. 2. To fall into a long pond.

dunnage bag See duffle bag.

duplex flyer 1. Also called duplex aerial yarding system. A logging method using a double slackline. 2. A specific donkey was used to power the system, having two sets of engines taking steam from one boiler. The skyline was operated by one engine, and the other ran the haulback, hauling line and the strawline. Also flyer. (P.N.F.)

duplex loader A donkey with two loading line drums. Each drum was operated by a separate engine so the lines could be moved independently. The engines ran in either direction. (P.C.F.)

duplex loading system A loading method using two drums, two engines and two sets of loading lines and blocks, ending in two sets of tongs on one log. The loading engines could be either independent, or if car-mounted, could get steam from the yarder boiler. In this case the combined yarder and loader were known as a unit. (P.C.F.)

dust a dam To fill the cracks or small holes between planks in the gate of a splash dam with earth or gravel. (N.W.) LL

dust bowler A newcomer to the lumber mills and woods from the dust bowl. (P.C.F.)

duster A dead standing yellow-pine tree with a sound heart. (S.F.)

Dutch coupling Also Russian coupling. See bucker's coupling.

Dutchman 1. A rig-up on a high lead show that consists of a piece of line anchored to a stump and tied to a block through which the main line passes at a spot near the spar tree. The purpose is to change the lead out from the bull block to overcome a bad curve, in order to maintain as nearly as possible a fair lead from the tree. A Dutchman is also used to pull the mainline to one side to keep logs off the track, or road, at the landing. 2. A prop put under a log to keep it from pinching the saw during bucking. 3. A prop used for any similar purpose, such as supporting the hitch of an arch while it is being hooked onto a tractor. 4. A pass block that is put around a tree to haul up the rigging. Thus the rigger will say, "We hung the Dutchman." 5. A short piece of rail on the railroad a few inches long to close a gap where rails are joined.

The term is also used in the following ways in the Northern Forest: 6. A splinter or slab of wood to be put under a wrapper chain to keep it tight, or to level up the load on a sleigh. 7. Wood left in the undercut to support the heavy side of a tree when it was falling on a swing. 8. A short stick placed transversely between the outer logs of a load to keep any logs from falling off. See grouser.

Dutchman show A setting where a Dutchman must be used to get away from a curve in the mainline in yarding. (P.C.F.)

Dutch oven 1. A crude campfire oven used by turning one frying pan upside down over another. 2. Heavy cast iron cooking pot with a tight fitting lid. LL

Dutch whistle A piece of bark lying on its side, waiting for the unwary logger.

dynamite 1. Snoose. 2. More than one dose of a laxative. 3. To slam on the emergency brake, usually breaking a train apart, and usually derailing a car. Also bang juice.

dynamiter A bad-order car on the railroad that had something wrong with the air. (P.C.F.)

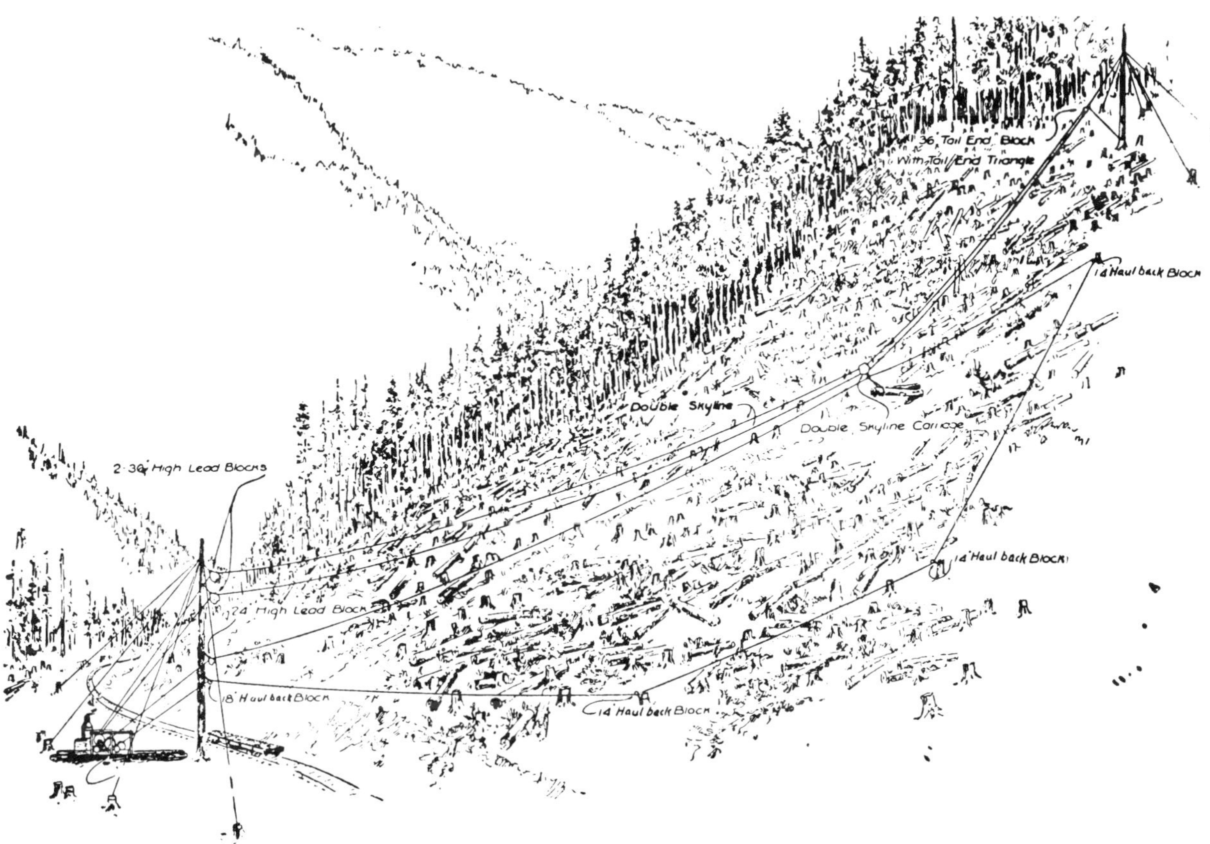

The **duplex flyer**. *From "Logging," Bryant, Wiley & Sons, 1923, Philip Schnell collection*

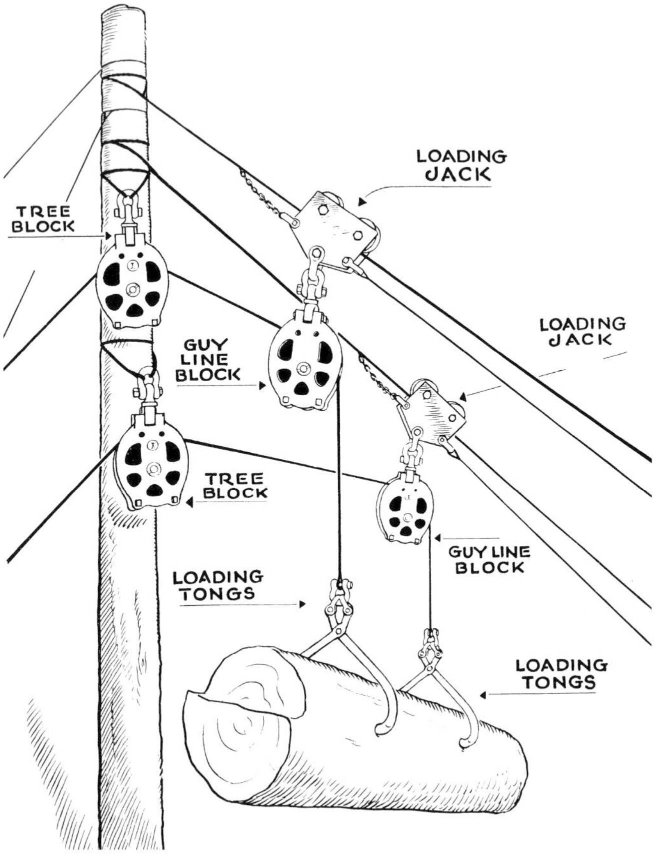

The **duplex loading system**. *From Young Iron Works catalog, Glen Comstock collection*

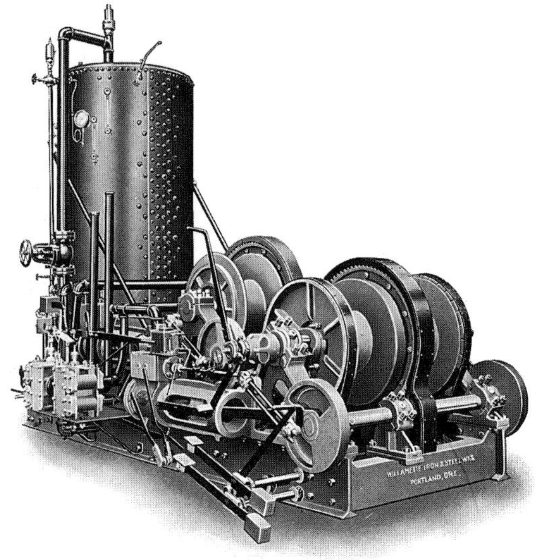

A **duplex loader**. *From Willamette Iron & Steel Company catalog, 1920, Glen Comstock collection*

A **Dutchman** in the track of the Deep River Logging Company at Deep River, Washington. *Courtesy of S. Lawrence*

E

eagle eye 1. The boss, the super, or the push. 2. A lookout. 3. A locomotive engineer. (P.C.F.)
early morning latchers Leather shoes with a metallic fastener in place of shoe strings. (App.)
ears Heavy iron brackets extending out from both sides of a single pole loading boom. The swinging lines are attached to the ears.
earth slide A furrow in the earth in which logs are dragged; sometimes iced in winter to facilitate skidding. (App.)
earwig The air-operated loading tongs that resemble the pinchers of an earwig. Invented by Lon Rees of Crown-Zellerbach. See air tongs.
east side East of the Cascade Mountains in Washington and Oregon. (P.N.F.)
easy as falling off a log Said of something very easy. This common expression of the log ponds and popular speech originated in the river drives of the North Woods and can be traced to the Maine river drives of the early 1800s.
easy runner 1. A free-pulling saw. 2. A locomotive that handled well. (P.C.F.)
eat dirt in the summer and fight mud in the winter A popular expression of the woodsmen of the Redwood Country.
eating soup with a fork Said of a man working hard without seeming to accomplish anything.
eats like a shagpoke (shitepoke) Said of a heavy eater. The shitepoke is a small heron. (P.C.F.)
elbow bender Said of a man who drinks too much.
electric haulage The cable that supplied electricity to the electric motors that were used to yard in the woods some years ago. (P.C.F.) See baloney.
elephant The delivery point or terminal of a log flume, widened to increase the ease of handling. (P.C.F.)
elephant ears Conk on Douglas fir. See conk. (P.C.F.)
elephant trunk hook A small hook with the tongue curled back upward to end inside the link from which the hook hangs, making it a semi-lock hook. (P.C.F.)
elkhorn A branchy moss-like plant found in the fir woods, used occasionally as Christmas greens. (P.N.F.)
Empire A skidding and loading unit used in British Columbia, built by Empire Manufacturing Company, Vancouver, British Columbia.
empties Any kind of empty logging cars. LL
empty as last year's bird nest Said of something very empty. (P.C.F.)
end dogs See end hooks.
end-haul To carry dirt to the end of the cut instead of sidecasting it in building a road.
end hooks The hooks, one at each end, used in loading logs. The loader passes a rope through an eye on the hook to guide the log; the rope is used to pull the hook free when the log is placed at the desired position. Also end dogs. See cup hook.

The **terminal** of a log **flume**, near the Deerlodge National Forest in Montana. This type of terminal was known as an **elephant**. *From "Logging," Bryant, Wiley & Sons, Philip Schnell collection*

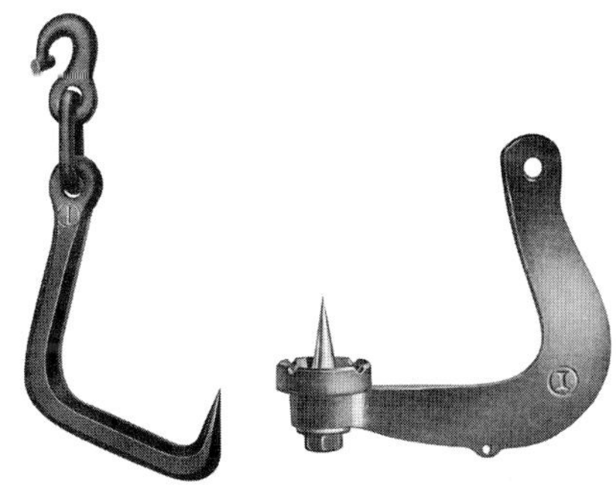

Two types of **end hooks**. The **cup hook** type of end hook (right) helped prevent **slabbing** of logs while loading, and were later required by state safety boards. *Young Iron Works catalog, Glen Comstock collection*

endless line roader A road engine designed to use a single endless line, such as Fout's grip wheel.
end mark A log mark on the end of a log. (L.S.) LL
endo To move anything endways. (P.N.F.)
engine 1. A locomotive. 2. A donkey in the late 1800s.
engineer 1. The head man on a donkey crew. 2. The man who drives a railroad locomotive. 3. The man who lays out a project to be constructed in a lumber mill. 4. In lumbering, a man skilled in some branch of engineering, such as a mechanical engineer, an electrical engineer or a safety engineer. 5. See logging engineer.
engineer's change The act of reversing lines on a high lead setup. (P.C.F.)
engine hill Any steep hill where a hoisting engine was set up to pull sleigh loads of logs up the steep grade. (L.S.)
equalizer 1. Any club or fighting object that makes a small man just as tough as a bigger one. 2. A whiffletree, with a long handled trip hook to carry a single chain, carried high on the rumps of the horses and used when loading logs with a cable. (L.S.)
estimate See cruise.
estimator See cruiser.
euchered out of To be cheated out of something. (P.C.F.)
Evans' skidroad system A skidroad, invented by David Evans of Eureka, California that had cast iron vertical spools on the turns in the road to keep a tow line in place. The hauling line was fastened to the first log, and a heavy line pulled the logs in. The logs were kept in the road around the curves by the many cast iron vertical spools along the skidway. (Red.)

The Excelsior Redwood Company of Eureka, California used the **Evans' skidroad system** for roading their logs. Note the vertical **road rollers** used to control the line through the curve. *Ericson photo, Pacific Lumber Company collection*

even-age management Forest management in which the crop trees are all of a similar age.
evener In horse logging, to even up the pull of the horses, a bar was attached to the singletree and placed between them and the log.
evenflow The production, from a national forest or other unit of land, of the same amount of timber annually for an indefinite, extended period of time.

A 13x14 Willamette steam donkey with an **extended fire box** being rigged up at Silver Falls Timber Company, near Silverton, Oregon. *Willamette Iron & Steel collection*

excelsior 1. Long string-like shavings made from wood and used as a stuffing for mattresses and as a packing material for many articles. (P.C.F.) 2. Noodles. (P.N.F.)

excited as a cat at a mouse show Said of a very excited man. (N.W.)

exhaust in the stack A method of forcing draft in a donkey engine or locomotive by passing exhaust steam into the smokestack, which forces sparks out and is dangerous in fire weather.

expansion joint On the railroad, a diagonal separation between two rails, that allowed for longitudinal movement without distorting the rails.

explosive shackle A shackle used in joining lines that have been reefed up tight in building a boom. The strain is so great that the pin cannot be released, so a hole is hollowed out behind it to take a small piece of dynamite that blows out the pin when the lines are ready to be released. (P.C.F.)

extension 1. A line spliced to the mainline for anchorage. 2. A line wound sharply around a block. 3. A rope added to the end of a mainline to reach logs in yarding. 4. Any new railroad construction.

extended fire box An extended fire box built on steam donkeys to take longer wood to increase production. Pioneered by Willamette, the rear of the fire box was extended and allowed the use of standard four-foot cordwood as fuel and made for better steaming. In later years, some manufacturers built boilers with both front and rear fire box extensions. Also extension fire box. (P.C.F.)

extension slash Unburned logging slash carried over from one year to the next. The logger who made the slash was responsible for any fires that started in it. (P.C.F.)

extension stake A long car stake permanently attached to a logging car. (P.C.F.)

extract wood Wood that is suitable for the manufacture of tannin extract. Chestnut was the principal native wood in the Northern forest.

eye A loop spliced into the end of a wire rope.

eyeballing a line To make a preliminary guess survey by eye, and not by instruments. (P.C.F.)

eye dog See boom dog.

eye splice A wire rope that is unwound and then interwoven to make an eye.

F

FB That portion of a tree (when the tree falls and breaks) from the stump to the break. FB refers to "falling and bucking." Fallers get paid a higher wage scale for bucking the FB than for bucking the AB. (Red.) See AB.
FS 1. The United States Forest Service. 2. A man working for the agency.
face The front edge of a stand of timber.
face log See head log. LL
fairlead (er) 1. A device consisting of three or four rollers arranged in the form of a U or hollow square, so that a cable passing through it can be carried out or reeled in from any direction. The purpose is to make the line spool evenly on the drums when moving, chunking out or yarding at an angle to the lead. Called fairleader in the late 1800s and early 1900s. 2. A cable guide on an arch (wire rope guide). 3. Any block or roller attached to the front of a donkey, or winch, to give the proper "lead" to the incoming line.

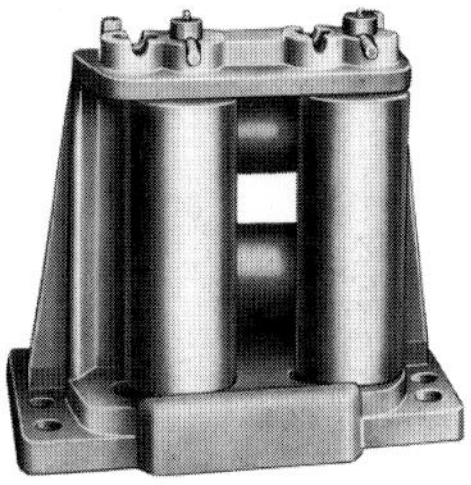

A four-roller **fairlead**. *Willamette Iron & Steel Company catalog, 1920, Glen Comstock collection*

fair lead In yarding, a wire rope almost in a straight line from the donkey to the woods. When the line is at right angles, it is called a square lead. If an attempt is made to reach behind the donkey for logs, it is called a diamond lead.
fair lead bunk See head block.
fairleading Bunching logs a short distance up to the arch with a line running from the winch through the fairlead. (P.C.F.)
fairlead timber 1. The cross timber on the front of the donkey sled to which the fairlead was fastened or on which it was mounted. Also fair lead bunk, head block. 2. A cross beam.
fair off To become clear weather. (Maine)
fake 1. A gas-powered donkey. See stink pot. (P.C.F.) 2. To do anything: "Fake up some biscuits."
fakealoo A machine, or some sort of setup that was cobbled together. (P.C.F.)
fall 1. A rope or cable used in hoisting, or used as a lifting line in hauling. 2. To cut timber. Fell is used in the past tense at times: "We fell three trees yesterday." Fall is the popular term used in the woods: "He is going to fall the tree;" "He is falling the tree." When the trees are down, then the popular phrase is "felled timber." Fell is the technical word. See fell.
fall a quarter To fell timber in any one quarter of a donkey setting. (P.C.F.)
fall block 1. A block to which lines are attached on either side for use in loading, or as a single sheave carriage on a tightline skyline. 2. A block used in the bight of a line, as in the main line of the North Bend and slackline systems. In both cases, a long narrow block with the sheave at the top; generally balanced so most of the weight is at the bottom. Also swinging block. (P.C.F.)

A **fall block**. *From Lamb Logging Equipment catalog, Glen Comstock collection*

fall block system Another term for a tight line method of logging. See North Bend system, South Bend system. (P.C.F.)
fall-burn To burn the slash after the first fall rains to reduce the danger of fire before the next dry season. (P.C.F.)
faller A woodsmen who fells trees. Also chopper, sawyer.
faller's strip A piece of timber given to a set of fallers and generally marked off by the bull buck. (P.C.F.)
falling ax The bits of the double-bitted falling ax were longer and

Two **fallers**, working for the C.A. Smith Timber Company, near Coos Bay, Oregon, make an **undercut** on a **Douglas fir** tree. *Philip Schnell collection*

narrower than the regular double-bitted ax, designed especially for chopping the undercut.
falling crew A crew of two or three fallers. (P.C.F.)
falling irons See falling plates.
falling on the square A method of paying timber fallers, not by scale but by the area of the cross-section of the stump. See big square and government square.
falling plates Thin, wide iron plates that are placed above and below falling wedges when the wood is so soft that the wedges cut into it. Also falling irons. (P.C.F.)
falling saw 1. A long two-man saw with a straight belly to saw the huge redwoods in the late 1800s and early 1900s. (Red.) 2. A crosscut saw that was used to fell timber. See crosscut saw.
falling set Normally a pair of loggers who worked together as partners. (P.C.F.)
falling to the lead Felling timber parallel to the direction in which it is to be skidded to the landing. (P.C.F.)
falling wedge A long tapered wedge (steel, wood or aluminum) to keep the saw free during the final cut on a falling tree. The wedge also helps to direct the fall of the tree.

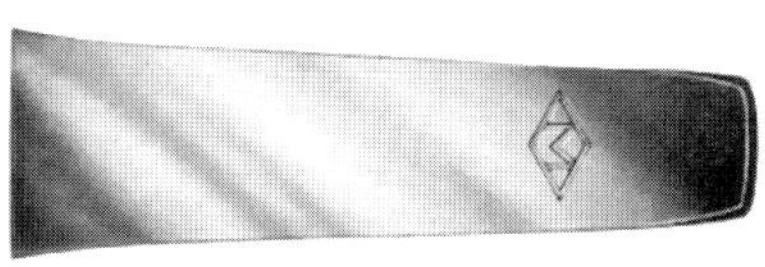

A **falling wedge**. *From Mallory Catalog No. 6, Glen Comstock collection*

fall in rounds To fell timber in two or more separate operations to reduce breakage. (P.C.F.)
false bunk 1. A large log used as a bunk for building a preload of logs for a truck haul. 2. Also applied to the timbers on either side of the swinging bunk on the old single axle truck trailers. They prevented the trailer from tipping forward or backward and breaking the reach when brakes were applied or when striking obstructions. Also dead bunk.
false gate dam This dam permitted a small cascade of water from a lifted splash board to strike a false gate hinged at the bottom. The weight of the water knocked down the false gate and allowed the main gate to pop up.
family camp A logging camp where there are homes for the married men. (P.C.F.)
family show A small outfit run by members of one family. (P.C.F.)
fan dodge Dysentery. (P.N.F.)
fantail 1. A bell with several sockets for chokers. 2. To lay out radial runs for pullboat logging, each main run having one or more branches. (S.F.)
farewell man The welfare man. (S.F.) (L.S.) LL
farm See turpentine orchard.
farmer 1. A green logger. 2. A person who is incompetent or stupid. (P.C.F.)
farmer eye An eye splice formed by stranding a cable and rerolling it to form an eye; not considered good by expert loggers. (P.C.F.)
farmer's laydown An easy logging show. (P.N.F.)
farmer splice A poor splice. (P.C.F.)
farmer wood 1. Small logs cut by a farmer or a small logger on a farmer's property. 2. Small logs taken from a farm. (P.C.F.)
farrier A blacksmith who shod the horses. (N.F.)
fast In saw filing, a saw, when raised up, is said to be "fast" in those places that come up to the straight edge.
fat pine The pitchy wood often found at the base of fir or pine trees. (P.C.F.)
feather edge An ax or saw edge filed so thin that it feathers or breaks off, leaving a dull area, evidence of poor filing.
feather merchant 1. A small logger. 2. Someone trying to sell a bill of goods. (P.C.F.)
feed bag 1. A lunch pail. (P.C.F.) 2. A nose bag as applied to horses.
feeder 1. A small trough that carries more water into a log flume. 2. A short spur flume that carries logs to the main flume. 3. The man who maintained the horses and kept the hovel cleaned out in a horse camp. (N.F.)
feel cold steel To get fired. (P.C.F.)
feister 1. A no-good person. 2. A feisty person, full of fight. (P.C.F.)
fell To cut timber. This is the technical or dictionary word. Fall is the popular word used in the woods. See fall.
felled Said of trees that have been cut down. (P.C.F.)
felled (fell) and bucked Timber that is down and bucked, ready to be yarded. (P.C.F.)
feller The technical word for faller. Feller is popular in the East, and faller is popular on the West Coast.
feller-buncher A mobile machine that cuts trees with giant shears or a saw and then stacks the trees in a pile.
felling The business of cutting down trees.
felling cut See back cut.
felling saw A crosscut saw that was used to fell timber. Alex Polson, the famous Washington logger, was supposed to have used the first one in the Douglas fir region. See crosscut saw.
felling to lead Felling a tree in the direction that it is to be skidded.
fell so far the bluebirds built a nest on his ass before he landed Said of a long fall off a tall spar tree. (P.C.F.)
female splice A polite term for cunt splice, a temporary splice in a wire rope. (P.C.F.)
fence boom A patent log-towing boom used at one time on the Great Lakes. (Eastern Canada)
fender To skid a log around an obstacle by using a short log.
fender boom A boom on either side of a gate to guide logs going through sluice gates. Also glancing boom. See sheer boom. LL
fender log 1. A log or heavy pole placed at the side of a skidding trail to prevent the skidding load from rolling off the road. 2. A log used in the bull team days to help sheer a turn of logs around a bend.
fender skid A log placed on the lower edge of a trail slide to hold the logs on the road. Also shear skid. See trail slide.
fernhopper A graduate from a school of forestry. See logging engineer. (P.C.F.)
ferrule A metal collar babbitted to the end of a wire rope to make a choker. See nubbin, butt hook, basket.
ferrule, butt hook The socket on the butt rigging into which the choker nubbin is fitted.
ferry traveler A simple type of skyline bicycle carriage.
fharrigan See wanigan. LL
fid 1. A short sharp steel tool used in splicing wire rope. 2. A marlin spike.
fiddle butts Large spruce butt logs suitable for the manufacture of musical instruments. (N.W.)
fid hook 1. A flat hook with a narrow slot and a movable tongue; it is used to tighten two ends of a chain and to lock the connection.

A **ferry traveler**. *Mallory Catalog No. 6, Glen Comstock collection*

A **fid hook**. *From Mallory Catalog No.48, 1948, Philip Schnell* collection

The tongue is used as a lever, and the hook is brought down into the slot, and a ring on the shank of the hook is slipped down over it to lock the tongue in place. 2. A flat hook without the tongue; a link of chain is grabbed in the slot to make the lock in cinching up a load.
fight hang-ups 1. To straighten out cables, chokers, blocks or any messed-up rigging from the many tangles that occur in yarding. 2. To fight a log that is hung up and resists the yarder.
fight the bell Said of a choker setter who tries to get ahead of the crew by holding the bell on the end of the choker wire. (P.C.F.)
figure eight (8) 1. A method of putting a choker around two logs lying in different directions so that they will skid together when yarded without fouling the choker. 2. A knot that will not slip on the end of a line to make a temporary eye. (P.C.F.)
figure eight (8) hook A hook that was once used to connect the haulback to the butt rigging. It stayed locked as long as there was a pull on the line. (P.C.F.)
figure head A bookkeeper. (P.C.F.)
filer A man who sharpens, fits and repairs saws. See fit, fit a saw.
filing shack A special shack with big windows and usually a sky-light to give the filer the best possible light on his work.
filler The core in the wire rope that is either iron or hemp.
fill-out To add enough cars to a train to make up the full load that a lokie can pull over a given stretch of track.
fillyloo bird A dirty bird that appeared in loggers' stories; eats red peppers and flies backwards for cooling purposes. (P.C.F.)
Filson 1. A brand name of woods clothes. 2. A cruiser coat. (P.C.F.)
fin A log fastened underwater to direct the course of logs floating down river. (N.F.) LL
fin boom A boom attached at one end, the other swinging free out in the stream. A winch controls a series of wooden fins attached to the downstream side of the boom. When logs are coming down river, the fins are pulled about at right angles to the boom, forcing it out into the stream when it shunts logs into a storage boom. When the fins are released, they drop back roughly parallel to the boom which then floats in line with the current, leaving the river passage open. A fin boom was still in use on the Coquille River in Oregon in 1950. Also rudder boom.

A **filer** at work in his **filing shack** is **fitting a saw** for use in the Northwest woods. The saw, which had four cutting teeth alternating with the **rakers**, was known as a **four cutter**, and was the most popular type of saw used in the woods. *From C.H. Swenson collection*

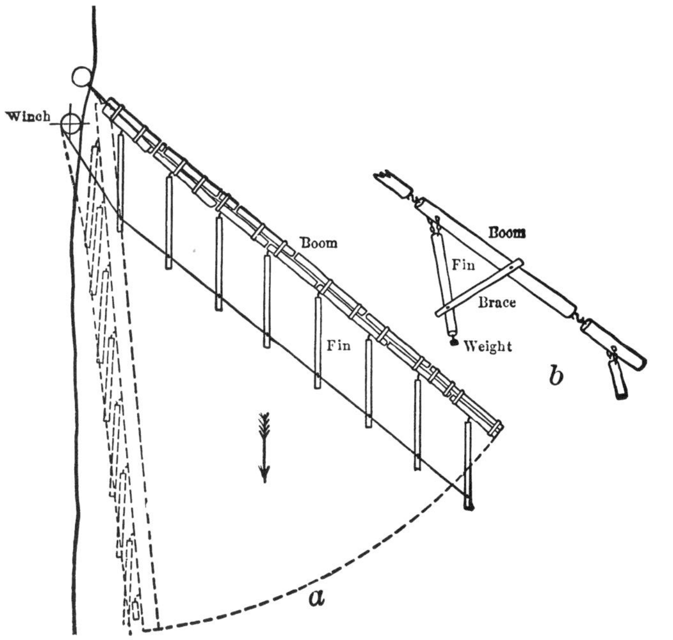

Diagram of a **fin boom**. *From "Logging," Bryant, Wiley & Sons, Philip Schnell collection*

fine grain Close-grained wood, usually found in old growth fir which grew very slowly. (P.N.F.)
finger link A light hook shaped like a pelican's beak. Also called pelican hook. (P.N.F.)

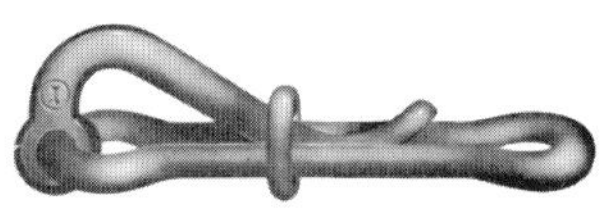
A **finger link**. *Young Iron Works catalog, Glen Comstock collection.*

fin(n) grub hoe A heavy hoe made from a shovel used for leveling tote roads. Also ground level. (L.S.) LL
fink A company guard or a stool pigeon or any other low-life double-crossing so-and-so.
fin(n) saw See swede saw.
fin(n) splice A wire rope splice in which the frayed-out end of the line sticks out from a tuck like a fish fin. (P.C.F.)
fir Usually Douglas fir.
fire boss The man who makes the decisions on forest fires.
fire box 1. Usually the stomach of one who has just returned from a weekend in town. 2. The area in connection with a steam boiler in which the fire is contained. (P.C.F.)
fire boy A fireman on a lokie or donkey.
firebreak Anything to stop a fire, such as a creek, ridge or any man-made stop.
fire bug 1. A man who intentionally sets forest fires. 2. A man who is careless with fire in the woods.
firecrackers Beans. (West Virginia) LL
fire danger Weather conditions that are good for working but extremely bad for fire and usually indicates a shutdown of logging.
fire hazard 1. Slash that is left on the ground. 2. Any situation that contributes to the risk of fire.
fire in the hole 1. A lighted fuse on a powder charge. 2. The warning shouted by the powder monkey after the fuse has been lit. (P.C.F.)
fire line A line or deep clearing around a fire to prevent it from reaching fresh fuels.
fireman 1. A fireman on a lokie, skidder or donkey. 2. A man fighting a forest fire. 3. A volunteer in the fire department of a company town. 4. A fire warden in a logging camp.
fire-out To burn out any fuel lying between the fire line and the edge of the fire so that the fire may not build up and break over the line.
fire patrol A forest protective association that is privately operated.
fire road A road built to get men and machines to a fire.
fire scar A burn on the trunk of a tree. LL
fire season 1. The dry season between June and October. 2. The period designated by the state or federal authorities during which prescribed precautions must be taken by any workers in the woods.
fire's out Said of a killed motor on a gas or diesel donkey. This popular expression is even applied to an automobile. (P.C.F.)
fire trail A cleared trail around a slash area to prevent a fire from escaping.
fire-up 1. To start a fire in a locomotive or steam donkey to get steam ready before the crew arrives. 2. To start almost anything of a mechanical nature.
fire warden 1. The man who has the authority to shut down a logging operation. 2. The man in charge of fire control for a company, an association or a state district.
fire watch The watchman who stays after a day's logging to watch for fires.
firm offer An offer that states price, quantity and other details, and the acceptance of which is not subject to confirmation by the offerer before he becomes bound.
firmwood The sound or solid wood in a log suitable either for chips or for solid wood products such as lumber or veneer. Char, voids and soft rots are not considered firmwood.
firmwood cubic The chipable wood content in a log expressed in cubic measure.
fir region An abbreviation for the Douglas fir region that includes Southwestern British Columbia, Oregon and Washington west of the Cascades, and Northwestern California.
first cook The pastry cook who is the boss in a large camp. (L.S.)
first faller 1. The lead man in the falling crew. 2. The bull buck on a small operation. (P.C.F.)
first growth The virgin timber.
first loader The head loader.
fish 1. Anyone good for a touch. 2. A person who is easily fooled. (N.F.) LL
fish-eyes (eggs) Tapioca pudding.
fish plates Plates used to hold the ends of two rails together and are bolted through the rails from each side.
fishskin 1. A poor waterproof slicker worn by a green hand that is not suitable for work in the woods. 2. A type of raincoat worn by commercial fishermen. (P.C.F.)
fishtail hook A hook, with the point flattened out, that is used for general purposes, such as holding the eyes of the straps in crotch line loading. (P.C.F.) Also used as an arch hook or cat hook. See cat hook.
fist choker hook An old flat choker hook shaped like a partly closed fist. (P.N.F.)
fit 1. To fit up or sharpen a saw. 2. To get along well on the job: "He fit good." 3. To notch a tree for felling, and after it is felled, to mark it into log lengths. (N.W.)
fit a saw To put a saw in good condition by joining, setting and filing the teeth. (N.F.) LL
fit hook A hook closed by a ring that could be pounded shut to tighten on one or more lines. Later it could be pounded loose.
fitter 1. The one who notches a tree for felling and who marks the log into lengths into which it is to be cut. 2. One who cuts limbs from felled trees. (N.F.) (Canada) LL
fixer The camp blacksmith. (P.N.F.)
flaggin' 1. To protect the rear end of a train by using a red flag, fusee or torpedoes or protecting any situation with a flag man. 2. To hurry, such as "flaggin' his ass down the road." 3. To flag for trucks entering a busy route from a blind access road.
flaggin's The dinner that was toted out to the woods on a sled by the cook. (N.F.) LL
flagman The person who transmits orders from the tong hooker to the steam skidder leverman. (S.F.)
flamethrower One of many varieties of oil-burning torches used to start fires in slash or to start backfires when fighting forest fires.
flange A projecting rim or collar attached to a donkey drum to narrow the drum.

flapper 1. A piece of wide belting attached to a handle that is used to slap down grass and to fight brush fires. 2. A wet sack also used for the same purpose. (P.C.F.)
flat 1. A flat car on a railroad. 2. A bench or flat place on a mountain. 3. A flat wheel on a logging car. 4. A flat tire. 5. A pancake.
flat bed Any truck with a platform body.
flat car See flat.
flathead A faller. (S.F.)
flat hook A choker hook similar to the boxing glove hook.
flatlander 1. A farmer. 2. Not a mountain logger. 3. A foreigner to the woods or area. (P.C.F.)
flat raft A log raft, one layer deep.
flats Hotcakes.
flat spot A flat spot on a railroad car wheel caused by the sticking of the brakes which makes the wheel slide. Also skid flat.
flatten-down To unscramble the logs in a jam so that they will move freely when hit by water from a splash dam upstream. (P.N.F.)
flat wheel 1. A lame man. 2. A wheel on a railroad car or locomotive with a flat spot. (P.C.F.)
flea-bag A flop house or cheap hotel.
fleet Several rafts of lumber fastened together to drift down a river with the current. (N.W.)
fleet angle The angle measured between a line from the center of the bull block sheave to the center of the main drum and a line from the center of the sheave to the rim of the drum. The fleet angle should be held to about two degrees to avoid undue wear on the cable.
flip the chip To do expert work with an ax, usually applied to undercutting. (P.N.F.)
float 1. A big raft to hold a donkey or a camp building. See donkey float. 2. See drive.
floatage Logs spread over the surface of a river or lake. LL
floaters 1. The logs floating in the log pond, as opposed to the sinkers. 2. Men engaged in towing and moving timber on bays, canals or the ocean. (App.)
floating camp 1. A camp built on rafts in the steep inlets of British Columbia where level ground was scarce. 2. Any camp built on rafts where level or solid ground was scarce.

A **floating camp** on a cypress operation in Louisiana. The dining room and office are on the ground floor and the sleeping quarters are in the upper story. The building on the left is the camp store. *From "Logging," Bryant, Wiley and Sons, Philip Schnell collection*

floating rear Logs of a river drive left stranded in still water which may be floated back into the current. (N.F.) LL
float out from under his hat Said of a man who falls into the river. (L.S.) LL
float road A channel cleared in a swamp and used to float cypress logs from the woods to the boom at the river or mill. (S.F.)
floggins Dinner in the woods. (L.S.) LL
flood See splash.
flood dam See splash dam.
flood gate The main gate on a splash dam.
flood logs To move logs down a river by flooding the stream with water from a splash dam. See splash.
Flora's chickens Ravens. Joe Flora was a famous "highballer" of whom it was claimed would stack anyone killed in a convenient spot until quitting time, rather than shut down in the middle of the day. Maybe the ravens took the place of buzzards! (P.N.F.)
flowage A head of water dammed up for fluming.
flowage rights The rights that permitted flooding of lands along the streams by water backed up by the dams and allowing river crews to cross the properties affected, remove logs and carry out other duties. LL
flow land A river bank that is covered with water during freshet time, or as a result of splashing.
flume A wooden water trough (usually V-shaped) on trestles or cribwork to carry logs or lumber to milling points or loading points. The famous Thielsen flume in the Blue Mountains of Oregon carried 50,000 board feet of lumber and 300 cords of firewood daily to Milton. See water slide, wet slide.

A **flume** serving the Fischer Bros. Lumber Company at Marcola, Oregon. *Steven R. Gatke collection*

flume box The trough, supported on a trestle, that carries the water in a flume system.
flume chaser, flume herder, flume walker A man who patrols the flume to keep the logs or lumber moving and to look for needed repairs.
flume tender 1. A repairman who maintains a flume. 2. See flume chaser.
flunky A helper of any kind.
flush dam See splash dam.
fly A large canvas to cover equipment temporarily or to cover a temporary woods camp.
fly-blowed Said of someone or something that is no good. (P.C.F.)
fly-boom Boom sticks floating in the middle of a log drive. If a log drive developed, the boom sticks were fastened to trees on the banks to stop the flow of logs. (L.S.) LL
fly bread Raisin bread. (P.N.F.)
flyer A donkey used for slackline logging. See duplex flyer. (P.C.F.)
flying chokers To yard with chokers, such as flying two chokers or flying three chokers, depending on how many chokers are used.
flying drive A fast drive in which the main part is quickly put through, without stopping to pick the rear. (L.S.) LL
flying Dutchman 1. A rigging hookup for high-leading parallel to the railroad. See Dutchman. 2. A slotted piece of steel to fit over the part of the sheave on a loading block, to prevent the tongs from jumping into the throat of the block. (P.C.F.)
flying machine 1. A fast donkey. 2. A fast skidder. (P.C.F.)
flying parson A minister who occasionally came to camp to ask for a donation to support a community church. (L.S.) LL
fly road See tote road.
fly-rollway A skidway or landing on a steep slope, from which logs

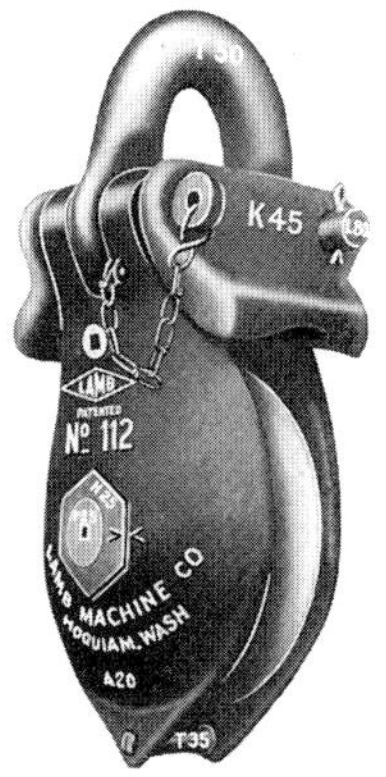

A **flying Dutchman** block. Note the bumpers attached to prevent **tongs** from jumping into the throat of the block. *From Lamb Machine Co. catalog, Glen Comstock collection*

are released quickly by removing the brace that holds them. (N.F.)

fog belt 1. The narrow 150-mile strip along the Northern California coast where the Redwood forest is now concentrated. 2. The Sitka spruce-hemlock strip along the coast in Oregon, Washington and British Columbia.

foggin' 1. To make a fast run on a logging railroad or a truck road. 2. To throw out large clouds of steam from a lokie or donkey. (P.C.F.)

fog horn A smoking pipe. (West Virginia) LL

fog so thick the camp carpenter shingled half a mile of it off the bunkhouse before the sun come up A medium fog. (P.C.F.)

fog so thick you could seine a salmon in it A medium fog. (P.C.F.)

fog so thick it has to be cut up in pieces like building blocks and laid to one side to make a passage through it A thick fog. (P.C.F.)

folded up his face Said of a person who quits talking. (P.C.F.)

folklore According to Professor Wayland Hand, Head of the University of California (Los Angeles) Center for the Study of Comparative Folklore and Mythology, the following superstitions or logger folk beliefs are the only ones that have been recorded in the United States: 1. It was bad luck to start a new job of hauling in logs (starting a log drive) on a Friday. 2. Some loggers would quit if a fat man came to work in the woods because three successive accidents would happen. 3. Many loggers would not report to work on Friday the thirteenth. 4. If a logger got killed in the woods, two more would soon die after in the same way. 5. Many loggers used to wear copper bands around their wrists or ankles to keep from getting rheumatism. 6. The best cure for a hangover is to have sexual intercourse in the morning before going to work. 7. A good blow-job (oral copulation by a whore) will clean out your tubes. 8. If you want to change your luck, have sexual intercourse with a colored woman. 9. One eye on the treetops will save two. 10. Lumberjacks never ground their axes on Sunday because it was bad luck. If they did so, they would probably cut their feet the next time they used their axes. LL

follow A chopping area where trees have been felled. (Maine)

fool killer 1. A bent pole or limb under tension, which can cause severe injury when released. 2. Any dangerous situation not likely to be recognized by a greenhorn. (P.C.F.)

footage 1. The quantity of board feet in logs or timber. 2. The amount of board feet in timber on an acre, or logged in a day, or dumped in a boom, etc.

foot block The rigging once used in connection with the loading

An early logging truck brings out a big cedar log over a **fore and aft pole road**. The **shear rails** prevent the wheels from slipping off the tracks. Such **truck shows** first proved their practicality during the First World War, growing in popularity through the 1920s until, by the 1930s, they had displaced all but the most well-established railroad operations. *D. Kinsey photo, Labbe collection*

boom and attached to the base of the tree.
footing 1. A place to stand while working, specifically while felling a tree. 2. A foundation for a building, bridge, skidder or other logging equipment.
footin's Heavy, thick stockings. (Maine)
foot plank The walkway alongside a flume, used by the flume chaser.
forbidden fruit Pork. (S.F.) Also grunt.
force burn Under wet conditions, a slash fire that is forced to burn with oil and which will not burn the marginal green timber.
fording Yarding logs across swampy ground which is like fording a river. (P.C.F.)
fore-and-after A man with a one track mind. (P.C.F.)
fore-and-aft plank road The usual type of plank road, with two or three heavy timbers on each track, laid on stout cross pieces like ties.
fore-and-aft pole road In the days of truck hauling (heavy wagons) in the woods, logs were laid lengthwise on cross ties, usually two or three logs on each track. Flat surfaces were hewed on top, and a small pole was laid as a curb so as to keep the truck wheels in the tread.
fore-and-aft road 1. A V-shaped trough made of logs, used for roading long distances or across swampy ground or unfavorable ground. 2. Small skids placed so as to make logs climb or turn in a desired direction in ground logging operations. 3. Logs placed on the inside bottom of a flume to keep large logs from rubbing on the bottom of the box. 4. A skidroad in steep country for sliding logs into water. (N.F.)
forest engineer A logging engineer from the company. (P.C.F.)
forester A man trained in forestry who marks the mature trees for cutting, supervises the tree farms, supervises the fire trails, makes growth studies of trees, etc.
forestier A French-Canadian lumberjack.
forestry A profession involving the science, business and art of creating, conserving, and managing forests and forest lands for the continuing use of their resources, material or other values.
fork The branching of a tree trunk, a road, a creek or a river.
fork lift A machine having two arms extending forward that can be raised or lowered. Used to handle lumber, logs, etc. around the mill yard.
forty A standard unit of forty acres in land surveying or measurement.
foul 1. To tangle the lines or the rigging. 2. To ruin a job. 3. To stop a car on a siding near the main line so that a passing train will hit it. 4. To run a line off the sheave of a block.
found This term was used to indicate board and a bunk to sleep in.
founders tree A 364-foot redwood tree at Dyerville in Humboldt County, California which was believed at one time to be the tallest tree in the world.
four-bunk log car A skeleton car with two extra bunks to handle relogged material and other short matter.

A **four-bunk log car**, built for Craig Mountain Lumber Company, near Winchester, Idaho, by Seattle Car & Foundry Company. *Seattle Car & Foundry catalog, Labbe collection*

four-cutter A crosscut saw that has four cutting teeth and which alternate with the rakers. This was an improvement over the early two-cutting teeth.
four-drum donkey A donkey that is used as a combination swing and loading donkey.

A **four-drum yarder** built by Willamette Iron & Steel Company. This example is a **compound-geared two-speed yarder.** The upper left drum is the **haulback**; upper right is the **strawline** drum; lower left is the **mainline** drum; and lower right is the **heel block drum**. *Willamette Iron & Steel catalog, 1925, Glen Comstock collection*

four-drum yarder A donkey equipped with four drums that control the mainline, haulback, straw line and heel block.
four-four The whistle signal on a donkey of four longs and four shorts, indicating that a man has been hurt.
four L The Loyal Legion of Loggers and Lumbermen, an association woods workers organized near the end of World War I and which was eventually outlawed by the Wagner Act during W.W. II. Its main objective was to offset the I.W.W. and at first used much patriotic propaganda. After the war, it was organized as an association of employers in the Northwest lumber industry.

The late Stewart Holbrook, popular logging writer who was a reporter and editor for the 4L periodicals, said that the 4L preserved the eight-hour day and greatly improved conditions in the camps. The 4L, which was used at times to connote lazy loggers and loafing lumbermen, was originally sponsored by the Spruce Production Division.
four-line skidder A double-end skidder which worked from two directions at once; used in the Pine Country.
four paws Double log grabs that were hooks for fastening logs together in ground lead yarding. See double couplers. (P.C.F.) (App.)
fourth of July A shutdown of about 10 days, until enough men sobered up to get the operation going again. (P.C.F.)
four-up A team of four mules. (S.F.)
Fouts' grip wheel A donkey on a car, moving itself on a railroad by means of a cable running over a big wheel equipped with grips. The original grip wheel, which was designed by George Fouts, a San Francisco man who patented the grip wheel, was not mounted on a car and was intended to be used as a yarder or roader. The basic idea was to operate an endless cable, thereby eliminating the time lost on the haulback. The grip wheel itself consisted of a big bull wheel, around the perimeter of which were mounted cable grips. Levers and cams operated the grips so that throughout two-thirds of the revolution the grips held the cable while they were allowed to open and release the cable during the remaining one-third of the arc. The first grip wheels, like Dolbeer's donkey design, had a horizontal wheel, and the grip wheels were designed to be mounted at the side

A Clyde **four-line skidder** in use somewhere in Michigan. *Clyde Equipment Company catalog, From the Larry Nyce collection*

of the skidroad, but proved impractical for yarding. The design proved more effective when used as a locomotive. Although the design remained the same, the power source moved, rather than the cable. Simpson Logging company of Shelton, Washington bought two of them in November, 1896. A year later the company rebuilt two of them into walking dudleys in their shops and put them to work in their camps.

frame The basic structure of any equipment, such as a locomotive, donkey, etc.
framing The frame, or skeleton of a building or structure, made up of windows, doors, diagonal braces and studs.
freak A new piece of gear or equipment.
free-end method An overhead cable system of logging in which the butt rigging is free to be pulled around so the crew can get logs easily. The butt rigging is not tied to a fall block. (P.C.F.)
free logs Unbranded logs that have been washed down a river, or cast ashore when an ocean raft breaks up.
free steamer A locomotive or donkey that is easily fired and makes a lot of steam.
freeze cat A polecat. (P.N.F.)
freeze the pin To tighten the slack on a train so that the brakeman cannot pull the uncoupling pin.
freight To move supplies by team to the logging camps in the 1800s and early 1900s.
freighting his pack Said of a man in the late 1800s and early 1900s when he quit camp with his pack on his back. (P.C.F.)
freshet A spring flood that would help to move logs in the stream to the mill in the early days of logging. LL
Fresno (scraper) An early horse-drawn scraper or dirt scoop. Also slusher.
friction The clutch on a donkey that is operated by hand, steam or air. It holds the drum against a turning wheel driven by gears connected by the crankshaft to the engine.
friction blocks Hardwood blocks used as clutch facing.
friction drum A gypsy spool on which a line was not permanently attached. Pulling power was obtained by taking several turns of line

The machine on the left is a **Fouts' grip wheel**, which was an advance over the little **Dolbeer** on the right. The grip wheel was intended to operate an endless cable to which the logs could be attached for the trip to the landing. The grips on the big drum would grab the line through two thirds the arc and release it through the remaining third. However, it proved less effective in practice than in theory. This was on the operations of the Simpson Logging Company at Shelton, Washington, and most of these grip wheels were mounted on cars and converted to **walking dudleys**. The photo was taken in 1898. *From the Pete Replinger collection*

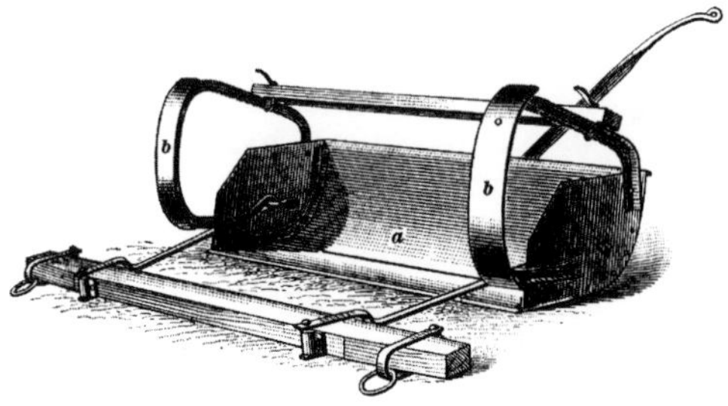

A **Fresno scraper**. *Steven R. Gatke collection*

A steam-operated **friction**. *Willamette Iron & Steel Company catalog, 1913, Glen Comstock collection*

around the drum and letting friction do the work as the drum turned. See nigger-head.
friction nigger See nigger-head.
friction pin The key that forces the drum into contact with the moving gear wheel in the friction assembly on a donkey.
froe A wedge-shaped cleaving tool with a handle set into the blade at right angles to the back; used to split shakes and shingles. Also froe-knife, shingle froe.
frog 1. The crossing piece on a railroad switch. Also rerailing frog. 2. The point where the log chutes come together. 3. A Y-shaped branch of wood once used as a rough sort of sled in skidding small logs. 4. The junction of two branches of a flume. 5. A timber placed at the mouth of a slide to direct the discharge of the logs.

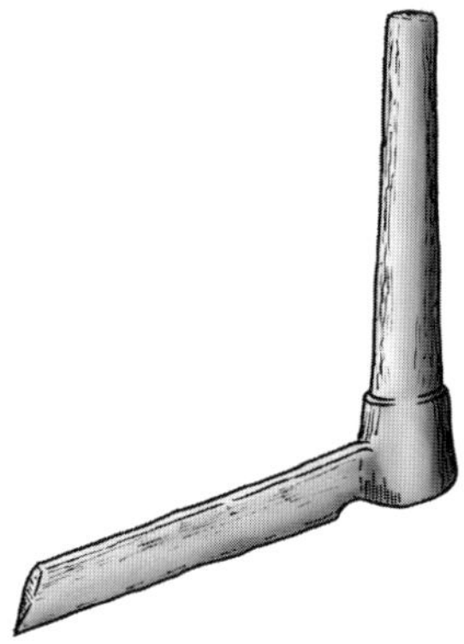

A **froe**. *Drawn by Philip Schnell*

frog eggs Tapioca. (West Virginia) LL
frogger See sled tender.
frog hair A very fine measurement in forest surveys. See cunt hair. (P.C.F.)
frog shoveler A member of a chute crew or a yarding crew who cleans out dirt and bark at the frogs. (California)
front 1. The cutting edge of the timber, such as the marginal trees on the edge of a clear-cut area. 2. That part of the forest fronting on drivable streams. 3. The point at which logging on a particular operation is being conducted.
front end The smoke box of a lokie where spark-arresting screens are supposed to be.
frost crack A split running up and down a trunk of a tree, caused by the extreme cold.
fry cake A mixture of dough and water cooked in a frying pan.
full of staybolts Said of a log full of knots.
full-rigged Said of a spar tree that is rigged and ready for yarding. (P.C.F.)
full scale A pay scale for fallers that gives the gross scale of a log without deduction for breakage or defect. Also bigness scale. LL
fusee 1. A red flare used for emergencies on logging railroads and on truck roads. 2. A back-firing flare for use in fighting forest fires.

G

gabboon A spittoon, sometimes found in the better bunkhouses. Also gobboon. (N.F.) LL
gabriel The long tin horn to announce that dinner was ready, before the gut hammer and iron triangle were used. Also gut horn. LL
gaff The steel point of a pike pole, consisting of a spike point and a spur, which was used to push logs in a log pond.
gaffer A foreman. (P.N.F.)
gall 1. A protrusion on anything. 2. A sore spot on an animal caused by a harness that does not fit properly. 3. A scar caused by friction, as where limbs rub together.
galloping goose 1. Usually a small steam or diesel lokie that does not run smoothly. 2. A three or four-wheeled old-time traction engine that was used to pull heavy log carts in the California Pine Country. 3. A loose or noisy crummy or speeder. 4. Any rough-riding piece of equipment.
gall stones 1. Rocks in the beans. 2. Pits in a cherry pie. (P.N.F.)
galluses Suspenders. A popular brand in the Lake States was Silver King. (N.F.) LL
gandy dancer 1. A section hand on a railroad, so-called because of the movement or "dance" in tamping ballast. 2. A name given to Hindus working on the dock at the end of a flume. 3. A pick and shovel man. (L.S.) Also gandy-hand.
gap 1. Any pass in the hills. 2. A log-sorting works on a stream. 3. The many kinds of gaps in standing booms, labeled according to their use, such as swing gap, closed by a log hinged at one end; a slide gap, opened by a log that slides to one side, and so on.
gap stick A log that is floated across the throat of a sorting gap to close it off.
gas engine Any rig in the woods or lumber mill that is powered by gasoline, such as a donkey, loader, locomotive, pile driver, pump or speeder. The first gas donkeys appeared about 1911, but they were no threat to steam. Effective gas yarders appeared in the woods in the mid-1920s.
gate boom A swing boom across a river that can be closed to catch logs or opened to let debris go by.
gate stem The upright timbers in the gate of a wooden dam. (N.F.)
gazabo 1. An itinerant worker. 2. A member of the Fellowship of Woods Workers. (L.S.) Also gaxabo. LL
geared engine Any locomotive powered through a gear train, such as a Climax, Heisler or Shay locomotive. These engines with a short wheel base and extra power were used on steep, crooked logging roads. See each name for more detail.
gear grinder A truck driver.
Geary high lead cap A patented rig to fit on the top of a spar tree that was supposed to save time in rigging up a tree. It had various connections built for attachments of blocks and guys. Also riggin' cap, tree cap. (P.C.F.)
gee A command given to horses or oxen, directing them to move to the right. (Haw was to turn left.) LL
gee throw A heavy, wooden lever, with a curved iron point, used to break out logging sleds. (N.F.)
geezler Tongs to handle pulpwood. More often called "son of a bitch." (N.F.) LL
Georgia pine A trade name for Southern yellow-pine timber from the Atlantic Coast region.
geraniums Cabbage. (P.N.F.)
Gerlinger spark arrestor A locomotive spark arrestor developed by Gerlinger of the Dallas Locomotive and Machine Works in Dallas, Oregon. Also known as the Pacific Spark Consumer.

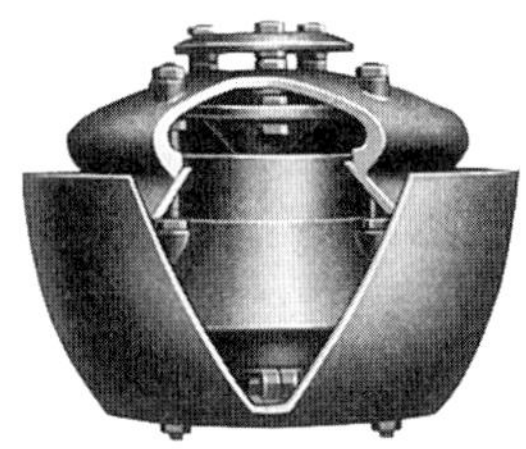

A Gerlinger spark arrestor. *West Coast Lumberman, June, 1935, Glen Comstock collection*

get a bulge on 1. To take advantage of a worker. 2. To get a head start. (P.C.F.)
get a fix on 1. To get a line of sight on a fire to plot its location on a map. 2. To zero in on almost anything.
get a holt on The popular phrase meaning to "get a hold of."
get a skate on To get drunk. (N.F.)
getaway road A path picked out ahead of time for quick travel away from a tree about to fall. (P.C.F.)
get in a jackpot To get into trouble.
get one's ashes hauled To have sexual intercourse, usually with a skidroad whore.
get rollers put under hill To get fired. (West Virginia)

Gin pole loading on the operations of the Bridal Veil Lumbering Company, a few miles east of Portland, Oregon. *Labbe collection*

get the book Advice for a green or ignorant logger or mill hand who does not know how to do his job.
gettin' lank inside Said of a tree before it falls. (P.N.F.)
get-up A general composition or structure of something: "It's a strange get-up."
get your time To be fired.
Gibson raft An ocean-going raft built on a floating mat, not in a cradle. It is square and not cigar-shaped like the Benson raft, and much smaller and more quickly built than the Benson. Generally about 160 to 210 feet long, containing some million feet of logs. Invented by James Gibson of Vancouver, B.C. See ocean-going raft.
gig To move up and down along the river's edge. (L.S.) LL
gig-back 1. To go back to the head of a stretch of rapids to run through again. 2. To go back to the beginning of a haul over an iced road by a short cut that might carry an empty sleigh but not a load. (N.F.) LL
gigging-trail A trail along the side of a stream used to follow log drives. (N.F.)
Gilchrist 1. A type of jack used by hand loggers to get logs into the water; the jack was also used in the early days to jack logs on and off cars. Also called Hercules. See crank jack. 2. A self-oiling high lead block.
gilflirt To get fouled up in horse logging. (P.N.F.)
gilhooley A crosshaul. (Oregon) See parbuckle.
gillagahike (jillaghike) 1. A rig used to do an odd or peculiar job. 2. Any machine to roll up heavy lines. 3. A motor speeder. 4. A small donkey.
gilleary A cant hook. (L.S.)
gilley A platform hung on the pile line, on which the crew could be raised and lowered along the face of a bent while spiking the braces in bridge building.

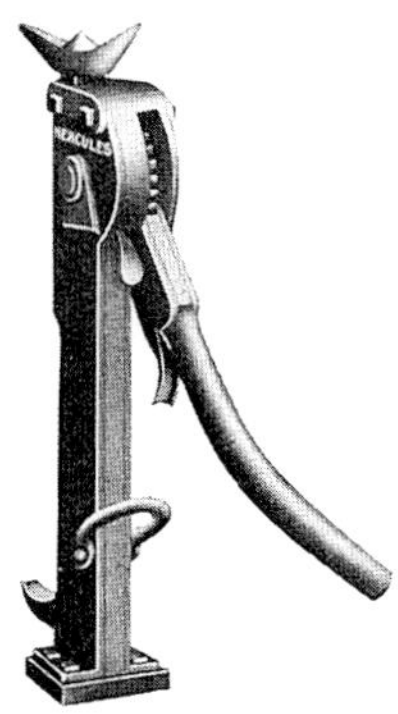

A **Gilchrist** logging jack. *From Mill & Mine Supply Co. catalog, 1916, Glen Comstock collection*

gilley engineer The engineer on a small donkey that is usually used for some purpose other than logging, such as pile driving. (P.C.F.)
gill guy A line that is tied to the tail tree and angled off to one side to relieve the strain on the tail tree when the tail hold is not directly behind the tree because of some obstruction.
gill-poke See jill-poke.
gilly g'lieu A hunting cabin or shack. (P.N.F.)
gimlet An officious, conceited man in camp. (App.)
gimpy 1. A lame man. 2. Said of a machine that is not operating correctly. (P.C.F.)
gin An abbreviation for gin pole.
gin pole An almost endless variety of rig-ups using a pole or two (or three) for loading logs. Also jin pole.

gin-pole side The opposite of the face side. See Canada side.
gin tree A gin pole used in raising a spar tree. (P.N.F.)
give a spread To push logs apart in the center of a boom in order to squeeze more logs in the middle. (P.N.F.)
give 'em a face (fist) full of knuckles 1. To knock a man down. 2. An encouraging expression to a fighter. (P.C.F.)
give 'em the ax 1. An expression which denotes impatience or disgust and directed to an obnoxious person. (Maine) 2. To get rid of a person; to fire him.
give'er snoose 1. To pour on the power. 2. To ask for more steam. 3. To hurry up. (A former tribute to the potency and popularity of snuff.)
gizzard 1. A man's stomach or other innards. 2. The inside of a machine.
glad hand 1. An airline coupling used between a truck and trailer. 2. The airline coupling on railroad equipment.
glancer A sheer skid, a short log used to sheer logs around a turn in the road or an obstacle of some kind. Also tender skid.

Roading logs on the operations of the Hammond Lumber Company in Humboldt County, California. Near the junction of the two **skid roads** a vertical **road spool** helped guide the line in the curve, while a **shear log**, also called a **slew skid** or **glancer**, prevented turns from the second skid road from cutting the corner. *Carranco collection*

glancing boom See fender boom and sheer boom.
glim A lamp. (L.S.) LL
glisse skids Freshly peeled skids on which logs are slid instead of rolled when being loaded. (N.F.) Also slip skids. LL
gloves The leather, canvas or rubber gloves used by loggers or mill men. Fallers, buckers and snipers used light leather gloves until about the middle 1920s. The rigging crews used heavy leather gloves reinforced with more leather.
glut 1. A hardwood wedge used with steel wedges for extra help in felling trees and splitting wood. 2. An iron wedge with a large hollow back designed to take a wooden insert. 3. A wooden wedge used in tie making. (S.F.)
goad stick The sharp stick that the bull whacker used on the oxen.
go a gumming To gather spruce gum. (Maine)
goat 1. A switch engine. 2. A rig-up donkey. 3. A small burner used to start a fire in an oil-burning donkey. See billy goat.
go-back road 1. The other end of a come-back road, the road that the empties take when going back to the woods from the log dump.

An early view of **bull team** logging on the operations of Puget Mill Company in Kitsap County, Washington. The **bull puncher** is holding his **goad stick**. The **fallers** are obviously posing for the photographer, since trees were not normally cut this high up the stump. *"The Timberman," October, 1949, Philip Schnell collection*

2. The road that cats take when going out to yard in more logs to the landing. 3. A road that unloaded logging sleighs can return to the skidways for reloading without meeting loaded sleighs on the way to the landing. (N.F.) (Canada) Also short-road.
gobbler A rubber-tired front end log loader. Also called a log stacker, stacker.

The Wagner Lumberjack was the first **log stacker**, or **gobbler**, ever built. This one is fishing logs salvaged from the **Tillamook Burn** from a Southern Pacific **gon**, to be turned into lumber by the Diamond Lumber Company of Tillamook, Oregon. The sawmill was located inside the ex-Navy blimp hangar in the background, which was the largest wooden structure in the world. *Philip Schnell collection*

gobboon See gabboon.
gobby The Canadian jay. (N.F. (N.E. Canada) See whiskey jack.
godendart A crosscut saw. (Quebec)
go-devil 1. A small sled, without a tongue, often made from the natural fork of a tree and used as an aid in skidding logs on stony or bare ground. (N.F.) (Canada) Also crotch and travoy. In Eastern Canada called a mogie. In the Southern Forest called alligator and crazy dray. 2. A two-wheel log cart or bummer. 3. A name for any unnamed vehicle. 4. A fast traveler.
go down the road To quit or to be fired.
go fishing 1. To lay off the job. 2. To go on strike. (P.C.F.)
go it blind To work at a job without knowing anything about the job.
gon An abbreviation for gondola, a railroad car with low sides used to haul logs and lumber.

A **go-devil** loaded with hardwood logs in Michigan. *"Logging," Bryant, Wiley & Sons, 1923, Philip Schnell collection*

A Washington Iron Works Trakloader loading logs for the Lewis River Logging Company of Cougar, Washington, in 1949. This machine was equipped with a **gooseneck boom** to aid in **heeling** the logs with one **tong**. *"The Timberman," October, 1949, Philip Schnell collection*

good corner A section corner that has been relocated legally. Sometimes difficult to locate in heavy timber.
good ground Usually ground that was easily logged. Also orchard show. (P.C.F.)
gooseneck 1. An early spark arrestor with a wire mesh tube about the size of a stove pipe, curving back down from the top of the stack to a washtub. (P.C.F.) 2. Short iron dogs placed into the bottom of a log chute to slow up logs just before a turn or arriving at the delivery end of a chute. Also called a scotch. (P.C.F.) (App.) (R.M.F.) 3. A brakeman's stick. (P.C.F.) 4. A wooden bar used to couple two logging trucks. (P.C.F.) 5. The yoke of a block. 6. A type of boom particularly a heel boom with a crook in it. See gooseneck boom. (P.C.F.) 7. A fishtail hook to hold chokers. See drawbar hook. 8. A hook which holds the whiffletrees in place. (L.S.) 9. The point of draft on a logging sled; it consists of a curved iron hook bolted to the roll. (N.F.) 10. A V-shaped pair of thills joining the forward and rear sets of runners of a logging sled. (N.W.) 11. A hook made of heavy iron used on the front of a sleigh to hold the evener onto the pole. (L.S.) 12. A bar used to couple two logging sleighs. (N.F.) Also bumper pole, dray hook, rooster, slipper.

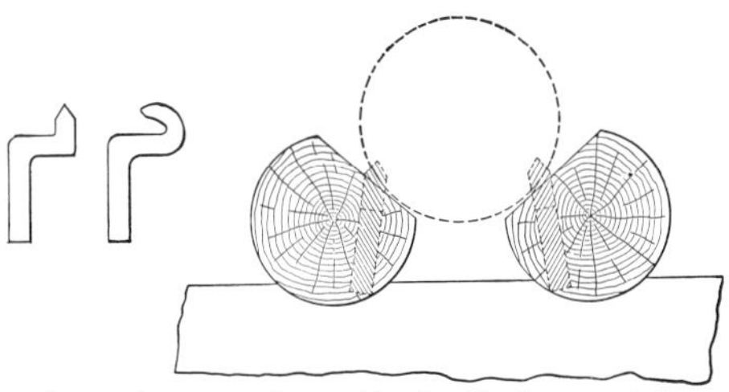

Two common forms of **goosenecks** used for checking the speed of logs on heavy grades, and the manner of placing them in the **skids**. *From "Logging," Bryant, Wiley & Sons, 1923, Philip Schnell collection*

gooseneck boom A boom on a loading shovel or crane that is bent for greater stability and ease of use. See gooseneck.
gooseneck hook A flat hook used to catch a link of chain in order to lock it.
goose pen A hole in the base of a large tree caused by fire burning out the rotted center. In the late 1800s the early Humboldt County settlers of Northern California used the hollow stumps of the Redwoods for pig and goose pens. (P.C.F.)
goose pen butt The base end of a log that has a place hollowed out by fire.
goose pen tree A tree with a goose pen.
goozlum Gravy. (P.N.F.)
gopher 1. A digging tool used to scrape a hole under logs which were hauled by big wheels. A chain was secured around the log and tied to the frame. 2. The person who made a hole under logs so that chains on a pair of logging wheels could be placed around them. 3. In power logging, the person who digs holes under the log so that a choker can be adjusted to it. 4. A small power shovel. (P.C.F.)
gopher hole 1. A hole that is blasted under a log to push a choker through. 2. Any hole that is dug for blasting powder.
gophering Prospecting in the area around camp when not working. (P.C.F.)
gopher man 1. A powder man, specifically one who blew holes under logs for chokers. 2. The man who dug under logs that were to be hauled by the big wheels. (P.C.F.)
gorilla A logger working in the brush. (brush ape) (P.C.F.)
gospel whanger A revivalist. (Maine)
got a half nelson on the payroll Said of a man who has relatives among the bosses of a company. (P.C.F.)
got'er by the ass 1. Said of a man who handles a job successfully. 2. Said of a job that is all but completed.
got'er by the face 1. Said of a man who is in charge of the show. 2. Said of a man who is handling a job successfully. 3. Said of a lumberjack who has saved his money and is now quitting his job. (L.S.)
got'er made 1. Said of a man who knows his job well. 2. Said of a man who has climbed to the top and who now gives orders. 3. Said of a man who quits his job after making a stake.
got'er on a down hill pull Said of a job when things are going smoothly.
got flies on it Said of something not very good.
go to logging To begin logging. (P.C.F.)
got the rollers put under him Said of a man when he got fired. (West Virginia) LL
got the wrinkles out of his belly Said of a logger who complains. At first a hungry man has no complaints, but after he puts on weight and gets the wrinkles out of his belly, he then finds things to gripe about. (P.C.F.)
gouging Scarring of the logs in a river drive from the pounding on the rocks and shoals. (P.N.F.)
government job A personal job that is done on company time. (P.C.F.)
government sale Stumpage bought from a federal timber agency, usually the Forest Service or Bureau of Land Management.
government square The actual area of the top of the stump, used in figuring a faller's pay. Also big square, falling on the square, small square. (P.C.F.)
grab 1. A hook. 2. A spike set in the bottom of a log chute to slow logs coming down a steep pitch. 3. A stop used on a hauling line.
grab driver 1. A dogger. 2. A grab hook man.
grab'er by the face To handle a difficult job in the right way.
grab hook A hook used to catch a link of chain to cinch it up, such as binding logs or lumber on a truck.

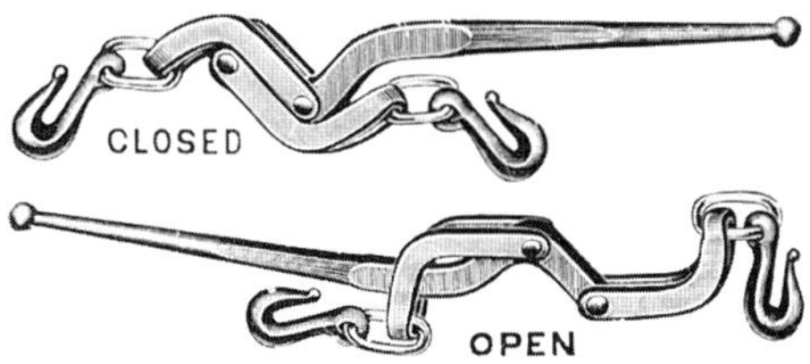

A catalog cut showing a **binding chain** in use. On each end of the **binders** is a **grab hook**. *Kloepfer Hardware Co. catalog, Glen Comstock collection*

grab hook man In the early animal-skidding days, the logger who set the grab hooks or tongs into logs to be skidded; the ancestor of the choker setter. Also grab hooker.
grab hooks 1. A pair of heavy tongs hung from a link, which was in turn attached by the tag line to the main cable of a roading operation. 2. The paired hooks used in grabbing onto, or in hooking up, logs to be skidded.

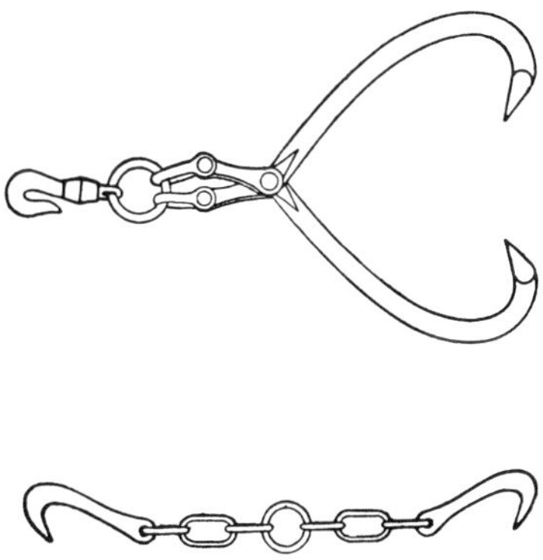

Two types of **grab hooks**. *From "Logging," Bryant, Wiley & Sons, 1923, Philip Schnell collection*

grab iron Any iron handle, that a man hangs onto or climbs onto, on a piece of equipment. Also grab.
grab link See slip grab.
grab man In a roading operation, the man who fastened logs together with grabs when making up a turn for a donkey. See grab driver.
grab maul A hammer for driving grabs. (West Virginia) LL
grabs 1. Dogs. 2. Hooks. 3. Skidding tongs. 4. Paired grab hooks. 5. Handholds.
grab-setter The person who attaches the grabs when logs are transported on logging wheels. (S.F.)
grab skipper A light sledge hammer with the peen ending in a sharp point, that is used in detaching log grabs. (S.F.) (App.) LL
grab the shitty end To take the worst part of a job.
grade 1. The quality of a log or piece. 2. A hill. 3. A truck or railroad route. 4. An abandoned road used as a fire road. 5. To determine the quality of a log or lumber. 6. To prepare a road or railroad grade.
grade destroyer car A flat car, built by Seattle Car & Foundry Co.,

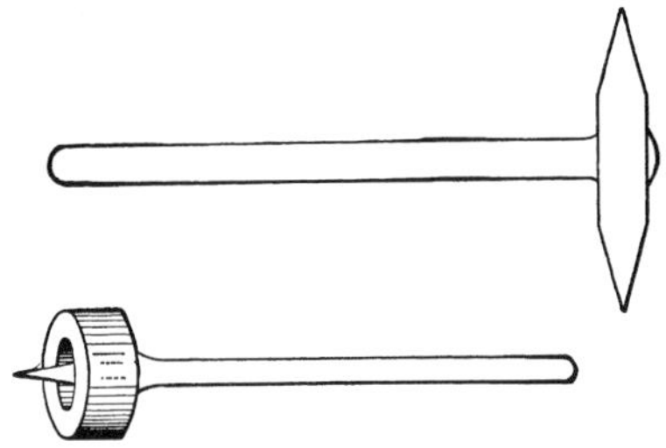

A **grab maul** (upper); a **grab skipper** (lower). *From "Logging," Bryant, Wiley & Sons, 1923, Philip Schnell collection*

with a heavy bulkhead on one end, and used on incline railroads. The bulkheads prevented logs from sliding off on steep grades.

In 1914 the Seattle Car & Foundry Company began delivery of these **grade destroyer cars** to the Yosemite Lumber Company of Merced Falls, California. The 80% grade of the Yosemite **incline** dictated the development. Note the **incline rollers** to hold the line in place on the curve. *Lyle Spears collection*

grader The machine used to keep the truck roads smooth and properly graded. Also inspector.
grading crew Men skilled with hoes, picks and dynamite for road and railroad building. (L.S.) LL
grain-fed Said of a buxom girl. (P.C.F.)
grandma 1. The low gear on a truck. 2. A slow logger.
grape picker A man who shines up to the boss. (P.N.F.)
grape stakes Split redwood stakes of different lengths that are used to hold up grape vines; now also used for fencing and novelty work in patios. Split cedar was used in the old days also.
grapple 1. The modern tong assembly, usually operated by air or hydraulic power. 2. To attach loading or skidding tongs to a log.
grapples 1. Two small iron dogs joined by a short chain and used to couple logs end to end when skidding on mountains so that several logs may be skidded by one horse at the same time. (N.F.) 2. Skidding tongs.
grass line A light line. See guinea line, haywire, strawline.
grauser The vertical traction cleat on the track pad of a tractor.
gravel a dam To cover with gravel or earth the upstream side of the

Using a **grapple** mounted onto a **cat** to load redwood logs onto a truck of the Simpson Timber Company, near Arcata, California. *David Swanlund photo, Carranco collection*

timber work of a dam to make it water tight. (N.F.) LL
gray backs Lice. LL
grease monkey 1. The logger who greased the skids on the skidroad so that the logs would skid better in the late 1800s and early 1900s. Also grease dauber and greaser. 2. A mechanic who usually greases the cats and trucks. (P.C.F.)
greaser A cook. (Maine)
grease swabber A poor cook. (P.C.F.)
grease the pig To grease the squealing wheel on a log truck pulled by horses.
grease the skids under 1. In the bull team days, the job of the grease dauber. 2. To put a man on a slippery footing with the boss, helping him to get fired. (P.N.F.)
greasy shirt A mechanic.
great guessing game from Maine In the early 1800s, an expression for cruising in the Lake States. LL
green back A forester.
green-cut Said of logs or poles cut from live timber, as opposed to dead timber.
greening-up Said of cutover areas in which vegetation is starting to grow after a heavy fire. (P.C.F.)
green strip An uncut strip of timber left along streams and roads. See buffer strip.
green timber 1. The standing forest that is not cut. 2. Green trees as opposed to dead trees.
green twig A reckoning cruiser often carried a green twig and would bend a kink in it every hundred paces to record his reckoning. (L.S.)
grid See apron.
griddle greaser A cook's helper. (P.C.F.)
grip hook A cup-type end hook used in loading logs. See cup hook.
grips Tongs used in skidding. Other terms used in the Northern Forest are chain-grapples, coupling-grabs, grabs, grips and skidding hooks.
grip wheel See Fouts' grip wheel, hill-climbing engine, walking dudley.
grizzly 1. A screen used to shake out gravel. 2. A man who is difficult to get along with. (P.C.F.)
groover A sleigh having chisel-like blades set to cut ruts in roads that logging sleighs would follow. (L.S.) LL
gross scale The volume of a log as determined from a log rule, usually expressed in board feet, in which no deduction is made for defect.
groove-faced Said of a donkey drum that is grooved so that the wire rope will wind smoothly onto it.
ground The type of area on which a logging operation is operating, such as a high lead ground, cat ground, good ground, etc. (P.C.F.)
ground apples Rocks. (S.F.)
ground haul Another term for ground-lead, hauling logs on the ground by a donkey as opposed to the high lead system of lifting logs up by one end.
ground haul engine A donkey used in ground leading. See road engine and road donkey.
ground hog 1. The second loader in a horse logging show. (P.C.F.) 2. A power skidder that skids logs on the ground. (P.C.F.) 3. The tie cutter who lives away from a main camp, back in the woods. (L.S.) (S.F.) 4. The man who sent the logs to the sky-loader at the skidway. (L.S.)
ground-lead In roading logs, the old method (after the bull teams) of using a donkey to yard logs on the ground from the woods to the landing. Machines replaced the oxen and were the first step toward the complicated rigging and highball operation of a modern camp. Also ground line, ground pull, low lead, low line.
ground level See finn grub hoe.
ground line See ground-lead.
ground loader One of the loading crew who attaches the tongs or loading hooks to the logs or who guides the logs up the skids. (N.F.) Also bottom-loader, hooker, hooker-on, sender and send-up man.
groundman A helper or handyman around a crane, shovel, loading rig, etc.
ground pig The man who works on the ground, fastening chains to logs to be raised. (P.C.F.)
ground plane A pick-ax, grub hoe or mattock. (L.S.) LL
ground pull See ground-lead.
ground rope system The use of manila rope in ground lead yarding.
ground skidder A skidder that is used as a high lead rig, without a skyline.
ground-slide To slide logs down a steep pitch without the use of a chute.
ground so rough it would brain a bird Said of very rough ground. (P.N.F.) One old-timer told the story of topping a tree in ground so steep he cut out the top and when the tree swayed back, he just undid his belt and stepped off!
ground yarder A donkey used in ground-lead yarding.
grouse ladder A hooter tree, a tree with many rough branches that serves as perches for grouse. See hooter tree.
grouser 1. A chunk or limb that is placed between the chains, or under a log, to even up a load of logs. 2. A man who complains. 3. A device that was let down into the sand to hold a tug boat in place while it was drawing in a log boom. (L.S.) 4. A knot, piece of wood, rock or anything used as a chink when loading. (L.S.) Also Dutchman. 5. A heavy chain wrapped around the lead log in a turn to slow it on a slope when ground logging with animals. (West Virginia)
grouse tree See hooter tree.
growth ring A layer of wood (as an annual ring) produced during a single period of growth.
growth The increase in diameter, basal area, height and volume of individual trees or stands during a given period of time.
grub 1. Sprouts of hardwood growth, such as grub oak or grub maple. 2. Food. 3. Junk. 4. To dig up trees or brush in clearing roads or land. (West Virginia)
grub hoe A heavy hoe used in fighting fire and road leveling, a mattock. LL
grub-hoe carpenter Usually a camp carpenter who did much work with a few tools. (P.C.F.)
grub-pile 1. A supply of food. 2. A call to come and get the food. (P.N.F.) LL
grubstake 1. Money for food. 2. To supply a man or a crew on shares, such as a prospecting trip.
grub worm A big worm that bores into wood, especially the dead trees.
grunt 1. An electrician's helper. (Red.) 2. Pork. (S.F.)
guard a hill To keep a steep downhill logging road in condition for a sleigh haul by the use of hay, straw or sand. (L.S.) LL
guard rail 1. On a sharp curve on a logging railroad, an extra rail was laid inside the running rail to prevent derailment. 2. A short length of rail inside a switch to prevent the flange of the wheel from pulling off the point of the switch when going on to a side track. 3. A rail used on plank roads to keep truck wheels from slipping off on

Ground-lead logging on the operations of Sunset Timber Company, near Raymond, Washington. *From Vera Jogtich collection*

curves.

gudgeon-pin See king pin. LL

guessing stick A cruiser's stick that is marked, gives the diameter of trees and the number of logs in a tree. See Biltmore stick.

guide roll 1. A roller used to guide the cable on an incline railroad. 2. A roller used as a block in ground lead operations. (P.C.F.)

guinea 1. A short stake used as a reference mark in logging engineering. 2. A person of Italian descent. (N.F.)

guinea line See strawline.

gulch 1. A deep, narrow ravine. 2. The trench made by yarding for a while on the same road.

gullet On a crosscut saw, the space between the cutting teeth and the rakers to allow the shavings to curl up and to be pushed out of the cut.

gully washer A heavy rain. (P.N.F.)

gum 1. Pitch. 2. Crude turpentine.

gumboot show 1. Logging in a swamp. 2. An extremely wet day in the woods. (P.N.F.)

gummed the works To foul things up.

gummer 1. A short-toothed saw used in the woods (felling or bucking) 2. A saw that has been filed so often that the teeth are getting short. 3. A machine to deepen the gullets and to reform the saw teeth.

gumming Failing to keep even the two ends of a rolling log.

gum up To get a saw stuck when making a cut, usually gummed up with pitch.

gun 1. An abbreviation for a splitting gun, a heavy pipe in which a charge of black powder was inserted to split logs that were too large to handle in the late 1800s and early 1900s. 2. An abbreviation for gun stick. 3. A term for a transit or for a level.

gun a log See cannon a log.

gun a tree To sight the direction in which a big redwood will fall. (Red.)

gun boar See catamaran.

gun boats Pieces of tire casing that fit over caulked shoes when walking on a good floor. (P.N.F.)

gundalow A flat-bottomed boat. (Maine)

gun'er 1. To apply power at once. 2. To give the engine full throttle.

gunned Said of failing to get a log on a car or a sleigh so that one end rests on the ground, resembling the barrel of a cannon. (L.S.)

gunning Using a sight gun to follow the line of fall of a tree. (P.C.F.) Also sighting.

gunning stick See gunstick.

gunny-sacked Said of anything fouled-up or broken. (P.C.F.)

gunny sack show A poor operation; a haywire show. (P.C.F.)

gunpowder Black pepper. (P.C.F.)

guns The action of a log that goes up endwise. (N.F.)

gunstick Two sticks attached by a hinge which is used to determine the direction of the fall of a tree. Also sight gun, timber compass.

gut 1. A narrow passage. 2. Bologna. (L.S.)

gut a show To remove only the best logs in a stand of timber. (P.C.F.) See high grade.

gut chain 1. When loading logs, a chain used to pull the wing logs closer together, making a better saddle for the log on top. 2. A chain around the belly of a load. See belly wrapper.

gut hammer The dinner gong, usually a triangle and bar to announce that dinner was ready; in some camps also used to awaken the men in the morning. The gut hammer was struck in a rotating fashion by another piece of steel. Also dinner triangle, hash hammer, iron triangle.

A **chopper** in Humboldt County, California, uses a **gunstick**, or **timber compass**, to **sight** the fall of a redwood tree. *David Swanlund photo, Carranco collection*

gut heater A shot of whiskey. (P.C.F.)
gut horn The long tin horn that was blown by a cook's helper to call the logging crew to eat immediately because the food was hot. Also gabriel.
gut log The center log in a load of logs.
gut robber 1. A camp cook. Also stew builder. 2. A very poor cook. Also can opener artist.
guts 1. The popular term for courage. 2. The insides of an engine or a human being.
guts, feathers and all The complete works. (P.C.F.)
gut-shot An ulcer or serious stomach trouble.
gutter man The man who cuts the limbs off a felled tree. (L.S.)
gutter road The path followed in skidding logs. Also skidding trail. See dray-road.
gut wagon The wagon that brought in the camp supplies. (P.C.F.)
gut wrapper A piece of chain or cable within a load of logs used to hold together an individual tier. (P.C.F.) See gut chain.
guy, guy line 1. A supporting line on a pole, spar tree, tail hold, etc. 2. A line used to hold raft timbers together. (N.W.)
guying drum A small drum on a skidder that was used to tighten guy lines. Used on early Clyde skidders and on recent portable loaders. (P.C.F.)
guy line loading A method of loading by using the guy lines that support the spar tree to carry the additional lines and blocks that are needed to load logs. This was the original crotch line loading system. The duplex loading system was also a "guy line" system.
guy spike 1. A railroad spike with a hole in the head that is used to fasten guy lines to stumps. The spike can be pried easily because of the hole. 2. Any railroad spike used for the same purpose.
gyppo 1. The term was originally used by Union men to refer to bushelers, or contract loggers, who gypped them out of jobs. Also chin-whiskered jobber, zippo. 2. A small outfit that works on a contract basis. 3. A popular name for a logger. (Red.) 4. To work the timber on a contract basis: "He is gyppoing the timber." (P.C.F.)
gyppo slackline A type of high lead rig-up in which a yarding block is hung on the haulback, and a short line from the block is attached to the butt rigging. A turn of logs can be lifted by adjusting the pressure on the haulback and mainline drums, thus reducing hang-ups. (P.C.F.)
gyp stick See cheat stick.
gypsy 1. A horizontal spool on the early donkeys. 2. A combination locomotive and donkey with two horizontal spools mounted on the engine where the pilot should be. The spools were powered by a crankshaft that could be thrown into operation by a clutch. See cogwheel drive, gypsy locomotive, Dolbeer locomotive. 3. A small gypsy

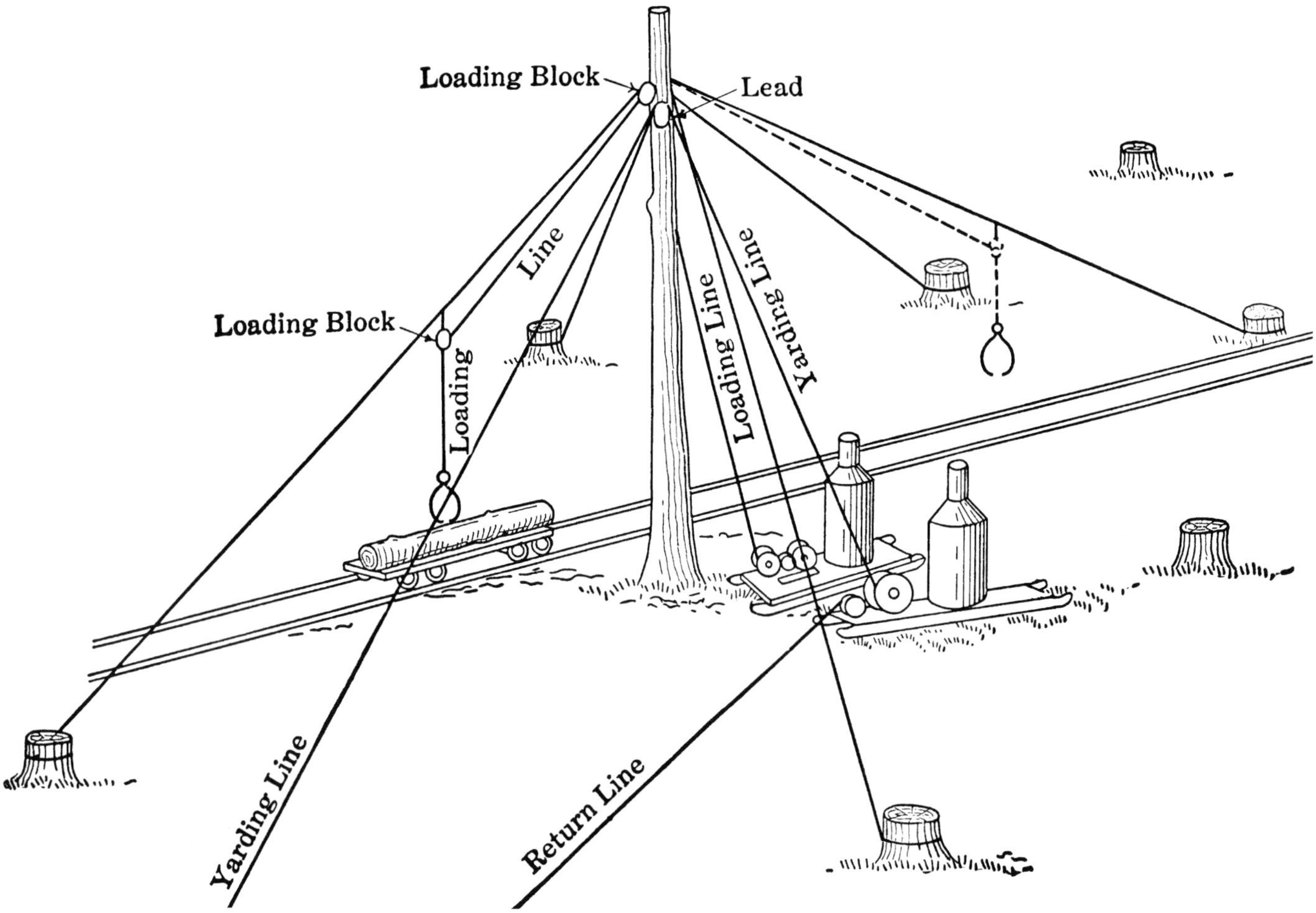

A single **guy line loading** system. *From "Logging," Bryant, Wiley & Sons, 1923, Philip Schnell collection*

on the main drum shaft of a donkey, used as a winch or friction loading drum to move cars, to parbuckle loads and other odd jobs.
gypsy engine See gypsy locomotive.
gypsy head A spool or drum mounted on the end of the shaft on a locomotive or donkey for loading and unloading logs. Probably borrowed from marine terminology along with the winch design. Also capstan spool, nigger-head, spool.
gypsy jammer A loading rig operated off a gypsy head.
gypsy line A line used on the gypsy head or spool to parbuckle logs.
gypsy locomotive A locomotive with a gypsy drum arrangement on the front deck. Also gypsy engine. See Dolbeer locomotive.

The **Dolbeer locomotive** was the most popular of the **gypsy locomotives**. The shaft driving the **gypsy** drums, or **nigger-heads**, was geared directly to the engines, while a sliding pinion gear on the axle could be shifted into neutral while the gypsy was in use. *From B.H. Ward collection*

gypsy tender The man who tended the gypsy: handled the line, wrapped it around the gypsy or drum and coiled it as the line was brought in. See spool-tender.

gypsy-type lowering engine A lowering engine using a very large gypsy-type wheel. A heavy duty roader was also used for the type of work, but the cable was reeled directly on and off the drum.

A **gypsy-type lowering engine**, showing its large **gypsy wheel**. *From Willamette Iron & Steel Works catalog, 1920, Glen Comstock collection*

gypsy wheel A large drum that held several wraps of cable to raise or lower a power car on inclines. See gypsy-type lowering engine. (P.C.F.)
gypsy yarder See Dolbeer.

H

habitant A French-Canadian logger. (P.N.F.)
hack 1. A caboose on a log train. (P.C.F.) 2. In turpentine orcharding, a strong gouge-shaped knife with a curved edge fastened to the end of a short handle, bearing on its lower end an iron ball weighing about four pounds. It is used to chip timber. (S.F.) 3. A bark mark or log mark. (L.S.) 4. To make ties. See tie hack. (P.C.F.)
hacker See tie hack.
hackin' A clearing made by girdling (or hacking) trees in order to kill them. These trees were later burned to help clear the ground. (West Virginia)
hackmatack The Algonquian word for the hemlock or larch tree. (N.F.)
haired-up Said of cutover land that has grown up with young trees. (P.N.F.)
hair pounder 1. An ox driver. 2. A horse teamster or mule skinner.
half-a-side The men and equipment required to rig up a side in advance of the regular crews. If an outfit operated two and one-half sides, it meant they had two sides, with one side rigged ahead. (P.C.F.)
half-breed 1. A combination ax for felling and swamping. 2. A donkey, between a roader and a yarder. Also breed. (P.C.F.) 3. A logger or mill worker who is half-white and half-Indian. Also breed.
half-burn To burn half the slash as a fire-prevention measure. It is difficult to burn half of anything. (P.C.F.)
half moon A piece of metal that is knicked out of an ax by striking it against a stone.
half-moon tie A tie made from a stick of timber yielding two ties, each with one smooth side. (S.F.)
half-side A cat yarding part of a setting to within reach of a donkey. (P.C.F.)
half-slewed Said of a man half drunk. (Maine)
half-sole a chew To spike chewing tobacco with snoose. (P.C.F.)
half way between the devil and a red pig Said of a person in a very dangerous situation. (Maine)
halo kumtux (kumtucks) A Chinook jargon term meaning "no understand, no savvy." (P.N.F.)
hamburger flats The skidroad in town. (P.N.F.)
hames That part of a horse harness that rests on the collar and to which the tugs are attached. LL
hammer 1. In pile-driving, the big, heavy iron weight used to drive piles. 2. The action of the pounding of an engine with worn bearings. LL
hammer a saw To pound a saw with a special type of hammer in order to adjust the tension within the metal and straighten a bent saw.
hammerman A pile driver engineer. (P.C.F.)

A U.S. Army **Spruce Production Division** railroad trestle under construction. The **hammer** on the **pile driver** is clearly visible. *From "Pictorial Review of World's War Activities Spruce Production Division, U.S. Army, Oregon & Washington," 1918, A.M. Prentiss photo, Philip Schnell collection*

hand A mill worker or logger. A popular term in the Pacific Coast forests since the influx of the mill workers from the South: "Do you need any hands?"
hand-bank To haul to the banking ground, with hand sleds, ties or other timbers that are to be floated. (R.M.F.)
hand-banker One who hauls ties on a hand sled from the stump to the landing. (R.M.F.)
handbarrow Two strong, light poles held in position by rungs, upon which bark or wood is carried by two men. (N.F.) Also ranking bar.
hand brakes Applied to disconnected trucks which had to have brakes set by hand, as opposed to air brakes on most equipment. Every air brake car also has a set of hand brakes.
hand dog See dog warp.
hand fallers In the late 1800s, the men who cut the timber with crosscut saws. (P.C.F.) (British Columbia)
hand-holt A grip or anchor to hold onto.
hand jack A small screw jack used in loading logs on cars before the days of the donkeys. See screw jack.
hand line Usually a light cable that the high rigger uses to haul up the pass line block with which he hauls up the heavy gear by running a line through the pass line block.
hand-log To move timber without the aid of animals or machines.
handlogger 1. A man who cuts trees, bucks logs and yards them into the water by hand. A common operation when the trees grew on steep slopes near the rivers or near the ocean. (P.C.F.) (Alaska) 2. Sometimes applied to loggers in the Pacific Northwest who use animals instead of power skidders. (P.N.F.)
hand logging Early British Columbia loggers, working without horses or machinery, rolled logs down the mountainsides into the sea by using hand jacks.
hand pike A short pike-pole, usually from 6 to 8 feet long, for handling logs piled in water. Also pike lever. LL
hand ring A ring on a swivel line hook, grasped to swing the rigging. (L.S.) LL

A **swivel line hook** equipped with a **hand ring**. *Kloepfer Hardware Co. catalog, Glen Comstock collection*

hand set A pair of hand fallers. (P.C.F.)
hand shoes Gloves. (P.N.F.)
hand-skid To move timber by hand to a point where it could be reached by horse. (R.M.F.)
hand skidder The man who ran along, throwing short skids under heavy logs as they were being ground-skidded. Later the skidroads were built. (P.C.F.)
hand-sluice To shoot logs down steep slopes on a crude slide made by felling timber down the slope, cutting off the top and arranging the boles so that a rough trough resulted. (Eastern Canada)
handspike A hardwood pole for prying logs before the use of the canthook and peavey. (N.F.) (Canada)
hand wheel friction A hand wheel-operated clutch on the haulback drum of a donkey. See friction.
hand winch Any winch operated with a hand crank.
handyman A good logger or mill worker who was skilled at many jobs and who could fill in when someone was absent.
handy with a hammer Said of a man who attempts to stamp his log brand on strayed logs that belong to others. (P.C.F.)
handy with the thumb Said of a scaler who cheats the cutting crew. (P.C.F.)
hang a block To put a block in position when rigging.
hang a boom 1. To rig up a loading tree. 2. To place a boom on a spar tree. 3. To put the boom in place around a bunch of floating logs. (N.F.)
hang an ax To fit a handle to an ax.
hang a set of sticks See hang sticks.
hanger A bracket used on a tree or A-frame to support a skyline or tram line. (P.C.F.)
hangfire A dynamite charge that fails to blow in the normal time required for the fuse to burn down.
hang fire To stop whatever is being done at once in order to figure out something.
hanging wedges In the drag saw days, the wide thin wedges that were driven into a log at right angles across the cut to prevent the log from breaking downward at the point of cut, and to keep the saw from being pinched as the log settled and closed on the saw cut.
hang sticks To lay out boom sticks in a pocket in order to make up a raft. Also hang a set of sticks. (P.N.F.)
hang the riggin' 1. To rig up a spar tree. 2. To prepare to fell a tree. (P.C.F.)
hang the throttle over the tank To pull the throttle all the way out on a steam locomotive, thus making a fast run.
hang tough tit To stick to a decision. (P.C.F.)
hang-up (noun) 1. Logs caught behind trees or stumps in a logging operation. 2. Tangled rigging. 3. Any operation that has failed. 4. A tree stuck against another. Also buckwheat or lodge in the Lake States. 5. A man who has a source of mental or emotional difficulty.
hang-up (verb) 1. To get logs caught behind trees or stumps in a logging operation. 2. To get the rigging stuck to prevent logging. 3. To get a tree stuck against another in felling. 4. To get fouled up in any operation. 5. To discontinue a river drive for some reason. (L.S.) 6. To get a load stuck in mud or behind a stump. (L.S.)
harbor (the) Grays Harbor, Washington, which has long been a famous logging center. (P.N.F.)
hard eye The filer. (S.F.)
hard face Any logger who looks, or is, tough. Particularly said of a man who gets the job done come hell or high water.
hard hat 1. An metal or plastic hat that loggers and some mill workers wear for safety. An old-timer said that in his day loggers wore soft hats and tin pants, but modern loggers wear soft pants and tin hats. Also tin hat. 2. A logger who wears an aluminum hat. 3. A construction worker. (P.C.F.)
hard-looking country Country with rough ground and difficult to log.
hardness The property of wood that is indicated by resistance to cutting, scratching, denting, pressure or wear.
hard-nosed Said of an ornery or bull-headed person.
hard rub 1. A heavy siwash in a line. 2. A tough job. 3. Bad luck. (P.C.F.)
hardtack Flat or stale bread.
hardtack outfit, hardtack show A camp with poor food. The name derived from the hard and cheap Swedish bread often served in camps in the Lake States. (L.S.) (P.C.F.) LL
hardwood 1. Lumber that is cut from dicotyledons. 2. A broad-leaved or dicotyledonous tree. Also broadleaf.
hardwoods The botanical group of trees that are broad-leaved. The term has no reference to the actual hardness of the wood. In the Idaho White Pine Country, for example, woods other than the soft white pine; in the Redwood Country, other species are known as whitewoods.
harness The rigging that is in place on a tree. (P.C.F.)
harp See crosscut saw.
harvesting Logging a tree farm.
hasher A waitress.
hash hammer See gut hammer.
hash house A cook shack.
has his head cut in An expression from railroad slang which refers to a man who is a clear thinker. Originally this expression meant to cut in on the air brakes of a train in order to have control. A man with his brains disconnected is as helpless as a train with no brakes. (P.C.F.)

has his tail up Said of a man not discouraged by hard luck.
hashslinger The cook. (L.S.)
haul 1. The distance in which logs are hauled in yarding. 2. The distance in which the logs are hauled from the woods landing to the dump. 3. The distance in which the logs are hauled from the dump to the log pond. 4. The distance and route logs were moved between points. (West Virginia)
haulage rope The mainline as it was called in ground lead yarding.
haulback 1. In yarding, a small wire rope traveling between a power skidder and a pulley set near the logs, used to return the main cable with tongs, chokers or hooks back to the next log. 2. The drum used to handle the haulback line. Also haulback drum. See return line, receding line, trip line.
haulback block A block on the spar tree through which the haulback line runs. Also trip line head block.

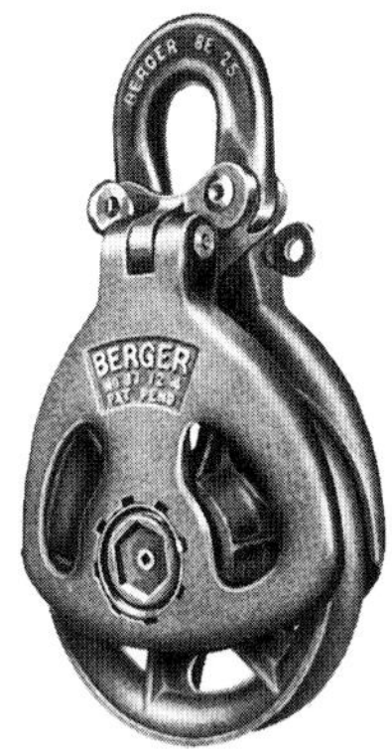

A Berger **haulback block**. *From Kloepfer Hardware Co. catalog, Glen Comstock collection*

haulback show An operation where logs are yarded away from the donkey, rather than to it. (P.C.F.)
haul drums The small drums on an electric donkey on which several turns of cable were wound before passing on to the main or winding drum.
hauled Said of a log jam which has been dynamited or broken free. (P.N.F.)
hauling drum The drum on a donkey that handles the mainline. Also mainline drum.
hauling-in line The mainline in the early ground lead system. Also haul-in line.
hauling line 1. The mainline in a cable system of lowering logs down a grade. 2. The skidding line in a skyline system. 3. The mainline on a high lead show.
hauling off The transporting of logs from the yards to the landing site. (Maine)
hauling piggyback To haul a trailer on the main truck. (P.C.F.)
haul-in line 1. In skidder logging, the line running from the donkey drum through the head block to another block hung from the carriage, and then down to the log. 2. A mainline on any donkey show. See hauling-in line and mainline.
haul it off the fire before it's cooked To leave a piece of work unfinished. (P.C.F.)
haul-up A straw line or light cable used in horse-hauling a heavy mainline into place in ground lead yarding.
have its valves lame Said of a steam locomotive, or any steam engine, that is out of time. When it is square, it is perfectly timed. When a steam engine is out of time, it is said to have a "limp" or to be "lame." (P.C.F.)
have its valves square Said of a locomotive in good working order, that is, it is perfectly timed. (P.C.F.)
have the deadwood on him 1. To have an advantage over a man. 2. To know the truth about a situation concerning a man.
haw A command given to a team to indicate to turn to the left. (Gee is to the right.)
hay 1. Money. 2. Money in a pay envelope.
hayburner 1. A horse in the 1800s. 2. A kerosene lantern.
hay chaser 1. A teamster. 2. A farmer.
hay-hill A hill on which hay or dirt has been scattered on the ice to act as a brake on a sleigh runner. (L.S.) LL
hayrack boom A type of McLean boom loader that has two parallel logs, which uses two pairs of tongs for McLeans' loading system, one if heeling the logs (which would make it a "heel boom"). (P.N.F.) See McLean boom loader, heel boom.

A **McLean boom loader**, or **hayrack boom**, loading logs. Each **tong line** hangs beneath a different leg of the boom, which keeps the log in line with the car to be loaded as the log is swung over the track. The outer end of the boom is supported by the **nose guy**, or **sail guy**. The **mainline** from the **donkey** operates the **loading tongs**, while the **haulback** is used to swing the boom. A counterweight, or **squirrel**, not shown here, serves to swing the boom back again. *Prentiss photo, Labbe collection*

hay road See tote road.
haywire 1. See strawline. 2. Said of anything and everything that is either poorly operated or poorly put together. Before the coming of the haybaler, with its haywire, the term used was "rawhide," since rawhide was used to patch things together before wire became available.
haywire from the cookhouse to the tail block Said of something that is extremely bad. (P.C.F.)
haywire outfit (show) Said of a poor logging outfit. See haywire.
haywire side 1. A rigging crew that has to handle a large number of cables. (P.N.F.) 2. The crew of a high lead camp that has poor equipment. See haywire outfit.
haywire the job To work in the cheapest way possible by using poor equipment and by taking chances. (P.C.F.)
hazel hoe A light grub hoe, used for tree planting or for building a fire line.
hazel hook See brush hook.
he Said of anything that is big and strong, such as a "he truck," or "he log," etc. (P.C.F.)
head 1. The top man or foreman. 2. A prefix used before other words to denote the foreman of a particular job or type of work. 3. A head of steam, to indicate operating pressure in the boiler.
headache rack A safety guard behind the truck cab.

head block 1. The main block on a high lead spar tree. 2. The big front cross timber on a donkey sled. Also fairlead bunk. 3. A huge diameter block used at the woods end of a fore-and-aft road when an endless cable and a grip wheel engine were used to road logs instead of an ordinary donkey. (P.C.F.) 4. A log placed under the front end of the skids in a skidway to raise them to the desired height. (N.F.) 5. A short piece of timber upon which the raft oar rests. (App.)

A large **steam donkey** owned by the U.S. Army **Spruce Production Division**. The **fairleads** are mounted to the **head block**, at the front end of the **sled**. The fairlead at the right is a **bullseye fairlead**, used to keep the **haulback line** in **lead**. *From "Pictorial Review of World's War Activities Spruce Production Division, U.S. Army, Oregon & Washington," 1918, A.M. Prentiss photo, Philip Schnell collection*

head brakeman The conductor on a logging train who usually rode the locomotive. (P.C.F.)
head bummer The boss or working foreman. (L.S.) LL
head bucker A bull buck. (P.C.F.)
head chopper The head faller.
head crane hooker The number two man, next to the head loader, working on a crane loader. (P.C.F.)
head dam A storage dam for holding water back until it is needed for driving logs downstream. (L.S.) LL
head driver An expert river driver when, during the drive, is stationed at a point where a jam is feared. Head drivers usually work in pairs. (L.S.) Also log-watch. LL
head-end brakeman The brakeman who rode on the locomotive. (Red.)
head faller The foreman in a felling crew.
head for the brush To return to the woods after a few days in town. (P.C.F.)
head grabs The grabs on the first log of a log turn; may be either crotch grabs or four-paw grabs. (N.W.)
head handyman The number one handyman or utility logger around camp. (P.C.F.)
head hook A hook with a latch that was used to catch the eye of the choker on a skidder show. A second hook held additional chokers. (P.C.F.)
head hooker 1. The foreman of the choker setters on a skidder crew. See hooktender. 2. The chief of a pullboat skidding crew. (S.F.)
head hook-on An early term for a rigging slinger or boss of the choker setters. (P.C.F.)
head loader The top man in charge of loading at the landing.
head log 1. The brow log at a landing. 2. The front bottom log on a skidway (rollway). (N.F.) 3. The front log in a turn of logs. Also face log. (N.F.)
head-long log loader The head loader in a camp where tree-length logs are loaded on disconnected trucks. (P.C.F.)
head mast An older term for the spar tree at a landing, as opposed to the tail tree. (P.N.F.)
head push 1. The number one man in the woods. 2. The boss driver in charge of a river drive. (L.S.) LL
headquarters 1. The distributing point for supplies, equipment and mail in a logging operation; not necessarily the administrative center. 2. Usually the site of the shops and engine facilities on a railroad show.
head rig A boss in a logging operation. (L.S.) LL
head rigger 1. The foreman of a rig-up crew. (P.N.F.) 2. The head rigging slinger, or foreman of the choker setters. 3. The man who was second in command to the hooktender in the early high lead or slackline shows in the Redwood Country. In the old days on skidder shows, the head rigger was the skidder foreman, in charge of the entire crew, and in this case was sometimes called the first rigger. In some places he may be a climber.
head rigging rustler The head rigging slinger. See head rigger.
head saw In the Pine Country sometimes means the head faller.
head short log loader 1. The head loader in a pine camp. 2. The head loader in a fir camp where 32- or 40-foot logs are loaded on skeleton cars or flats.
head skidder A foreman on a skidder operation.
head slinger In the high lead or slackline shows this man, who was foreman of the choker setters, directed or signaled the donkey engineer when to stop the mainline so that the logs with chokers could be attached to the line. (P.C.F.)
head stick The one head stick that is square across the front end of a raft, making a blunt bow. Formerly flat rafts were headed by two head sticks angled to make a pointed bow on a raft. The towing line was attached to the point of the two sticks. When moving, the logs often pushed into the angle and rode over the edge of the boomsticks. To overcome this, one head stick was used for a blunt bow. Logs are kept in place with a swifter laid across the top, not far back of the head stick.
head train loader In the days of railroad logging, the top man in a loading crew. (P.C.F.)
head tree A spar tree at the railroad or truck landing; opposed to the tail tree back in the woods on a swing operation.
head trip block See haulback block.
headworks A general term applied to all or most of any part of the following: 1. The lowering engine at the top of an incline. 2. The main block, its housing and rollers for the main line at the top of the hill. 3. Includes both 1 and 2, plus any other rigging deadmen, trestle work, housings, etc. (P.C.F.) 4. The intake of a flume, or the point at which logs enter a big sorting works on a river. 5. The operating machinery on top of a splash dam. 6. A platform or raft, with windlass or capstan, that is attached to the front of a log raft or boom of logs for warping, kedging or winding it through lakes and still water, by hand or horse power. (N.F.) (Canada) 7. In river driving, the buoy, a small raft anchored in midstream. (N.W.)

A **headworks** used to tow log rafts across small lakes. The winch is hand-operated. *From "Logging," Bryant, Wiley & Sons, 1923, Philip Schnell collection*

heart The mature wood at the center of a tree.
heater boom A small round boom of logs. (P.N.F.)
heat test A test to see if the train brakes are working. The car wheels are felt at the bottom of the hill: a cool wheel means no brakes; a burned hand means a stuck brake.
heaver 1. A fireman on a wood-burning steam engine. (P.C.F.) 2. The man who fires coal. 3. The fireman on a logging locomotive. (L.S.)
heaves A lung disease of horses that prevented proper breathing. (N.F.) LL
heavy Said of a man or beast with the heaves, asthma or any breathing difficulty. (L.S.) LL
heavy as a dead preacher Said of something very heavy. (P.N.F.)
hedge A fence made by felling trees. (Maine)
heel 1. To load logs with a heel boom. 2. To heel logs, meaning to settle the heel of the log against the boom. See heel boom. 3. To tighten up the heel tackle when rigging up a steel spar skidder.
heel block A small block on the heel of a loading boom.
heel block drum A small drum on a yarder or a skidder which is used to raise or lower the skyline by reefing on the heel tackle.
heel boom Any loading boom that uses a single tong and heels the short end of the log against the underside of the boom. Also tracked loader.
heel line A light line for tightening the heel tackle attached to a skyline.
heel loading block The block on the tip of a heel boom through which the tong line passes. Also called tong line block.
heel rails Iron or old railroad rails used to shoe the underside of a heel boom to take the sharp blows of logs being bumped up against the boom; however, a large part of their purpose was to keep the heel of the log from slipping out from under the boom. (P.C.F.)
heel strap See boom strap.
heel tackle 1. Gear used to tighten a guy line on a skidder. 2. A system of lines and blocks used to tighten the skyline.
heel tackle rig 1. A skidder. 2. The heel tackle setup as used on a skidder.
heft 1. The weight of any object. 2. To lift an object.
hefty 1. Heavy. 2. A strong man. 3. Anything well-built.
he has a long-handled ax Said of someone cutting over the line on someone else's property or stealing timber. (L.S.) LL
heifer dust Snuff. (L.S.) LL

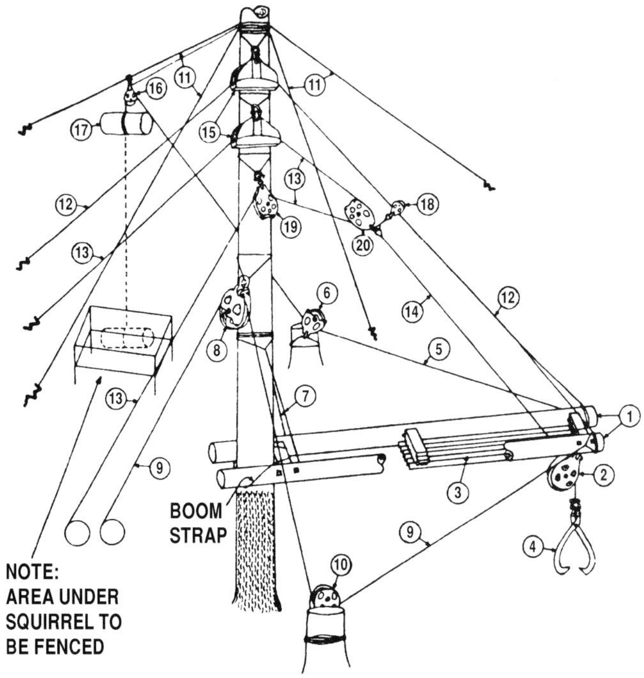

The **heel boom** loading system. 1) loading boom; 2) tong line block; 3) heeling irons; 4) tongs; 5) boom swing line; 6) squirrel line swing block; 7) boom hold-up straps; 8) haulback lead block; 9) haulback line; 10) haulback swing block; 11) buckle guys; 12) sail guy; 13) loading line; 14) tong line; 15) tree shoe; 16) squirrel block; 17) squirrel; 18) sail block; 19) load line lead block; 20) loading or fall block. *From "Safety Standards for Logging Operations," Washington State Department of Labor, 1961, Philip Schnell collection*

heifer gear A small spur gear that runs on the bull wheel of a donkey.
height-of-land line A height-of-land is the top, crest or highest point of land between two watersheds. Logs could not be hauled uphill, so the height-of-land where a lumbering operation was going on was marked by a spotted line (i.e., a blazed trail) to help the loggers

In June, 1957, the Long-Bell Lumber Company operated this diesel **side** above Keasey, Oregon. The logs were **heeled** on the **heel boom** for loading. Note the small log or **squirrel**, running up and down the **guy line**, which acted as a counterweight to swing the boom back. *Labbe photo*

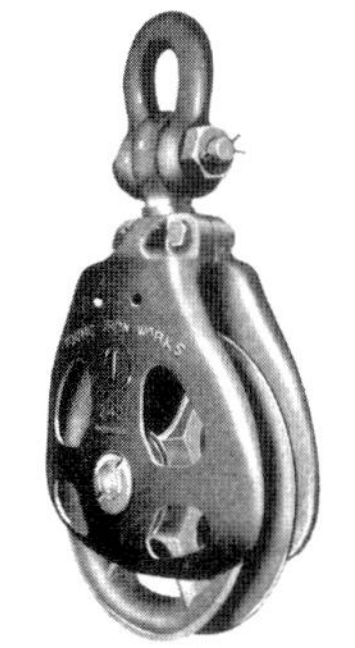

A **heel loading block**. *From Young Iron Works catalog, Glen Comstock collection*

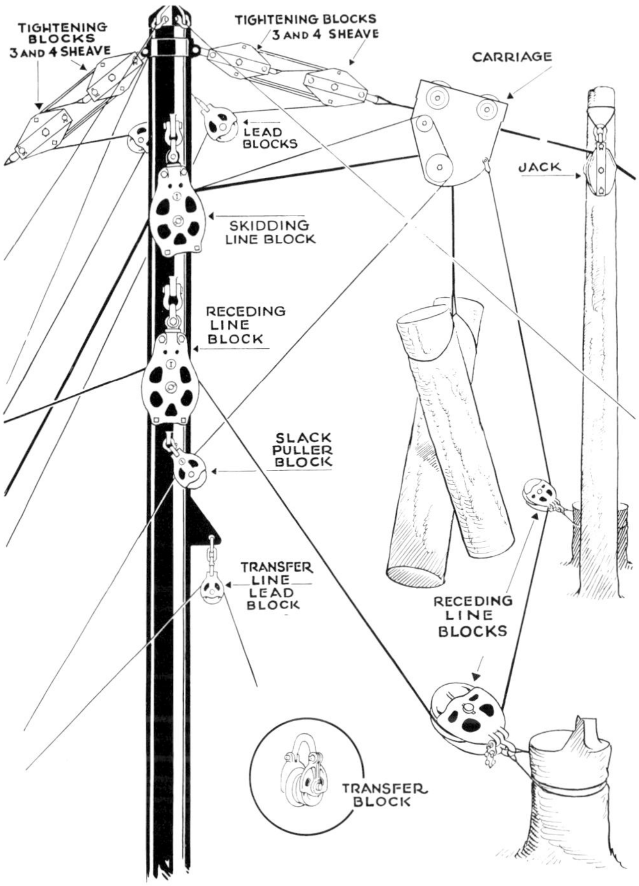

A steel-tower **skidder** system. At the top is shown the **heel tackle**, used to tighten the **skyline.** *From Young Iron Works catalog, 1952, Philip Schnell collection*

avoid cutting over the top. To distinguish it from other spotted lines, the spots were often painted red. (Robert E. Pike, "Tall Trees, Tough Men")

Heisler A geared locomotive that was used on the logging railroads. The cylinders were slanted at 45 degrees and set at right angles to the engine frame. The crankshaft and drive shaft were geared to one axle in each truck, and the second axle was connected by a siderod.

he is wearing a wood kimona Said of a lumberjack who is dead and in his coffin. (L.S.) LL

helicopter logging In recent decades, helicopters have been used in logging or hauling lumber from or to inaccessible places.

hell for Used as an intensifier, for example, "hell for strong," or "hell for stout." (P.C.F.)

hell town in the pines Said of Seney, Michigan. (L.S.)

he logs on section 37 Said of timber being stolen from someone else. There are only 36 sections in a township; therefore, section 37 is a mythical description. See rubber forty. LL

helper 1. A cat or truck that is used to help another truck, etc. on a rough road or steep pitch. 2. In railroad logging, another engine used to help push on a steep grade. 3. See second faller.

helve The wooden handle of an ax. LL

hemlock West Coast hemlock is fine textured, straight-grained, stiff and easily worked; it adapts itself readily to many different uses. It is particularly distinguished by its strength, freedom from pitch, freedom from splintering and its ease of working. It dries well, either by air-seasoning or kiln-drying, and when properly seasoned and stored is subject to a minimum of shrinking or checking. It is used primarily as pulpwood and preferred in making paper. Also Western hemlock or Pacific Coast hemlock.

hemlock show 1. In the 1800s and early 1900s, hemlock logs were not wanted, and any poor show in scrubby timber was called a hemlock show. 2. The logging of hemlock. 3. An operation designed to harvest hemlock primarily.

hemlock side 1. The poorest side in any logging operation. 2. A side logging primarily hemlock.

hemlock special A long train of sleighs loaded with logs being pulled by a steam hauler. (L.S.) LL

hemp The core of a wire rope.

hen fruit Eggs. (S.F.) (West Virginia) (N.W.) LL

hen house show A poor logging outfit. (P.C.F.)

henskin 1. Thin soled shoes that cannot take caulks. 2. Light underwear.

Hercules A brand name for a type of jack, for a donkey, for a disconnected truck, for a logging bunk and for an unloading boom.

herder One who patrols a lumber or log flume to prevent jams. (California)

herkin Snags or obstacles that hinder log-handling in river water. (P.N.F.)

hermit 1. Any logger who lives alone. 2. A logger who works alone. (P.C.F.)

herring chokers 1. Loggers from New Brunswick. 2. The term was occasionally applied to loggers from Prince Edward Island. (Professor David C. Smith, Department of History, University of Maine at Orono.)

hew 1. To cut a piece of wood with an ax or adz. 2. To shape a piece of wood or timber.

Many loggers were artists with the ax and **adz**, as these hand-**hewn** redwood timbers testify. *From Kramer Adams collection*

hew a ride To cut off a flat area onto a log so that it would ride on that side, preventing it from rolling, in the early days of animal skidding. (P.C.F.)

hewer The broad-ax man who hewed tree trunks into masts, square timber and railway ties. (N.W.)

hewing dog A metal bar with spiked ends to hold a log in place while hewing. (L.F.) LL

A small narrow gauge **Heisler** locomotive of the Swayne Lumber Company, at Oroville, California, photographed in June, 1936. *Guy Dunscomb photo, from M. Koch collection*

hibernate To leave the woods and shack up for the winter. (P.C.F.)
hick 1. A lumberjack. 2. An uncouth person. (P.C.F.)
hickey An iron bar or lever used for tightening brakes on a logging train. See brake hickey.
hide The outside of anything.
hidebehind A stump or anything to hide behind during blasting operations or when a tree goes down.
hide gravel To force gravel under the ties in raising the track.
highball 1. A signal sent from the woods to the donkey operator that means "go ahead" or "go fast." 2. Used the same way on the trains. 3. To work fast.
highball outfit, highballin' outfit 1. A company that makes its men hustle—no loitering on the job. 2. A company that is after production first of all, with little consideration for the comfort and safety of the men.
highballin' To go fast.
high behind A mythical monster high in the behind. Human flesh was a necessary part of its food. It ran backward over its victim whose body it had a way of hiding so that it was rarely ever found. If the body was found, the head, which to the high behind was the most desirable part, was always gone. The effect of this tale was startling to the greenhorn. (Agnes M. Larson, History of the White Pine Industry in Minnesota.)
high climber A person who tops and rigs spar trees or poles; in some camps he also rigs the tree. (P.C.F.)
high climber block A light block that is used by the rigger at the top of a spar tree. Also climber's pass line block. See pass block.
high climber's elevator 1. A light wire rope that is used to pull up the gear to the high climber when he is rigging a tree. 2. A pass line. Also jack line, haywire, strawline.
high-deck To cold-deck logs by skidding into small piles parallel to the railroad track for loading on cars.
high grade 1. Good quality timber 2. To cut only the best trees in a stand of timber. Also gut a show. 3. To pick out the top grade logs and leave the lower grade in the woods, a practice frowned on as wasteful and causing a fire hazard, but sometimes resorted to in a desperate effort to make ends meet. Also gut a show.
high hat skinner Said of a horse skinner because many of them wore a tall black hat with the top not dented in to keep water from settling on it. (P.C.F.)
high hitch A logging arch in which the arch was a part of the cat itself for greater lift and more tractive effort; one stage in the development of the logging arch design.
high lead (system) The popular system of logging using a spar tree that carried the mainline and haulback cables through blocks high above the ground. This was an improvement over the low lead or ground logging that dragged the logs over the ground. Invented in Michigan in the mid-1880s. Also high lead logging, outhaul system.
high lead block The block on the mainline on a high lead show.
high lead logging See high lead.
high lead show Said of the steep or rough terrain that was best logged by the high lead system.
high lead tree See spar tree. LL
high lead unit Combination yarder and loader, both usually mounted on a railroad car like a skidder. (P.C.F.) Also unit.
high line 1. The high lead. 2. The skyline. 3. A cable tram. 4. The upper of two logging roads in the mountains.
highline flyer Another term for Lawson's skidder. See Lawson system.
high pole A tall spar tree. (P.C.F.)
high rigger 1. A logger who tops and rigs a spar tree. 2. In some camps, the high rigger rigs only after the climber has topped the tree. (P.C.F.)
high spar A steel spar skidder. (P.C.F.)
high stump A stump that is higher than a specified standard.
high-water pants Pants that are very short on the leg. If they are cut off, they are stagged.
highwheelers The loggers behind the big wheels. (P.C.F.)

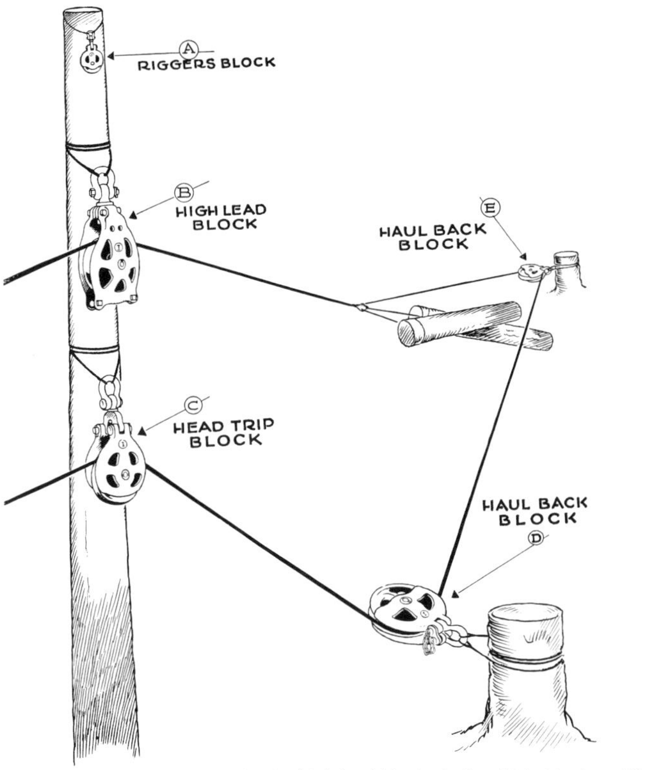

Basic **high lead system**, showing the **high lead block**, the **head trip block**, and the **haulback blocks**. *From Young Iron Works catalog, 1952, Philip Schnell collection*

high wheels Two-wheeled carts designed to straddle the load and raise one end for easy yarding. Wheels were as high as 15 feet; the forerunner of the logging arch. (P.C.F.) See big wheels.

hill-climbing engine A self-powered car that moved loads, or turns of logs, on an incline by pulling itself along a fixed cable. See Fouts' grip wheel and walking dudley.

hillside salmon Deer killed out of season. (P.C.F.)

hillside show A logging operation requiring the use of an incline car or a switchback, in railroad logging. (P.C.F.)

hind-end shack The brakeman who rides on the caboose end of a train. (P.C.F.)

hindu, hondu 1. A cable, approximately 2 feet long and doubled to 1 foot, an eye in one end and a hook in the other, used to connect two sections of strawline. 2. A short strap to fasten lines on a skidder show.

hip deep to a tall Indian Said of a measure of snow or mud. (P.N.F.)

hire-out To go to work for a specific company: "He hired-out to Simpsons."

hiring-wage A uniform wage scale throughout a region for new employees that was once followed in the logging camps.

his nibs An important person; usually a demanding person who is in authority: "Besides the men's living quarters, there was the office where 'his nibs,' the foreman, had his headquarters." Always used with mockery.

histe the turkey To take one's personal belongings and to leave camp. (L.S.) (West Virginia) LL

hit a gill-poke To go ashore, on a log drive, on the limb of a tree that overhangs the river. (L.S.) LL

hitch 1. A hookup in rigging. 2. A hold of some kind. 3. The drawbar of a cat. 4. The trailer hitch on a truck. 5. To put a choker around a log. 6. To put tongs into a log. 7. To tie up lines or rigging.

A Willamette **high lead unit**, built for the Booth-Kelly Lumber Company in 1920. The **unit**, which was usually mounted on a car but sometimes on a **sled**, combined both the **yarding** and **loading** engines in a single unit. In this photo, the newly-delivered car-mounted unit was being broken in on the operations at Wendling, Oregon, northeast of Eugene. *Willamette Iron & Steel Company photo, Labbe collection*

A Sheboygan, Michigan, shingle mill in the 1880s. Note the **high wheels** in the background and the **drag cart**, or **bummer**, in the foreground. *M. Clark photo, Labbe collection*

hitchhiker An entangled log, not choked with the turn, which has been hauled all or part way to the landing. (P.C.F.)
Hitler The early German-designed chain saw. The best chain saw patents were held by the Germans, and with the start of World War II, the patents were released to Western manufacturers. (P.N.F.)
hit the brush 1. To go out on a cruising job. 2. To go to work in the woods. 3. To jump off a runaway truck or train. Also join the birds.
hit the pike 1. To head for town. 2. To quit the job and head for town.
hit the steel 1. To be fired. 2. To quit a job. In the early days of railroad logging this expression meant to walk down the track away from camp.
hitting the high places (spots) 1. Said of a man who quits his job and leaves camp. (West Virginia) 2. Said of a man in an unusual hurry. (P.N.F.) LL
hi-yu Much, plenty or abundance, from the Chinook jargon. (P.N.F.)
hoedag 1. Any grubbing tool, such as a grub hoe, planting hoe or fire hoe. 2. Anything seen in the woods that is unknown and unnamed.
hoffler, hoofler A man who fastened a hook during loading operations in the Pine Country. LL
hog A locomotive.
hogan boys The lumberjacks who had the D.T.s (delirium tremens).
hogan's alley A wooden walkway between the bunkhouse and the wash house. (P.N.F.) (British Columbia)
hog around Said of a machine that is slipping around on the ground. (P.C.F.)
hogback Any sharp ridge. LL
hogger, hoghead A locomotive engineer.
hoist 1. A winch or power device to reel in a cable. 2. An incline railroad. 3. A signal to the loading engineer to lift a log. 4. To lift anything.
hoister A loading engineer, specifically a jammer operator in the Pine Country.
hoisting beam A heavy square timber placed above the sluice gate of a dam and used as a fulcrum for the pry. See sluice gate. (L.S.) LL
hoisting block The biggest block (actually a large frame with several small sheaves at the corners) used in logging which almost covers a railroad car, used in Sessom's system of incline railroad log-hauling. (P.C.F.)
hoisting car See block car.
hoisting line See loading line.
hold, holt An anchor to which a block is fastened. (P.C.F.)
holdback The haulback line, specifically when under tension to keep the turn partly off the ground in yarding.
holdback block A chunk of log that was slotted into the bottom of a chute and placed ahead of a turn of logs; it was held back by a cable to a snubber, thus acting as a brake on the logs going down the chute. (P.N.F.)
holdback line A haulback used as a brake on a downhill haul in the O'Gorman system of cable logging. (P.C.F.)
holder See shank.
holding boom Floating logs held in storage. See storage boom.
holding ground A place in the water where logs or pulpwood were held in a boom and released as needed. (N.F.) LL
holding wedge The same as a hanging wedge, but with a small eye at one corner to which a piece of rope could be attached; used by buckers to drive across a cut, preventing the logs from rolling or falling down while the cut is in its final stages.
holding wood That part of a tree that is left uncut when felling a tree in order to hold the fall of the tree in the desired direction.
holdover fire A fire that secretly keeps burning in a snag or wood pile, and that breaks out later when all fire is thought to be out.
holdown line 1. A wire rope tied to a loader or yarder and anchored to a rock or stump to prevent the sled from lifting on a pull. 2. See anchor line.
hold-up strap A short cable that supports the heel of the loading boom on a spar tree.

hole A term used by brakemen to indicate the area between the loads where they stood to apply the brakes. (P.C.F.)
hole in his boiler Said of a logger with an ulcer or stomach trouble. (P.C.F.)
hole in the sky What tall trees stick through; a measure of height.
hole in the water The spot where the man fell into the river. (App.)
hole ringer A teamster. (P.C.F.)
holes in the doughnuts are too big Said of a camp that offered poor food or the pay was poor. (P.C.F.)
hole up 1. To wait until bad weather is over. 2. To quit work. 3. To retire.
holiday The open area in the timber. (Red.)
hollow butt 1. A hollow log with questionable floating ability because of defects, such as knots or rot. 2. A log with an open center due to rot. (L.S.) LL
holt See hold.
homeguard 1. The local workers who have been on the job for many years as compared to the boomers who come and go. 2. Often local farmers or laborers who work in the woods when they can, but who are permanent residents of the area.
homesteader A logger without a family who stays in a camp for a long time.
homestead the joint To work a long time in one place. (P.N.F.)
home tree A spar tree at the landing, as opposed to the tail tree. (P.C.F.)
hondu See hindu.
honeycomb A type of rot usually found in fir.
hoofler See hoffler.
hook 1. One of the endless varieties of hooks that were and are used in logging to make quick couplings. In the animal skidding or ground lead days, the loggers needed a narrow throat to keep the line from falling out when the line slackened. Locking-type hooks were invented, such as the butterfly or Dennison. At the present time a ferrule is babbitted on the end of the choker, and this knob sockets into the hook on the butt rigging, thus eliminating the locking-type hook. 2. To steal.
hookaroon Another term for pickaroon, the sharp, one-tanged tool for moving timbers when loading or unloading, and for construction work.
hooker 1. The hooktender or foreman of a yarding crew. In the late 1800s he tended the grab hooks used in hooking logs together for a pull to the landing; or he directed the other men to the logs to hook. 2. The head man of the choker crew on a skidder. 3. A whore on the skidroad. The terms hook and hooker were transferred to the skidroads of Western cities and were very popular. A whore house was a "hook shop" or "hook joint," and a whore was a "hooker," and the men usually got "hooked," meaning a poor exchange for the money.
hook'er down a notch To run a machine one gear lower.
hooker-on A tongsman or early choker setter. Also ground loader.
hook'er down, hook'er up Said to indicate the position of the Johnson Bar in the locomotive, which controls the valve setting. Once the train was rolling, one could "hook'er up," or bring the bar back a few notches, which gave a quicker cutoff and saved fuel and steam and made the fireman's job much easier. The opposite would be to drop it in the corner, or open the valves up for a heavy pull. The term was carried over by engineers and truck drivers to cover the same use of gears.
hooking 1. Said of a man who handles the job of hook-tending. 2. Said of a man who handles the tongs at a landing.
hooking-on point The switch where a logging spur runs into the mainline of a logging railroad, and where loads are hooked on to trains headed for the log dump or for shipping points.
hook line A line used as a slack pulling rig on a cableway system of logging. It ran along the ground from the donkey and hooked into the butt rigging.
hookman One who works with a cant hook or peavey. (L.S.) (R.M.F.)
hook-on The procedure of setting a choker on a log in early-day yarding.
hook on 1. To hook tongs on logs to be loaded. 2. To set chokers behind a cat. 3. To hook tongs on logs to be skidded in the early high lead or skidder shows. 4. To go to work for a company.
hook-on man 1. A second loader. 2. A choker setter. 3. The man who unhooks logs at the yarder and attaches chokers for the swing to the landing.
hook plate An iron device for attaching straps to a spar tree. See tree plate.
hooks Climbing irons used by high climbers and which are strapped to the legs. The sharp spurs are used to dig into the bark.
hook shop See hooker.
hooktender The foreman of a yarding crew on a logging side. In the 1800s, he was in charge of skidding, and he either hooked on the logs or told the teamsters what logs to hook on. Also head hooker or hooker in some camps.
hooktender's home A state mental hospital. (P.C.F.)
hooktender's special Snoose mixed with a chew of tobacco. (P.C.F.)
hookup Any arrangement that involves tieing together a number of pieces of rigging.
hooligan's hitch A raftman's knot for tying a rope in a ring of a boom chain. (P.N.F.)
hoosier 1. A beginner in the woods. (P.C.F.) 2. A worker indifferent to the interests of the laboring classes. (L.S.) See scissor bill. The use of this word as a term of contempt started when a Pacific Coast company recruited Indiana farm boys to come west as lumberwoods apprentices. In Bunyan lingo, "hoosier" meant, first, a logger who did not know his trade; then, one who could never master the necessary skills; and finally one who regularly slights his job. (Elrick B. Davis, "Paul Bunyan Talk," American Speech, Vol. XVII.)
hoosier up 1. To malinger. 2. To slow work purposely or to foul up a job. (L.S.) LL
hooter tree An old unmerchantable tree with limbs to the ground, usually found growing in the open, on which the hooters, or male grouse, perch and hoot during the mating season. Also grouse tree.
hoot-nanny 1. An undercutter or type of bracket to support the back side of the saw while cutting a log from the underside. 2. A name for almost any unnamed gadget.
hoot owl To start work as early as possible in order to get in a full day's work before the heat of the day or before of low humidity shut the job down.
hoot-owl shift An early morning job or shift.
Hoover holes Choker holes that were dug by hand during the depression because labor was cheaper than powder. (P.C.F.)
hop toad A derailer on a railroad switch to prevent cars from running out on the main track.
horn The stub of a broken branch left on a log.
hornbeam Any of a genus (Carpinus) of trees of the birch family having a smooth gray bark and hard white wood.
hornbeam lever A piece of wood made from an iron wood sapling that became very strong and durable; used before the cant hook was perfected. (N.F.) LL
hornet's nest 1. Bunched-up logs that make the job of setting chokers dangerous. 2. A ball of stranded cable on a line. (P.C.F.)
horn knot See spike knot.
horse 1. A very good worker. 2. To handle a job with bull strength and awkwardness.
horse around To fool around.
horsebacks Ridges, which are known to geologists as eskers and are formed by subglacial streams during the ice age and are noted for their abundance of gravel. (N.W.)
horse blankets 1. Heavy wool underwear. 2. Flapjacks.
horse bucker A stave bucker operated by a sweep horse power. (S.F.)
horse camp A logging camp where the skidding was done with horses.
horse dam A temporary dam made by placing large logs across a stream in order to raise the water behind it, so as to float the rear. (L.S.) LL
horse'er over To reverse an engine on a steam locomotive.
horse jammer A light gin pole guyed back to its runners, and used

for loading small logs with horses.
horse logs To salvage logs stranded on a drive, pulling them into the water with horses or a peavey.
horse outhaul In ground skidding, the operation of sending the horse back to the woods with the line.
horseshoe end A horseshoe hung on a loading skid pole on which logs were rested in a one-man loading operation. (L.S.) LL
horse skinner A teamster. (L.S.) LL
horse team In the woods there were usually six, eight or ten horses in a team, according to the need. They could make more trips in one day than the ox teams but often broke their legs. Horses were also used in the mill yard to move lumber.
horse wheel logging The operation of yarding logs with big wheels and horses; this was the standard method in the Pine Country for many years.
hospital agent A man or sometimes a woman from a hospital who visited camps selling hospital tickets. (L.S.) LL
hospital tickets The tickets sold to a lumberjack, usually for ten dollars, entitling the holder to hospitalization and care for a year in the event of injury or sickness. This system was the Blue Cross of the early logging days. (L.S.) LL
hostler 1. The man who helped the barn boss in the early days. 2. The man who lit fires and got steam up on the lokies. 3. A machine tender. 4. A mechanic. 5. A roundhouse man.
hot box 1. A hot bearing on a railroad car. 2. Machinery that is hot. 3. A difficult situation.
hotcakes are too round An excuse for quitting the job. (P.N.F.)
hot deck 1. A landing at a swing tree from which logs are swung to the head tree as fast as they are yarded; the logic is that the logs are hot and do not get a chance to cool off. . 2. The logs that are hauled immediately to the mill, not stored in a pile. (L.S.) LL
hot gravel The dry gravel spread over a grade in a thick road to serve as a brake. LL
hot-head An early semi-diesel motor started by heating a plug in the head with a blow torch until it was red hot. (P.C.F.)
hot-log To move the logs directly from the woods to the mill, without storing the logs at any point along the way.
hot logging Moving the logs from the stump to the mill without stopping. Also stump logging. LL
hot shot 1. A on a powder charge with a very short fuse. 2. A fast worker who was not quite as good as he thought he was.
hot skidway A skidway from which logs are loaded immediately. (N.W.) LL
hot stuff The logs that are to be hauled as soon as they are cut down. (L.S.) LL
hotter'n love in hayin' time Said of a very intoxicated logger. (Maine)
hot yarding The complete operation of twitching the logs onto the skidway and then rolling them directly onto the sled. (N.F.)
Houghton tree jack A tree jack with a bearing surface made of the idler wheel section of a cat track instead of a number of sheaves. (P.N.F.)
hound A type of guard through which the tongue slid on a slip tongue big wheel.
house The heavily-built timber structure over a donkey.
hovel A structure to house the draft animals in the 1800s. (N.F.)
how's she loggin'? The same as "How are things going?"
hub A survey point.
hub tree The head tree at the landing, the hub of logging activity. (P.C.F.)
Hudson's Bay ax A small single bit ax with a wide cutting face, used by cruisers and others for light work until about 1970. Also cruiser's ax. (P.C.F.)
hugag An imaginary fierce animal that lived in Northern Minnesota and Western Wisconsin. It was as big as a moose, and its legs were jointless. It had a long upper lip, and its head and neck were hairless. At night it rested by leaning against a tree, and when a human being came along, the hugag fell upon him.
Hughie God. (P.N.F.)
hula dozer A bulldozer blade that can be wiggled up and down and sideways.
Humboldt 1. An undercut that is made upside-down. This cut was designed in Humboldt County, California to let the base of a redwood tree slide off the stump before the top of the tree hit the ground to keep the tree from breaking. 2. The flat chunk of wood taken out in making the undercut. (Red.)
Humboldt log rule, Humboldt scale The rule used to denote the volume of old-growth redwood logs, seventy percent of the Spaulding Scale to allow for defect (the Spaulding log scale less 30 percent). See Spaulding rule. (Red.)
Humboldt stroke In bucking before the introduction of the drag saw, a method of pushing a crosscut saw from the chest forward, as opposed to the pine stroke of bringing the arms along the side and pushing forward. (Red.)
Humboldt yarder A style of compound-geared donkey made by the Willamette Iron and Steel Works of Portland, Oregon. On these yarders, the main, haulback and strawline drums all revolved in the same direction.

An 11 x 13 **Humboldt yarder**. *Willamette Iron & Steel Co. catalog, 1913, Glen Comstock collection*

humidity The amount or degree of moisture in the air which is important to a logger. Fire laws require the woods operations to shut down if the humidity drops too low.
hump 1. A hill on a logging railroad. 2. A knoll that obstructs yarding.
humpback 1. A logging truck with its trailer on its back, returning to the woods. Also piggyback. 2. A saddle tank lokie. 3. See rerailing frog.
humped up like a chicken in a rainstorm Said of a very cold wet logger. (P.C.F.)
hung tree, hung-up tree In felling, a tree that jams against another in falling, which is dangerous to release.
hung up 1. To be stuck or stalled. 2. To get a log stuck behind a stump in yarding. 3. A log that sags in the middle, making it difficult to buck. (P.N.F.)
hungrier than a woodpecker with a hangover Said of a logger who is hungry but too tired to hunt for food. (P.N.F.)
hung up by one corner Said of steep ground. (P.C.F.)
hunks The foreign-born laborers, usually Hungarians or Southern Slavs. LL
hurling down the pine Cutting down the pine trees. (N.F.) LL
hydramatic logger A logger who is shiftless. (P.C.F.)
hydraulic brake In addition to the regular brakes on a donkey, a cylinder filled with water used as a brake to lower cars down an incline. The water was forced back and forth through a small port by a piston geared to the lowering drum. (P.C.F.)
Hyster 1. A trade name for a lift truck that lifts units of lumber onto piles or trucks or railroad cars. 2. A winch made by Hyster. 3. A logging arch made by Hyster.

A wood-fired steam **donkey** equipped with a **house**, somewhere near Astoria, Oregon about 1930. Note the **spring poles** attatched to the house, used to maintain tension on the **jerk-wire**. *From the Steven R. Gatke collection*

A **Hyster** logging **arch** pulled by a Caterpillar **tractor**. *Willamette-Hyster Co. photo, Philip Schnell collection*

I

I.W.A. The International Woodworkers of America, a former C.I.O. loggers' union.
I.W.W. The Industrial Workers of the World, whose members were known as wobblies, is still active. The group was organized in Chicago in 1905 and did some good in cleaning up some of the bad camp situations. Later, radicals infiltrated the organization, and all came to a head in gunfire and murder at Centralia, Washington in 1919. In that year a series of strikes brought legal action against the I.W.W. The abbreviation "I.W.W." was paraphrased by opponents as "I won't work."
ice a road To sprinkle water on a logging road so that a coating of ice may form. Ruts are cut in the ice for sleigh runners in hauling logs. See sprinkler, rut cutter. (N.F.) LL
ice box A large water box on a sleigh. It sprinkled the logging roads with water to make an ice road. (N.F.) LL See sprinkler.
ice breakers Ice guards, deflectors to protect bridges and piers during log drives. (L.S.) LL
ice guards Heavy timbers fastened fan-shaped about a cluster of boom piles at an angle of approximately 30 degrees to the surface of the water. The timbers force the ice to mount the guards and to break, thus preventing the destruction of the boom. (N.F.) LL
icer Same as ice box. (L.S.) See sprinkler.
ice tank See sprinkler.
ickter pooch A mechanical lubricator (pump) that oils the cylinder on a steam engine. The term is supposedly derived from the sound made by the pump. (P.C.F.)

Manzel was a popular brand of mechanical lubricator, or **ickter pooch**, used to lubricate the cylinders of steam **donkeys** as well as locomotives. The operating lever was actuated by the donkey's valve motion. *Mallory Catalog No. 12, Glen Comstock collection*

Idaho cork pine See cork pine.
Idaho white pine See white pine.
Ideal choker hook A deeply notched choker hook, designed to prevent the choker from escaping from the hook; invented by Frank Lamb of Grays Harbor, Washington. (P.C.F.)
idiot powder See snoose.
idiot stick 1. A peeling bar. See barking bar and spud. 2. A large scoop shovel. (P.C.F.)
idler 1. An empty car that is coupled to another car having a load of logs longer than the car, thus permitting the load to be coupled into a train. Used also with a skidder. 2. The idler wheels that support the cat tracks.

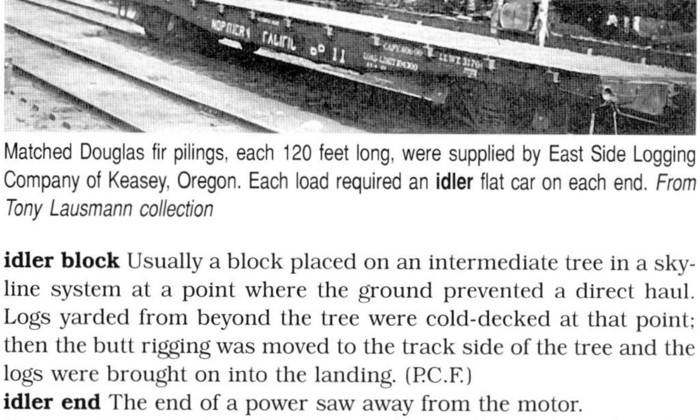

Matched Douglas fir pilings, each 120 feet long, were supplied by East Side Logging Company of Keasey, Oregon. Each load required an **idler** flat car on each end. *From Tony Lausmann collection*

idler block Usually a block placed on an intermediate tree in a skyline system at a point where the ground prevented a direct haul. Logs yarded from beyond the tree were cold-decked at that point; then the butt rigging was moved to the track side of the tree and the logs were brought on into the landing. (P.C.F.)
idler end The end of a power saw away from the motor.
if we can't repair it, we'll paint it yellow and send it back to the woods The shop foreman's explanation of how to handle a difficult repair job. (P.C.F.)
if you know, you don't go hungry A way of saying that a good logger can always get a job. (P.C.F.)

iktas Clothes, goods, things, any personal possessions. (From the Chinook jargon.) (P.N.F.)
I'll cut my suspenders (shoe strings) and go straight up Said by an angry logger who is working hard and getting nowhere. (P.C.F.)
illegal slash Slash labeled "to be burned" by state forestry inspectors, but left unburned by the operator because of weather or other reasons.
improve a stream To remove all obstacles in making a good driving stream out of a poor one.
improvement company A company in charge of building dams and keeping them in repair to supply water for log drives. (L.S.) LL
in a bind 1. Said of a pinched saw in cutting a tree. 2. Said of anything that is stuck. 3. To be in trouble of any kind.
in a jam 1. Specifically said of a wire rope that has run off the sheave and jammed up against the side of the block. 2. Trouble of any kind. 3. Logs jammed up in a river.
inch-and-a-half Time and a half for overtime. (P.C.F.)
incline A system of lowering logs down a very steep railroad grade. On some inclines the loads were counterbalanced by the empties, on others a lowering engine let down the loads and pulled up the empties. On a dead incline the cars were lowered by cable from a stationary engine. On a live incline the power plant itself was reefed up and down the hill, either carrying logs on cars or skidding them on the ties. In the latter case the power car was called a Barney, Dudler or Dudley. There are records of inclines operated on slopes up to 87 percent. John Yeon used an incline prior to 1900 on the Lower Columbia River, with a hydraulic lowering rig and a Mundy donkey for hauling up the empties. In the mid-twenties there were over 50 inclines in use in the West. The last one in Oregon, which was torn up in 1943, was operated by Bill Dyche at the Algoma camp north of Klamath Falls. It dropped off the big rim straight down to the lake shore level.

The **incline** of the Yosemite Lumber Company near El Portal, California, was one of the most spectacular. It was 8,000 feet in length with a drop of 3,000 feet. The grade at one point reached a maximum of 82%. *J. Stoddard photo, courtesy of Lyle Spears.*

incline engine See lowering engine, snubber.
incline rollers The many different types of rollers and spools, horizontal or vertical, to guide the lines used on inclines.

A catalog cut showing an **incline** using the **Sessoms system**. This system utilized a **block car** to gain a block purchase on a load. The line passed from the drum of the lowering engine to the block car and then back up to an anchor adjacent to the lowering engine. Notice the **incline rollers** at left, which would support and guide the lines. *Willamette Iron & Steel Co. catalog, 1925, Glen Comstock collection*

increment The increase in diameter, basal area, height, volume, quality or value of individual trees or stands during a given period.
independent air The separate air brakes on a locomotive, not connected to the train line brakes.
independent logger 1. One who logs and sells his output on the open market. 2. A logger in the 1850s who had his own team and his own claim. (Red.) 3. All loggers were independent.
independent log truck A railroad log truck that is disconnected.
Indian compass The moss on the north side of a tree. (P.C.F.)
Indian's devil An imaginary creature with terrible claws. It would tear the flesh of its victim into ribbons and suck his last drop of blood. At dusk the men could sometimes hear the Indian's devil moaning in the woods, a sign that it was hungry for human blood. Then the tenderfoot was warned to be on his guard. (Agnes M. Larson, History of the White Pine Industry in Minnesota)
Indian silks Overalls. (L.S.) LL
in good kilter Said of anything in good working order. (P.C.F.)
in-haul line The mainline of a high lead or skidder system.
inkslinger The camp timekeeper.
in-lead The wire rope coming off a donkey drum in line with the block through which it runs, indicating that the line pull is in lead with the spar tree.
inside seven load Two big logs against the chocks, two small ones inside and three on top. (P.C.F.)
inspector 1. A man from a state forestry department who inspects logging operations for compliance with the forest laws. 2. A logger who stays in a camp just for a short time. 3. The logging engineer. 4. See grader. 5. The man inspecting and measuring rafts at the market places. (App.) 6. The inspector for the Industrial Accident Commission. (Oregon)
instantaneous change donkey A yarding donkey with a low gear ratio for starting and heavy pulling, and a high gear ratio for speed once the load had gathered momentum. The gear train could be shifted from low speed to high instantaneously, even "on the fly." High gear was used on the haulback to return the butt rigging to the woods. (P.C.F.) See two-speed.
integrated logging A method of logging designed to make the best use of all timber products. It removes in one cutting all timber that should be cut, and distributes the various timber products to the industries that can use them to best advantage. (N.W.)
intensive management The entire complex of industrial tree-farming techniques: genetic selection, site preparation, planting, wood control, pest control, fertilization, thinning and harvesting.
interlocking gear On a skidder, meshing the main gears of the skidding and receding line drums. For returning the skidding line to the woods, the receding line was meshed to a high speed gear, and the

skidding line was kicked out of gear to run free. In hauling in, both drums were interlocked to provide equal line speed on the lines both coming and going.
interlocking skidder A skidder that uses an interlocking gear.

Willamette Iron & Steel Company built this huge **interlocking skidder** unit for Inman-Poulson Logging Company in November, 1923. The interlocking skidder engine and a **duplex loading** engine were mounted on a **skidder car** built by Pacific Car & Foundry Company of Renton, Washington. *Willamette Iron & Steel Co. photo, Jack Taylor collection.*

interval A narrow river valley. (Maine)
in the bark 1. Said of log rafts sent down the large rivers with bark on the logs. (L.S.) 2. Said of unpeeled poles and piling. LL
in the bight 1. A bight is a sage or bend in a wire rope that is on the ground or is flopping around. When straightened out the wire rope will snap inside the bend. Many loggers have been killed by being "in the bight." 2. A married logger. (P.C.F.)
in the boondocks Said of someone or something back in the woods. (P.C.F.)
in the brush Said of someone back in the woods.
in the clear 1. Out of danger. 2. Railroad cars on a side track that are clear of the trains that go by on the main line. 3. Clear of the rigging during yarding.
in (down) the corner Said of the position of the Johnson bar in the steam locomotives when shoved ahead to its last notch, in the right hand corner of the cab. In that position the engine was pulling the hardest, giving the job the best effort.
in the hole 1. Said of a train in a passing track waiting for another train. 2. A logging show at the bottom of a canyon. 3. An operation that cannot break even. 4. The bottom gear in a truck or machine.
in the rip track Said of a man in the hospital. (P.C.F.)
in the sock Said of a man who has a stake put away. (P.C.F.)
in the swing Said of a log that is hung up on both ends and swinging in the middle, making it difficult to buck.
in the tules 1. Said of a man out in the brush. 2. Said of a man who is hiding out from the law. (P.C.F.)
in the water Said of logs delivered to a purchaser at a booming ground or in a log pond.
in-turning Said of a cat coming into the landing with a turn of logs.
iodine 1. The first aid man. 2. Strong coffee. (P.C.F.)
Irish apples Potatoes in a cookhouse.
Irish baby buggy A wheelbarrow. (L.S.) LL
Irish coupling See bucker's coupling.
Irish snowstorm The act of putting snow on bare spots of logging roads to get out the last of the logs in the spring breakup. (L.S.) LL
iron 1. Specifically, all the rods, bolts and fittings used for a job such as bolting up a donkey sled. 2. Any kind of metal to make something around a mill or logging camp. 3. A railroad rail.
iron bender 1. A railroad brakeman. 2. A camp blacksmith.
iron burner 1. A blacksmith. 2. A welder.
iron duke 1. An iron stove in the cookshack. (L.S.) 2. A trade name for an early camp stove.
iron mule A dirt-moving machine.
iron ox The machine used in trailing logs, skidding logs down the ties.
irons See climbing irons.
iron triangle See gut hammer, dinner triangle.
Isaac's haulback Said of a steep logging show in which the butt rigging runs back to the woods by gravity. (P.N.F.)
Isaac's in the cab Said of a good grade on which a loaded truck moves by gravity alone. (P.N.F.)
it bites on both sides Said of a double-bitted ax.
it don't take him long to roll his blanket Said of a man who will quit for any reason.
it's a fit Similar to "it's a deal" or a job well done. (P.C.F.)

J

jabbing rod The piston rod on a steam engine.
jack 1. A lumberjack. 2. A tool of various sizes used for lifting. 3. A locomotive. 4. A hangar to support a line. See tree jack. 5. Money. 6. Bad (from jack salmon). (P.N.F.) 7. A steel structure used to support the tongue end of a log trailer at a preload. See preload.
jackass flu A bad hangover.
jack guy A guy line that carries the loading jack in a power-loading system. See nose guy, sail guy.
jack-hammer 1. A device used to make holes in rocks for dynamiting. 2. A busy woodpecker.
jack in the dark, where are you? A roughhouse game played in the bunkhouse. (L.S.) LL
jackknife 1. To straighten out crossed logs in making up a boom. 2. To pile up logs at the bottom of a chute. 3. To get a truck jammed against the side of a truck, or crossways to the line of travel.
jackknife boomsticks To prelay a set of boomsticks so they will fold out into place when a tug takes off with the head stick.
jackknife butt hook An early butt hook with a locking latch that worked on a spring. (P.C.F.)
jackknife dray A dray loaded so that logs could balance. The driver would step on the back of the logs going downhill to act as a brake, and step on the front of the load going uphill to make it easier to handle. LL
jack legs The outriggers and jacks that block up a skidder.
jack line 1. A light cable used in hauling gear or the rigger up to the top of the spar tree when rigging it for yarding. Also high climber's elevator. See pass line and strawline. 2. A line used on a guy line at a fixed distance from the tree.
jackpot 1. A mess of piled-up logs, either on land or when dumped in the water. 2. Logs felled haphazardly by a green faller. 3. Trouble ("to get into a jackpot"). 4. A bad slash. (N.F.) 5. A contemptuous term applied to any unskilled piece of work in logging. (N.F.) LL
jack screw In the 1800s, jacks were used to jack large logs on and off railroad cars, to move logs downhill into the water, and to move a skidder car off the railroad track. The jack was operated by a rod inserted through a hole in the screw in the base. (P.C.F.)
jackshaft 1. The shaft that transfers power from the bull wheel to other wheels on a donkey. 2. Any short shaft used to transfer power from one source to another.
jackstraw cold deck A cold deck on which logs are not piled carefully, a mixed-up mess.
jack up 1. To force a man to do better work. 2. To discipline a man.
jack works A loading dock built parallel to a railroad spur at car height. The dock had a dogged chain in the bottom of a trough which pulled logs alongside the empty cars. The logs were then rolled or parbuckled onto the railroad cars with cant hooks.
Jacob's staff 1. A strong bar that was attached to the axle of a big-wheel log carrier at right angles to the tongue. When the driver pulled on the staff, the logs were lowered to the ground which acted as a brake. 2. A pointed steel staff used as a support for a staff compass. 3. The wooden handle on a peavey or cant hook. Also jake staff, jig staff, jog staff, Johnson bar, Johnson board.
jag Drunk. (L.S.) LL
jagger(s) 1. The frayed, ragged strand or strands that break off in the wire rope, and that are especially troublesome to the choker setter. Also jigger(s), burrs, whiskers. 2. To fray a line.
jail hook In the early days, a heavy hook with a partly closed throat, making it difficult for a cable to slip out.
jake Said of anything that is satisfactory: "Everything's jake."
jake staff See Jacob's staff.
jam 1. Logs hung up on a river drive. 2. Any kind of mess, difficult situation or predicament. 3. To form an obstruction of logs in a stream. (N.F.) (Eastern Canada)
jam at Gerry's rock The popular ballad which relates a true story that happened during a lumber operation on the Penobscot River in Maine. The logs in the river had jammed, and "six brave shanty boys" and their foreman, young Monroe, volunteered to break the jam. This was perilous work, in which the men lost their lives. Only the mangled body of young Monroe was recovered. His lover, Clara Vernon, died of grief, and her last request, to be buried by young Monroe, was granted. (Roland Palmer Gray, Songs and Ballads of the Maine Lumberjacks)
jam boom Any boom hastily set behind the start of a log jam that prevented more logs from creating a larger jam. (N.F.) LL
jam cracker A riverman who is an expert in breaking log jams. (L.S.) LL
jam crew Rivermen who were skilled in breaking log jams, to see that the channels were kept open so that the logs could float down. Every jam crew had at least one man who knew how to break a jam. His job was to find the key log. The moment the peavey yanked at the vital spot, the whole mass began to tumble, and men moved with speed, for life was in the balance then. To move the key log was a daredevil job, and the man who did it was the hero of the crew.
jam hook See swamp hook.

Sunset Timber Company, out of Raymond, Washington, used this **donkey**, converted into a **jammer**, to load logs onto **skeleton cars**. *Courtesy of H. Samuels*

jammer 1. Originally a huge machine used in railroad logging in the Pine Country of the Southern forests. At a new landing, the jammer was usually jacked up on hydraulic supports, straddling the track. The empty cars were pushed under it and stored up the track. One car would be pulled back for loading, either by a cable from the jammer, or by a locomotive. Other types slide along on tracks on top of the cars, skid along the cars on runners or sit on a side track while the locomotive pushes cars to them. 2. A type of loader. 3. See McGiffert, decker. 4. A derrick for loading logs onto a sleigh or railroad car, operated by horses or steam. Also arsneau. (N.F.) (Canada) 5. An A-frame mounted on a sled for use in loading logs. (N.W.) 6. A river driver who helped to break up log jams. (L.S.)
jammer logging A cable logging system generally restricted to one skidding line and used for winching logs from distances up to approximately 300 feet (from the cutting area to a log collection point). Also short line skidding.
jammer puncher An engineer on a jammer.
jamming logs Loading logs on any vehicle, such as a railroad flatcar. (L.S.)
jam piers Heavy piers to hold back logs in river driving. (N.F.)
jam pike A heavy pike with an eight-foot handle and an eight-inch spike at the front end; used in the early white pine camps by rivermen, it was supplanted by the peavey. (N.F.)
jam pole See pike pole.
Japanese skidder See track layer.
Jap camp Designated camps for Japanese laborers who were imported during World War I. Before World War II, there were Japanese logging camps along the British Columbia Coast.
java Coffee. Also jerk water. LL
jawbone 1. Talk. 2. An unwritten contract. 3. Said of a deal made by too much talk and little common sense.
jay-grab, jay-hook A hook, with a recurved head, to each end of which a grab is attached by a short chain. The jay-hook is fastened to the top of the forward log of a turn on a skipper road and serves as the point of attachment for the draft. If the logs start to run, the draft animals can be automatically freed by turning them at right angles to the road. (App.)
jayhawk 1. To peel bark from a small tan oak as high as can be reached by using a long-handled bark spud, without felling the tree. (P.C.F.) 2. To strip only one four-foot length of bark from a tanbark oak, leaving the tree standing. (N.W.)
jay hole A space provided where the team can step out of the way while the logs are taken by. (West Virginia)
jay-off To step aside and let the logs go by. (West Virginia)
jemmy john The handcar on a railroad track. Also jimmy john and demi john.
jerk the hash Serve the food. (L.S.) LL
jerk water Coffee. (West Virginia) Also java. LL
jerk-wire 1. The out-of-date signal system of the whistle punk which was a wire attached to the donkey whistle and flipped by the punk. The electrically-operated whistles appeared around 1914. 2. To work as a whistle punk.
jerry gang A section crew. (P.C.F.) (British Columbia)
jersey cream 1. A good logging show. 2. Extra good timber. (P.N.F.)
Jesus Bill The evangelist Billy Sunday. (L.S.) LL
Jethro iron A large metal piece used on railroad ties that was invented by a man named Jethro. (L.S.) LL
jewelry 1. The odds and ends of small logging gear, specifically a small block. 2. The various kinds of gadgets put on logging engines to save coal or make more steam at less cost. (P.C.F.)
Jew's harp 1. A ring shaped like a Jew's harp in an ox yoke, used to shorten or lengthen the skidding chain. Also bitch chain or bitch link. In the North Woods called a slip grab. 2. The musical instrument once played in the bunkhouses. Now called jaw harp or mouth harp.
jibbo To remove a dog from a log. (N.F.)
jigger 1. In river driving, a small raft that the checker stood on to sort the logs. (N.W.) 2. A big wagon pulled by a pair of horses. (N.W.) 3. To pull a log by horse power over a level place in a slide. (N.F.) 4. See jagger(s).

A **jerk-wire** whistle system can be seen on this small donkey belonging to Yeon & Pelton, of Rainier, Oregon, on the Lower Columbia River. The **spring pole**, mounted to the **house**, acted as a spring, to allow the whistle lever to close when the wire was slacked. *J.F. Ford photo, from Jim Brady collection*

jig staff See Jacob's staff.
jig team A team of horses used to jigger logs. (App.)
jillagahike See gillagahike.
jill-poke 1. According to the late Frank Fraser, veteran redwood logger, the original jill-poke originated in the building of the skidroads. A notch was cut on the opposite ends of a log or skid. A post of jill-poke was driven alongside the notch with a wooden maul to keep the log or skid from moving. 2. A horizontal pole or lever used to dump logs from flat cars into a log pond. 3. A pole or plank used to shove something. 4. A loose plank on a plank road or bridge that flies up and spears a car or truck. 5. A branch lying in a road that flies up and spears a car or truck. 6. Any prop or timber used as a lever that relies on the momentum of the object to be moved to change its course. 7. A short spar, used instead of a jammer for loading and unloading logs. (N.F.) 8. Any of the three poles of a hoisting gin. (N.F.) 9. A single pole held in a vertical position by guys, which support a block and tackle used for lifting logs. (N.F.) Also gin pole. 10. In a stream, a log with one end stuck in the bank, where it may cause a jam. (N.F.) 11. An awkward or lazy person who fouls up an outfit. (N.F.) 12. To push logs off the flat cars. 13. To be hit or shoved off an object. 14. To puncture a cat radiator with a sharp stick or branch. 15. To be treated unfairly: "The mill owners have been jill-poking him." 16. To be stuck or hung up.

A **jill-poke** unloader used by the Wisconsin Logging & Timber Company at Oak Point, Washington, on the Lower Columbia River. *Courtesy University of Oregon*

jill-poke log In river driving, the log used to close a gap so that other logs could not float through. (N.W.)
jill-pole See jill-poke.
jim-binder A springy pole used for tightening a chain that binds a load of logs. (L.S.) See binder. LL

Logs on this train of the Laurelton Lumber Company at Laurel Park, Pennsylvania, were fastened with chains and then tightened by twisting a stick or small pole in the chains. This was known as a **jim-binder**. *Kline photo, courtesy of W. Casler*

Jim Crow 1. A type of rail bender used for bending or straightening steel rails. 2. See brake hickey. LL
Jim Crow load 1. A load of one big log. 2. A big log on top of two small logs on the outside bottom for the chocks to bear against or to secure the big log on the bunks. (P.C.F.)

A large **yellow fir** log leaves the woods behind **Shay** No. 1 of the English Lumber Company, near Mount Vernon, Washington, in 1908. A load of this type was sometimes called a **Jim Crow load**. *D. Kinsey photo, Labbe collection*

Jim Crow men The men who make ties. (S.F.)
Jim Crows The small unsalable ties. (S.F.)
Jim Hill A railroad spike used in the woods to hold heavy lines in place while splicing. Jim Hill was the builder of the Great Northern Railroad. (P.C.F.)
jinnie See travois.
jin pole See gin pole.
jobber A logging contractor or subcontractor.
jobber's sun 1. Said of the moon in a jobber's or contractor's logging camp; refers to the early and late hours of beginning and ending work. In the 1800s, a day's work was considered done when the moon, the jobber's sun came up. 2. A kerosene torch used at night so that the loggers and lumberjacks could put in longer hours. (N.F.)
job hog 1. A worker who took another man's job. 2. A man who was very eager to get work. (N.F.) LL

job shark An employment agent who sold jobs to lumberjacks.
job simple 1. Said of one who was afraid of losing his job. 2. A stupid person. (L.S.) LL
jocky box 1. The storage compartment in a truck cab. (P.C.F.) 2. A wooden box on the truck frame in which the chains, binders, etc. were carried on the return run. 3. A tool box for a donkey or locomotive.
joe hog A sled for skidding logs. (L.S.) LL
jog staff See Jacob's staff.
John D. An early name for the logger's oil bottle. (P.C.F.)
John is up the river Said of logs that will soon be coming down the river. (L.S.) LL
Johnny Dyer An air hose connection that was cut into the train line in the middle of a log train. If the swing brakeman rode here, he could apply the air brakes in an emergency. (P.C.F.)
Johnny newcome A newcomer. (L.S.) LL
Johnson bar 1. A long, heavy lever that was pinned on a notched steel arc and connected to the valve motion on a steam locomotive. It was used to reverse the direction of the engine, as well as to change the timing and amount of steam admitted to the cylinders through ports in the steam chest. 2. Any big heavy lever or pry.
Johnson board 1. A Jacob's staff. 2. A heavy bar attached to the axle of the big wheels to help raise logs.
Johnson pants Pants made by the Johnson Company in Vermont. (N.W.)
Johnson-Willamette A skidder that used a carriage operated by a two-drum donkey on a standing line. (P.N.F.)
joint 1. The place where two boom logs meet, connected by a boom chain. 2. A section of a raft. (Eastern Canada) 3. To tap the teeth or rakers of a saw to make them of equal length, after which they are filed sharply.
joints In river driving, logs bearing the same log mark and fastened to each other along the shore in rafts. (N.W.) See dropping off, swings.
jointer A gauge to measure the height of saw teeth.
join the birds To jump off any moving equipment when danger occurs. (P.N.F.) Also hit the brush, take to the bushes.
journey, journey cake A corn pone bread baked in a reflector oven. (L.S.) Also Johnny cake.
judy coon A raft with a capstan for floating logs across a lake. (L.S.)
juggler See log rolling.
juicery A saloon. (L.S.) LL
juice side A logging side where the yarder is powered by electricity. (P.N.F.)
jumbo A type of tongueless double sled used for short distance hauling. (N.F.)
jumbo block An extra heavy bull block with a wide throat and a wide sheave with a flattened groove.
jump See jump a tree.
jump a frog, jump a log To place a log to fit into a certain spot on a load. (L.F.)
jump a tree 1. To move a spar tree a short distance in a vertical position by means of rigging. Also slip, spar jumping. 2. To jump a tree off a stump in felling to prevent as much damage as possible. (P.C.F.)
jumper 1. A small wooden sled used to haul short length pulpwood or supplies. (N.F.) (Eastern Canada) 2. A lumberjack who accepts a lumber company ticket and gets off the train before reaching camp. (L.S.) 3. A plaid wool jacket. LL
jumper sled A log sled having a high cross piece on which one end of a log is supported when it is being dragged out of the woods or along a tote road. (L.S.) LL
jumping jack A logger who rested all week and worked on Sunday. (L.S.) LL
jump-up landing The area where loading is done by means of a gin pole raised or jumped up on the spot. (P.C.F.)
jump-up skids Logs laid from inside a rollway up against the brow log to keep logs from rolling onto the track. They could also be raised up and used to roll logs up higher on a load. (P.C.F.)
june bug A lantern set on the boom while loading logs at night. (L.S.) LL

Jumping a tree at Tidewater Timber Company, near Jewell, Oregon. *Labbe collection*

Winter logging on the operations of the Hallack & Howard Lumber Company at La Madera, New Mexico, using a **jumper sled**. *W.J. Duteau photo, from J. Hereford, Jr. collection*

jungle 1. The logging camp and woods. 2. The uncut timber in the back country. 3. The skidroad in town. 4. A tramp's hangout. 5. A thicket.
juniper juice Gin. (P.C.F.)
junk 1. A chunk or large portion: "A junk of wood for the stove." (Maine) 2. To cut up logs, which are left after a piece of land has been burned over, into convenient lengths for piling and burning. (Maine)
junk wagon The sled on which the bull cook and the cookee carried the noon dinner to the men in the woods. (N.F.) LL

K

kahallek special A slide loader used in the Idaho White Pine Country. See slide loader.

A **kahallek special** at work loading **flats** in Southern Idaho. *Labbe collection*

kantuckies Loggers or mill workers from Kentucky. (S.F.)
kaso A tall pail made from birch bark with a small top and broad square base. (L.S.) LL
katydid An early name for a set of big wheels. (L.S.) LL
kedge See warp.
keel A marking crayon used by the bull buck in the woods; in the lumber mill used by graders on the green chain and in the planing mill.
keeper A bolt-like device used to hold the bow of an ox yoke in place. LL
keeping food on the skids too long Said of stale food, especially bakery goods that were kept in storage too long. (L.S.) LL
keg-up 1. To collect material in order to be ready for a job. 2. To hide things in anticipation of a rainy day. 3. To stay in one place.
kegging up Getting drunk. (L.S.) LL
keister A packsack. (L.S.) LL
Kennebecker A knapsack or packsack. (N.W.) LL
Kentucks The people imported to Forest County, Wisconsin from Kentucky. They settled throughout the woods and were a great help in logging. LL
Kentucky windage Compensation for the side lean of a tree. (P.C.F.)
kerf 1. The width of the cut made by a saw. 2. A notch made by the axman to govern the direction of the fall of a tree. LL
kettle 1. A leaky donkey. 2. A leaky locomotive. 3. Any donkey or locomotive. 4. A tea kettle.
key log In river driving, the log that gets stuck first, causing a jam by other logs piling up behind it.
KG-ing The practice of using a KG blade (dozer blade shear) to clear unwanted vegetation in preparation for planting tree seedlings.
kick 1. Cork shoes. (West Virginia) 2. Said of power saws that jerk or kick backwards into a bucker, causing injuries; usually caused by a twig falling into the kerf. 3. To siwash, or change the direction of the lead by using a rub tree or Dutchman.
kick back Said of a falling tree in springing back off the stump as it comes down.
kicker 1. The boom of a power unloader that kicks logs off cars. 2. A light gas engine, specifically a small starting motor on a big diesel cat. 3. In felling, a block placed in position to hold the tree to the desired line of fall.
kicking Said of a falling tree in springing back off the stump as it comes down. (West Virginia) LL See jump a tree.
kicking pole A pole hung between two horses in a stall to protect them if one of them started kicking the other. (Maine)
killapie The mess that happens when the dipper stick of a shovel flops back over the cab or the boom of a cherry picker flies back if not guyed. (P.N.F.)
kill-dad An empty tin pail where all the lumberjacks threw their old or odd pieces of chewing or smoking tobacco. Anyone could borrow the tobacco for his pipe. (L.S.) LL
killig pole 1. A long pole used sometimes as a booster with a peavey. 2. A short, stout pole used in felling to push the tree in the desired direction. (N.W.) See jill-poke, sampson.
kimble A short logging car.
king pin 1. The big center pin on which a bunk swivels on a logging car or truck; that connects the front axle of a wagon; that controls the pivoting of a monorail; that connects the truck of a railroad car with the body and acts as a pivot when the vehicle turns; and many other uses. Also gudgeon-pin, king bolt. 2. The head man.
king snipe 1. The head man. 2. The foreman of a steel gang on a logging railroad. 3. A switch crew conductor. (P.C.F.)
king swamper The head swamper. (S.F.)
kink 1. A sharp bend in a wire rope. 2. A bend in the track.

A **killig pole**, or **sampson**, used in directing the fall of a tree. *From "Logging," Bryant, Wiley & Sons, 1923, Philip Schnell collection*

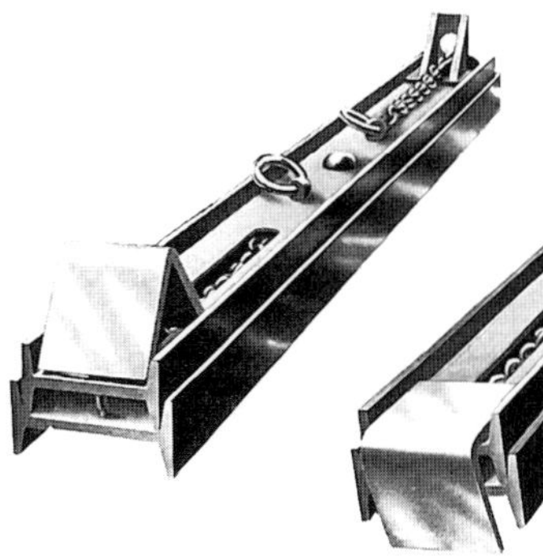

A **Hercules** log bunk equipped with **Knight chocks**, showing the method of releasing the **chock** for unloading. *From Seattle Car & Foundry catalog, Labbe collection*

kitchen car The cook car in a railroad camp. (P.C.F.)
kitchen mechanic Usually the dishwasher. (P.C.F.)
kitchen sweats Dances held at a lumber camp. (L.S.) (S.F.) Also shindig, wrastle.
kitten A small cat (tractor). Also kitty. See turn-out cat.
klatawa To go, travel, move away, migrate, start or leave. (From the Chinook jargon.) (P.N.F.)
klondike 1. A heavy sheet iron pipe of various lengths, that was placed over the chimney or stove pipe in winter to provide an extension that would stick out above the deep snows on the cabin roof. (P.N.F.) 2. A famous whore house near Manistique, Michigan, but the name was used to refer to any whore house. (L.S.)
klondike spuds Raw potatoes fried in a pan. (P.N.F.)
klootch, klootchman A female Indian in the Chinook jargon. (P.N.F.)
Knight chock A patented chock, invented by Jesse Knight, which holds logs and used on Seattle-made bunks. (P.N.F.)
knit To splice, specifically wire rope.
knob A ferrule on a choker that is babbitted to the end of the choker.
knock To cut.
knockdown block A block that can be opened up to lay in a line.
knocker See car knocker.
knock his ears down To thrash a person.
knock it down To fell timber. (P.N.F.)
knock off his hames To fire a man. (P.N.F.)
knock-out shackle A shackle with a driven pin, one not screwed in, for quick moving. This shackle was used on overhead systems where the skyline was moved frequently, to connect the skyline to the extension. The pin was quickly knocked out in order to unpin the eye of the skyline and permit the line to be moved quickly to a new tail tree.
knock out the bud To cut the top off a tree to be used as a spar tree. (P.N.F.)
knock trees down To fell timber. (P.C.F.)
knot 1. The portion of a branch or limb that has become incorporated in the body of a tree that may appear in sawn lumber. 2. To knock the knots or limbs off logs.
knot bumper 1. The man who knocks the knots off logs before they are loaded on a car or truck at the landing. 2. The man who cuts limbs from a felled tree. (West Virginia) Also limber.
knotcher See notcher.
knothead A stupid person (one who is useless as a knot in a board or tree). (P.C.F.)
knotting Smoothing the branch stubs close to the trunk of a fallen tree. (L.S.) LL
knowledge box A schoolbox. (S.F.)
knuckle 1. On the railroad, that part of a coupler that engages the coupler on another car or engine. 2. A joint on a piece of machinery.
knurl A live growth, like a huge wart, that grew on some hardwood trees. (L.S.) See burl. LL
kortick A fire tool that is half hoe and half rake.
kumtucks See halo kumtux.

L

L.S.W.U. Abbreviation for Lumber and Sawmill Workers Union, a union of the American Federation of Labor in the lumber mills.

lag a boiler To insulate steam boilers on locomotives and sometimes on donkeys.

lag a drum To bolt on metal or wooden slats to build up the diameter of a donkey drum and thus get greater speed in line handling.

lagging The insulation placed around steam boilers on locomotives and sometimes on donkeys.

lags In horsepower, the crosspieces for the horse to walk on. (Maine)

laker An expert log driver. (L.S.) LL

Lake States The states bordering on the Great Lakes, but the big lumbering states were those on Lake Michigan and Lake Superior.

Lamb cableway system An overhead cable skidder system with a special carriage, invented by Frank Lamb of Hoquiam, Washington. (P.N.F.)

land To bring logs to a landing or banking ground where they are cold decked or banked for later movement.

landing 1. Usually a flat area where logs are yarded to or assembled for transportation, to be loaded on cars or trucks; the center of operations on a logging show. 2. A platform where logs were collected and loaded on cars. 3. A cribwork of logs, constituting a platform alongside the railroad track, onto which logs are hauled by a donkey, ready for loading onto cars or trucks. (P.C.F.)

landing crew 1. A crew that constructs landings. 2. A crew that works a landing. (P.C.F.)

landing hooker Usually the head loader at the landing.

landing man 1. A chaser, the one who unhooks logs coming into the landing. 2. One who unloads logging sleighs at the landing. (L.S.)

landing show A designated spot on a map of timber to be logged, where a good landing will be located.

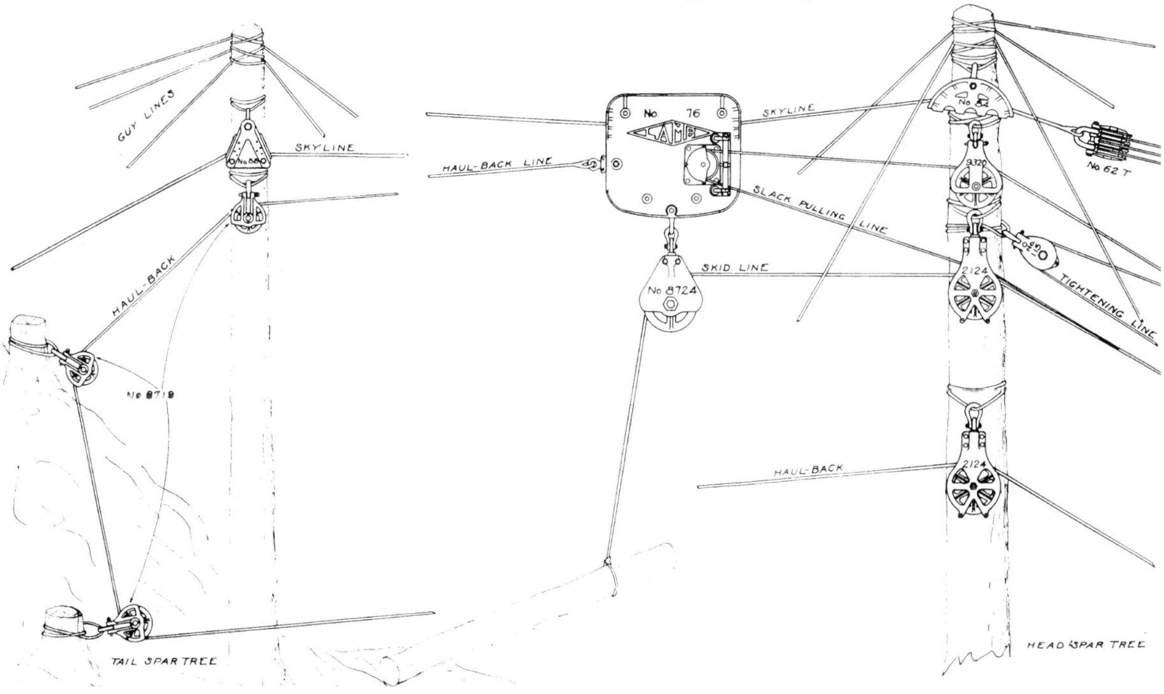

The **Lamb cableway system**. *From Lamb Logging Equipment catalog, Glen Comstock collection*

A railroad **landing**. This was C.C. Masten's operation near Svensen, Oregon. *Nelson photo, from Mr. Nordlund collection*

landing team In a horse camp, the team that was used for rolling, decking and loading around a landing.
landing tree A spar tree at the landing or loading point, as opposed to a yarder tree out in the woods.
land inspector A timber cruiser. Also scenery inspector.
land-looker 1. An appraiser. 2. A real estate man buying timberland on commission. 3. A cruiser.
land plane A mattock. (West Virginia)
land shark A timber speculator.
Lange hook tender The head cook in camp; from the Lange ranges which were very popular in the logging camps. (P.C.F.)
lang lay A method of laying, or weaving, wire rope. See lay.
lank inside Said of one who is hungry. (L.S.) LL
lap 1. The tops of trees, or slash, left in the woods after logging. 2. One position of the train line or engine brake handle in air brake use; all ports closed. 3. To place a long log across the joint where two boom sticks butt together.
larch 1. Western larch, tamarack. 2. In Western Oregon and Washington, the larch trees are sometimes called noble fir. See piss fir.
lard Butter or oleo, same as salve. (L.S.) LL
larrigan See shoe pack pie.
larrigans The oil-tanned moccasins with legs used by the lumberjacks. (L.S.) (Eastern Canada) LL
lash pole A cross pole that holds logs together in a raft. LL See bows.
latch A locking device that can be tripped open. Some skyline carriages had fancy latches which governed the rise and fall of the butt rigging.
latch hook One of the many old choker hooks that had a spring latch to lock and hold the choker in place.
Lawson system A popular double mainline system of slackline logging used in California. Chokers were placed on each end of the log that was carried overhead and loaded directly on the railroad car from the mainline without touching the ground. A four-drum donkey (usually a Washington duplex flyer yarder) was used. Developed by D. P. Lawson at Elk in Mendocino County, California. Also highline flyer.
lay 1. The path along which a tree is to be felled. 2. The method in which a wire rope is woven: in regular lay the single wires go to the left and the strands to the right; in left hand lay the wires go to the right and strands to the left; in Lang lay the wires and strands twist in the same direction, either right or left, making a more flexible rope than the other types.
lay-off A shut down.
lay of the land The estimate of a piece of timber from a logging viewpoint: location of roads, landings, etc.
layout 1. The rigging assembly that is set up and ready to log. 2. The company or outfit for whom a man works. 3. The leveling of ground or bed to break the fall of a redwood tree. (Red.) See bedding. 4. A logging plan. 5. Personal gear.
lay out 1. To hide out with a truck load of overweight logs until the state highway department scales are shut down for the day. 2. To sleep in the open with no blankets.
lay the wood down To fell a lot of timber. (P.N.F.)
lay up 1. To stop work and take it easy. 2. To store a machine. 3. To live with a woman.

A **stand** of **larch** timber in Northwest Oregon about 1904. *Kiser photo, courtesy J. Graves*

lazy haul To pull a log by horse power over a level place in a slide. (L.S.) Also jigger. LL

lazy man fire The procedure of pushing a long pole gradually into a fire, the fire tender being too lazy to chop it up into chunks.

lead 1. The direction in which the lines run out from the donkey, pointing in the direction in which the logs are to be yarded. 2. An older term for a snatch block used on a haulback where roads were frequently changed. 3. To hang a lead block.

lead block 1. A large block through which the mainline passes on the head tree. The position of the block directs, or leads, the line in the desired direction. In ground leading, the lead block was a wide-mounted contraption big enough to allow the butt chain to pass through. 2. Lead blocks were also used as tree rollers, and were thus called. 3. Also used on donkeys as a fairlead in earlier times.

A **lead block**. *Young Iron Works catalog, Glen Comstock collection*

lead chains See cross chains.

lead dog In skidding, any big dog driven into a log so that a chain could be attached.

leader 1. The growing tip of a tree. 2. An older term for a butt chain.

leaders In an ox or horse team, the forward pair. LL

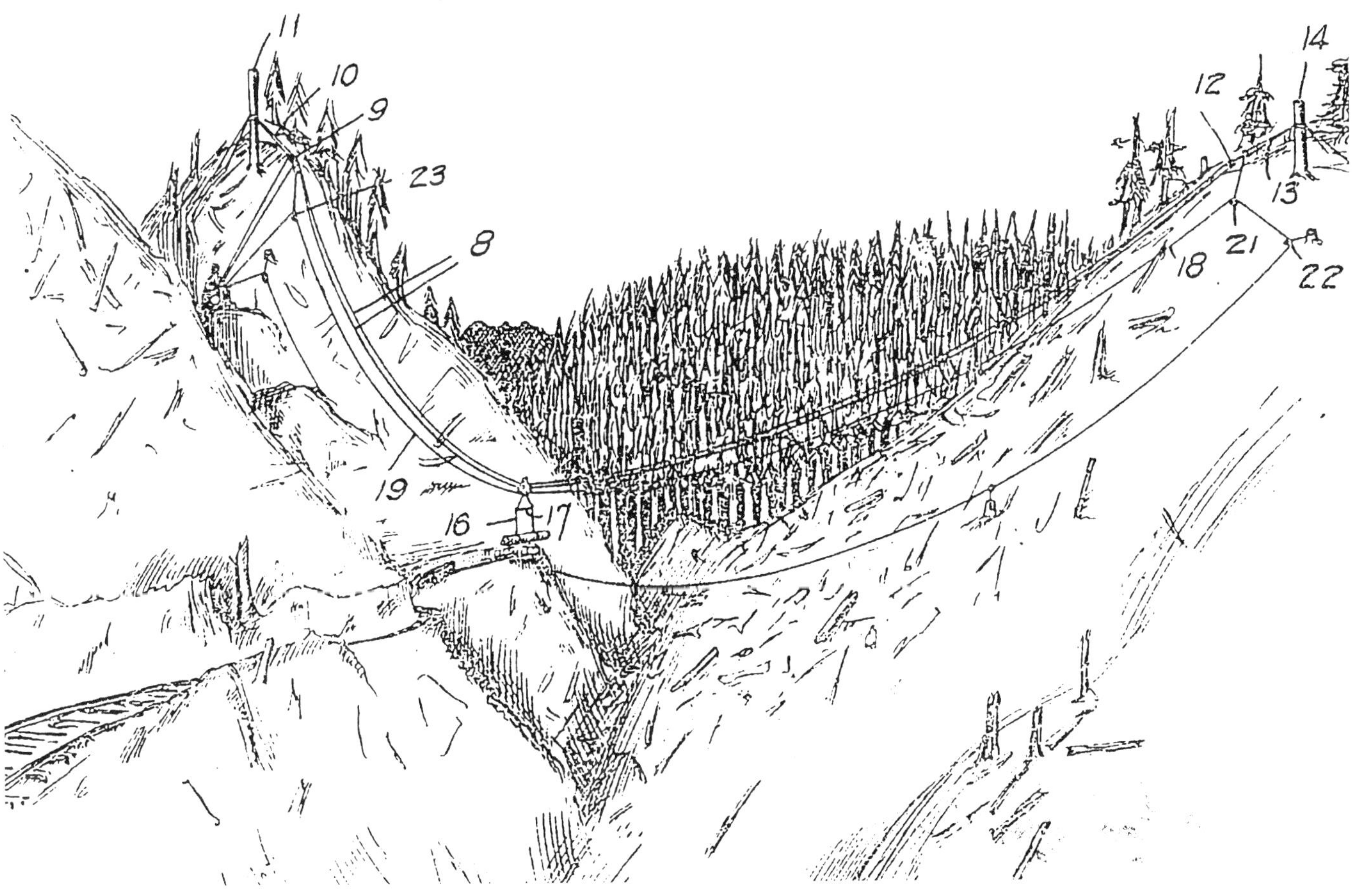

The **Lawson system**. *From the U.S. Patent Office*

leadfoot 1. A slow or clumsy logger. 2. A truck driver with a heavy foot on the throttle. (P.C.F.)
lead line An early name for a strip that was used to tie a tail block to a stump or anchor a snatch block when laying out or changing the lead in yarding.
leadman A boss or foreman.
leads fair 1. Said of a cable that pays off a drum and out to the woods at right angles to the axle of the drum or close to it, preventing undue wear on the block and line. 2. Said of a cable that lies flat in the groove of the sheave and does not ride or cut the side of the block.
lead strap A wire rope, with an eye at each end, used to anchor the block in setting a lead. (P.C.F.)
lead team 1. The front team where two or more teams were hitched to a sleigh. 2. The first team over the road in the morning; usually the best team in camp. (N.F.) LL
lead tree An older term for a spar tree. (P.C.F.)
leadway end The end of the log nearest the spar tree as it is yarded in to the tree. (P.C.F.)
lean The degree that a tree leans from a vertical line.
leaner 1. A tree that leans to a considerable degree. 2. A drunk holding up a pole or building.
leaning tree A leaning spar tree that can be used also as a loading tree. (P.C.F.)
learnt by hand To get experience by working and not from books.
leave-strip An area of uncut timber between two clear-cut areas. See buffer strip. Also green strip.
leave-tree A tree that is marked, not to be cut by the fallers in a stand of timber. (P.C.F.)
leg A prop used under the end of a log to prevent it from splitting while being bucked.
leg man A field man for a timber outfit who has to do a lot of travelling by foot.
leg up To prop up a log while bucking, to prevent it from dropping.
let in the daylight To fell trees.
level drum A donkey with both drums at the same level.
leverman 1. The operator of a skidder engine unit: the skidding leverman operates the skidding engine and the loading leverman the loading engine. 2. The operator of a duplex loading engine. 3. Sometimes any donkey puncher.
Lewis loading system A crotch line loading method that used a donkey with two main drums and a car line drum. (P.N.F.)
L-hook An L-shaped hook with a long cable, chain or rope attached. The hook is fastened to the rear of a turn of logs in the trailing portion of a slide and the draft animals to the cable. When the turn starts to run on a steep portion of the slide the hook is automatically released and prevents the logs from dragging the draft animals. (App.)

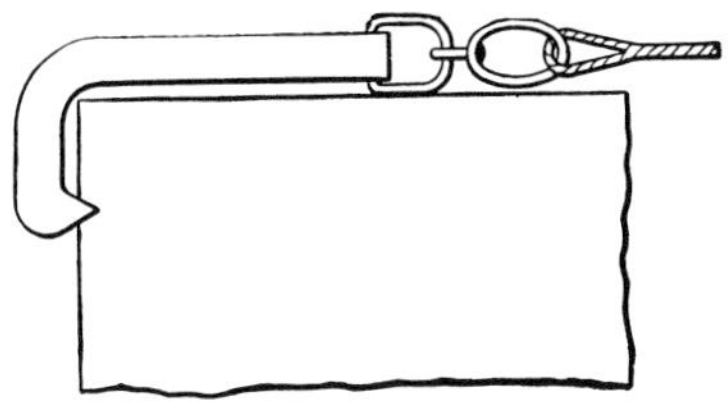

An **L-hook**. *From "Logging," Bryant, Wiley & Sons, 1923, Philip Schnell collection*

L-hooks The end hooks used in loading logs. See end hook.
license A timber license or a permit to cut public timber in British Columbia.
lid 1. The roof built over a donkey. 2. The steel roof guard required to protect the operator of a tractor. 3. A hat or cap.
Lidgerwood 1. The popular big skidder that was a large, powerful machine used for both skidding and loading. Some types carried a folding steel tower and others were tree-rigged. The machine was usually mounted on a big railroad car with swivel trucks so the skidder could be set off on short tracks at right angles to the mainline, to keep clear of traffic. The first skidder was built by the company in Michigan in 1882. It was improved by Dickinson, an employee, who set up the first skidder in Louisiana in 1891. About 1883, Horace Butters developed a skidder from a three-drum pile driver at Ludington, Michigan. A Lidgerwood was first used in the Pacific Northwest in 1904. 2. A logging donkey also built by

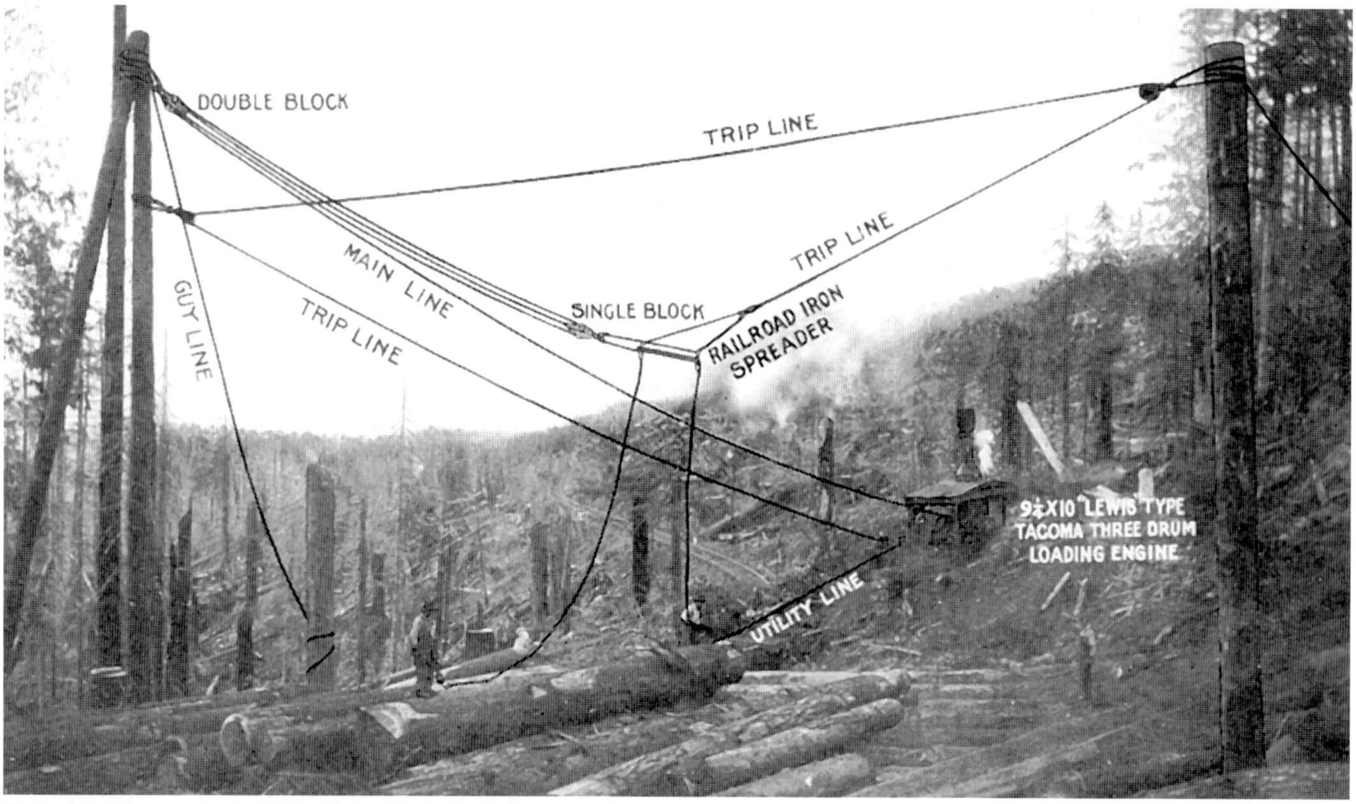

The **Lewis loading system**. *From Mallory Catalog No. 6, Glen Comstock collection*

The J.B. Miller Logging Company, near Toledo, Oregon, loading a car with **L-hooks**. *Labbe collection*

An unidentified **Lidgerwood** tower skidder at work in the Pacific Northwest woods. The **jack legs** supported and steadied the skidder during operation, and the running gear could be removed if needed, which allowed the skidder to be used as a **tunnel loader**. *From Mallory Catalog No. 12,1925, Glen Comstock collection*

Lidgerwood. 3. A plow for dumping ballast off railroad cars. 4. All skidders were known as Lidgerwoods.

Lidgerwood carriage A carriage used with the early skidders that consisted of two two-wheel sheaves connected by a stout iron pipe about 5 feet long. The hauling line ran through a block hung below the forward unit, and through a double sheave fixed below the back unit. The slack-pulling line ran over the second sheave in the back unit.

life preserver A peavey. (L.S.) LL

lifesaver A hospital ticket. (L.S.) LL

lift To raise and ballast the railroad track. (P.C.F.)

lift a tree 1. To set up a spar tree instead of using a natural tree. See raise a tree. 2. To wedge a tree being felled. (P.C.F.)

lift-gate In a logging sluiceway, a gate that may be moved up or down.

lift his plates To suspend a log hauler's permit to haul logs.

lifting line 1. An important part of the Tyler system of logging. The lifting line runs from the donkey drum through the carriage, down through a fall block, back to the carriage, out to the tail tree, and is then anchored to a stump. By paying the line cut, the fall block, the butt rigging hung from it, and the chokers, are lowered down to the logs. Reeling in on the lifting line lifts the gear up toward the tight skyline clear of the ground, where it rides in to the landing. (P.N.F.) 2. The loading line used in an overhead log loading system.

light-foot A man who is light on his feet, especially in loading operations.

lighthouse cat A tractor on which a steel tower is mounted for use in loading, or light yarding over short distances.

lightning landing 1. A landing having such an incline that the logs may roll upon the cars without too much assistance. (L.S.) LL 2. A very steep landing where logs might get away once the first logs are started.

light side The light side of a log that floats up.

lightwood Pine wood that is heavily impregnated with a resinous substance. (S.F.)

like a cow pissin' on a flat rock Said of a drenching rain. (P.C.F.)

like a duck takes to water Said of anything working smoothly.

like a handful of ants Said of a confused situation of any kind. (P.C.F.)

This tractor equipped with a loading tower and boom could be called a **lighthouse cat**. *From Pacific Logging Congress "Loggers Handbook," 1948, Philip Schnell collection*

like two oxen on a narrow bridge Close together. (P.C.F.)
Lima 1. A gear-driven locomotive especially suitable for work on heavy grades, sharp curves and uneven tracks of logging railroads. The Lima Locomotive Company called it a Shay. See Shay. 2. Any locomotive, gear- or rod-driven, made by Lima. 3. A power shovel or logging crane made by Lima. LL
limb To cut branches off trees or logs.
limber 1. One who trims off the limbs of felled trees. 2. Anything that bends readily. 3. See knot bumper.
limber boom A flexible boom, the sticks of which are usually joined to each other by means of short chains or short pieces of manila rope or wire cable.
limbing out Cutting off the limbs from a fallen tree. (L.S.) LL
limb wood Firewood cut from branches.
limit An area of 24 square miles of standing timber.
line 1. A wire rope. 2. A boundary. 3. A section line in public land surveys.
line car A car equipped with a large drum, usually operated by steam from the locomotive, used to pick up and store the various lines used in conjunction with a donkey logging operation.
line-changing Changing the rigging from one road to another on a donkey show. (P.C.F.)
line guard A series of loops along the leading leg of an arch, through which the line was threaded as a guard against the line whipping if it broke under a strain. (P.C.F.)
line guide A special type of spool equipped with wide wings used to guide cables on incline systems. See incline rollers.
line hook 1. An open hook used in making a loop in a cable or line for any general work. 2. See butt hook and choker hook.
line horse There was no haulback line when the early steam donkeys were introduced in the woods in 1882, and a horse was used to pull the line from the donkey back out to the woods.
line-log 1. To haul logs by cable attached to a winch on a cat where the ground does not permit yarding directly to the tractor. 2. To bunch logs up to an arch with a line run through a fairlead. (P.N.F.)

The **line car** was used to gather up the various pieces of line when a **side** was finished and carry them to the site of the next operation. A large drum and engine powered by steam from the locomotive here serves the purpose. *Labbe photo*

lineman A man in charge of hauling logs in a chute. (P.C.F.)
liner 1. The lumberjack who removed the bark from the log and drew a line by snapping a chalk line against it in the process of hewing timbers. (N.F.) 2. In surveying, a tree exactly on line which is usually given two blazes on each side, one above the other. LL
line retainer A roller with an arm that was used to keep a cable in place on inclines.
line runner A poor logging engineer because he could not do anything more than run a line in flat country. (P.N.F.)
line running The locating, tracing and marking of land ownership lines.
line tree A tree that happens to be on a survey line.
lining bar A heavy steel bar used for lining track and for other general work around a logging show.
link-and-pin coupler The early type couplers used on logging trains that were dangerous for the crew in trying to hook up cars on rough track. Guiding the link into a coupler and dropping the pin at the right moment was hard on the fingers. These couplers lasted on some woods lines until World War II.

A **link-and-pin coupler** at the West Side Lumber Company at Tuolumne, California. In this instance, it is coupled to a car with a knuckle coupler by removing the knuckle and retaining only the pin that held it. *Vernon Goe photo, Labbe collection*

link saw An early name for a chain saw. See Wolf link saw.
Linn tractor A make of logging tractor having a truck front end and tracks in place of rear wheels. Designed to haul logs on sleds in the snow.
lionite Anything that is strong; taken from a brand name for an old-time blasting powder. (P.N.F.)
little bull A forman not yet old enough or experienced enough to be called a bull of the woods. (P.C.F.)

During the 1920s, when roads were seldom ideal for the motor trucks and hauls were often too long for the cats, the **Linn tractor** proved a useful compromise. This operation was that of the Homestake Mining Company of Nemo, South Dakota, about 1930. *From "The Timberman" collection*

little casino (The 2 of spades.) A hanger-on. (P.C.F.)
little ditch The headwater of the West branch of the Susquehanna. (Pennsylvania)
live-burn To pile and burn slash as it is made.
live chute A chute on steep ground where logs will move by gravity.
live line Any line that is moving, or might be put into motion at any time.
live reel On tank trucks, a reel from which a hose is connected to the pump so that water will immediately run through when needed for fire.
liver pads Flapjacks. (P.C.F.)
live skyline A system designed to log uphill out of canyons. A two-drum donkey is used in which the skyline is handled by one drum and is fastened to a tail hold across the canyon. The carriage returns by gravity and is brought in by the mainline. Also shot gun system. See duplex flyer.
livestock Body lice. (L.S.) LL
lizard 1. A rough wooden bummer or dolly to keep the head end of the log up off the ground in skidding. 2. A tree crotch used as a crude skid under a log. 3. A sled for skidding logs. (L.S.) LL
load 1. A railroad car or motor truck loaded with logs. 2. The stress placed on a wire rope or the equipment. 3. The power output of an engine. 4. To load logs. 5. To load flat cars or box cars. 6. To load freighters. 7. To tamp dynamite in a hole before blasting.
loader 1. A person who supervises the loading of logs onto a truck or flat car. 2. Any kind of rig that loads logs or lumber.
loader leverman One who operates the levers controlling the drums on a power loading device. (S.F.)
loading bar A device used in place of a doubletree. It was a wood bar that hung on the horse's harness to which a long-handled grab hook was attached; used on a cross-haul team in decking or loading logs. (N.F.) LL
loading bitch A loading machine, usually a small donkey.
loading block A block hung on a guy line on various types of booms, and used in loading logs. Also loading lead block, pick-up block.

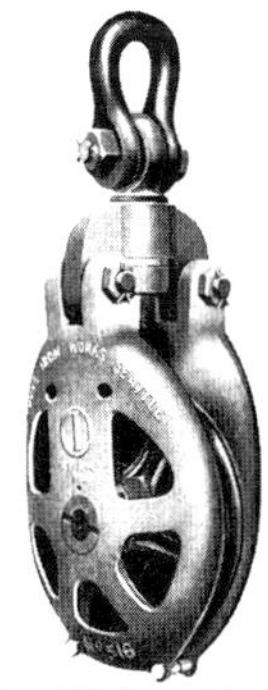

A **loading block**. *From Young Iron Works catalog, Glen Comstock collection*

loading boom 1. A heavy boom or arm used to swing logs from the landing to the truck or car. 2. The projecting arm of a log-loading machine that supports the log during loading; may be the swinging or rigid type.
loading chain A long chain used in loading or piling logs with horses. Also decking-chain, rolling chain.
loading hoist rope An older term for a loading line. (P.N.F.)
loading hook Any of many different types of hooks used in a loading operation, such as the various jammer hooks. LL
loading jack 1. A two-sheave carriage that rides on a guy line and supports the loading block. 2. A platformed framework upon which

In this landing scene of the Yosemite Lumber Company near El Portal, California, the various components of the loading setup are illustrated. The **loading block** hangs from the **loading jack**, which in turn hangs from the **jack guy**. *Courtesy of Lyle Spears*

logs are hoisted from the water for loading upon cars. (N.W.) LL 3. A dead bunk set up for loading logs or lumber in the absence of trucks. The truck would back under the prepared load, and the jack would allow the load to be lowered onto the truck bunk. Then the truck knocked the jack down, and the lumber rolls on the truck bed allowed the truck to back under the load. (P.C.F.)

A **loading jack**. *From Young Iron Works catalog, Glen Comstock collection*

loading lead block See loading block.
loading leverman The man who operates a loading boom and its long tongs, the first with a swing line and the second with a loading line.
loading line A wire rope that runs from the donkey through the loading line blocks to the boom. Also hoisting line.
loading machine The early steam-powered boom loading machines were the American and Barnhart which worked on rails laid on the log cars, and the Browning, Decker and McGiffert which loaded directly from the track.
loading pot A loading donkey.
loading punk The donkey puncher on a loader. (P.C.F.)
loading shoe A loading jack that has an ironbark shoe instead of two sheaves.

A **loading shoe**. *From Young Iron Works catalog, Glen Comstock collection*

loading spool A small spool or a gypsy on a donkey that is used to load logs and to move cars at the landing.
loading tongs The heavy pinchers for picking up logs in a loading operation.
loading tree 1. A small spar tree that is rigged only for loading. 2. A stripped-down tree left after yarding for use in loading out a cold deck. (P.C.F.)
loading tripod Three long timbers joined at their tops in the shape of a tripod for holding a pulley block in proper position to load logs on cars from a lake stream or deck. (L.S.) LL
loading trolley See loading jack.
loads 1. The number of car loads on a train. LL 2. The number of truck loads hauled.
lobby The recreation area in a logging camp. (App.)
lobby hog The bull cook. (App.)

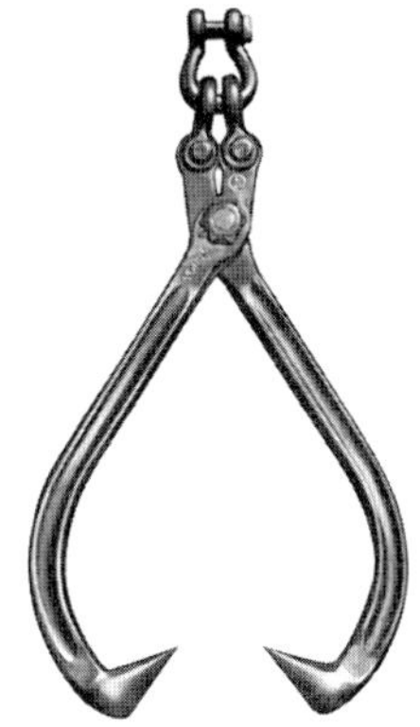

Loading tongs. *From Young Iron Works catalog, Glen Comstock collection*

locator 1. A man who located timber tracts for distant buyers in the days when the public domain was still open for purchase. See public domain. 2. A logging engineer specializing in road location. (P.C.F.)
location The rough line that is first laid out for a logging railroad or logging truck road.
locie, lockie See lokie.
lockage Sections of a log raft designed to be separated for passing through locks.
lock a raft To tow log rafts through a set of locks. (P.N.F.)
lock-down A strip of tough wood with holes, that is laid across a raft of logs. Rafting pins are driven through the holes into the outside logs, thus holding the raft together. (N.F.) LL
lock hook One of many types of hooks with a lock or latch to hold in the eye of the choker used in yarding.
locomotive cord In the days of the wood burners, the wood for locomotive fuel that was cut in 2-foot lengths and piled 2x4x8 feet to make a cord.
lodge To hang up a tree against another in felling.
log 1. The downed timber that is cut into sections. 2. Specifically to cut down trees and to transport the logs to a mill. 3. To carry out any part of a logging operation.
log agent A man who counted the logs as they came down the Penobscot River past the Argyle Boom. (Maine)
logan 1. A wet, marshy piece of land. (Maine) 2. See pokelogan.
log barge A barge for hauling logs across rough water on the West Coast. The shell of an old sailing vessel was frequently used. Later barges were so arranged that by flooding compartments on one side, the logs can be unloaded by tipping the barge on its side.
log berries Prunes.
log boar A short tongueless sled with wooden runners, used to haul logs. No part of the load is dragged on the ground. (N.F.)
log brand A mark of ownership stamped into the end of a log, with a stamping hammer, for identification in a log pond; registered by law. In the Lake States also called, brand, log mark, side mark and water mark. See bark hack.
log buyer Any man who purchases logs on the open market for a lumber company.
log chain Any chain used in logging.
log chair See logging dog.
log chute 1. A flume to transport logs down steep hills or around waterfalls. 2. A trough made of timbers and used for sliding logs downhill, either dry or by aid of water. Also timber chute, timber slide.
log deck A pile of logs where all logs are stacked horizontally to the ground surface in a uniform manner. See cold deck.
log drive The act of floating logs down a river or stream to an ultimate boom site. The early mills on the rivers used this system when the timber was cut near the rivers, but it is now a thing of the past. In the Lake States, the logs that would float were the pines. Hemlock, a softwood, and basswood, a hardwood, would float. All other hardwoods would not float and were not driven down the stream. Tamarack, considered a softwood, would float only well enough to be driven on short drives. The last log drive in Wisconsin was on the Flambeau River at Ladysmith in 1926. The last log drive in Minnesota was in 1937. The last drive in Maine was on the Big Machias River (Old Stream) in 1968. LL

The Caspar Lumber Company used this **log chute** for many years to get their redwood logs down the steep bank to the millpond at Caspar, California, on the Mendocino Coast. *Gregory Maxwell collection*

log drivers The men who keep the logs moving in the stream and follow to release and start all stranded logs. See drivers.
log driving Moving logs by water.
log dump 1. This was the end of a railroad where the logs were dumped into the water. 2. Any pond or storage place where logs are brought in from the woods and dumped.
log fixer The man who chopped a "ride" or slightly beveled one side of the logs so that they would travel down the bull road and not turn over and break a train of logs. In the Lake States also called a rosser or slipper.
log for chips To log pine and other hardwoods (other than redwood) for chips to be used in the manufacture of paper. (Red.)
log gatherer An older name for a skidder. (P.C.F.)
logger 1. Any man who works in the operating area of a logging show other than the Great Lakes and Canada. Also ape, brush ape, brush cat, brush monkey, rosin-belly. See lumberjack. 2. The owner of a logging outfit. 3. A good worker.
logger's can opener An ax. (P.C.F.)
logger's delight A tour of the skidroad bars and whore houses. (P.C.F.)
logger's dream 1. A trade name for a light portable yarding outfit built on a truck. 2. Any homemade rig of the same kind. (P.C.F.)
logger's physic Said of a big meal which will act as a physic. (P.N.F.)
logger's powder Dynamite or strong stumping powder used in blasting stumps.
logger's queen A female on the skidroad. (P.C.F.)
logger's smallpox Said of a man who was stomped on by a man wearing caulked shoes.

Eastern & Western Lumber Company used this homemade **unloader** at their **log dump** near Stella, Washington. *From Willis Gulker collection*

log-getter 1. A highball logging boss. 2. A highball logging machine.
logging On the West or in Maine, all or any part of the operation of cutting trees into logs and delivering them to an unloading point.
logging berries 1. Prunes which made better apple pies than peaches. (L.S.) LL 2. Beans.
logging block One of many blocks used in a cable logging system.
logging camp 1. A shelter for workers in the woods, in the East this was generally a log building. 2. A camp or headquarters which contained the bunkhouse, cook shack, stables and blacksmith shop.
logging car Any railroad car used for carrying logs.

Taylor Machine Works of Louisville, Mississippi manufactured the **logger's dream**. This one was owned by Cleve Robertson of Vernonia, Oregon. *From Pacific Logging Congress "Loggers Handbook," 1948, Philip Schnell collection*

logging chance A logging operation.
logging dog A short metal stake, sharp on one end with an eye on the other, which was driven into a log so that another log could be tied to it. There were many types of dogs used as a temporary anchor for a line.
logging engineer A professional man, a combination logger, forester, surveyor, cruiser, mapper, road locator, construction man, road builder, cost estimator, etc. Also fernhopper. See engineer.
logging horses Heavy horses used in logging. Many weighed one ton. They were generally purchased from Iowa or Illinois farms. The average life of horses on sleigh hauls and skidding was three to five years. (L.S.) LL
logging on the rise Getting logs into the water in time to ride downstream with the freshet. (N.F.) LL
logging railroad Any railroad designed primarily to haul logs, whether a private or common carrier.
logging recovery The volume or weight of logs which can be harvested from a given volume of standing timber.
logging setting See setting.
logging show A general term for a logging operation.
logging side The activity connected with any one landing. That part of a logging operation in which the logs are brought from the woods to the place to be loaded.
logging sled See sleigh.
logging sled road, logging sleigh road A road leading from the skidway to the landing. (N.F.)
logging timber Trees big enough to cut; merchantable timber. (P.C.F.)
logging timekeeper In the days of the logging camps, the versatile logger who took care of the scale for choppers, checked all logging crews on the job each day, scaled logs on cars, ordered supplies for the cookhouse, took care of the empties, etc. (P.C.F.)
logging tramp A traveling logger. (P.C.F.)
logging truck (log truck) 1. Any one of many different kinds and types of disconnected trucks used in railroad logging. 2. Any type of gas or diesel road truck designed to carry logs.
logging under the wire Loading ships offshore by wire chute. (Red.)

A **logging camp** of Webb Logging & Timber Company, near Brinnan, in Kitsap County, Washington. *From C. Wirkkala collection*

logging uphill Working hard and getting nowhere. (Red.)
logging wheels See big wheels.
logging woods The area where there is logging.
log grab An iron dog.
log grades Quality standards in logs, ranging from No. 1 peelers to the sawmill grades to wood logs and culls.
log handler An overhead cable method of decking logs. (P.N.F.)
log haul 1. A small one-drum yarder in the ground lead days. 2. Another term for bull chain or log chain.
log hauler 1. A teamster. 2. An engineer on a logging train. 3. A log truck and trailer. 4. A logging truck driver. 5. A steam or gasoline engine.
log-hungry 1. Said of a logging boss who drives his men, wanting to get a higher production of logs. 2. Said of a logger who will work long hours to earn a big check. 3. Said of an outfit short of logs that wishes to buy more stumpage.
log-hungry and money mad Said of a highballin' outfit that pushed for more than its quota of logs. (P.C.F.)
log it To work in the lumber woods. (Maine)
log jack 1. A screw-jack. 2. A lever-operated ratchet jack for moving logs.
log jam 1. In the early days of river driving, logs caught on any obstruction and piling up during a river drive. Also side jam. 2. A mixed-up mess of logs on a cold deck or landing or train wreck.
log jammer A loading rig used primarily in the Pine Country, but also common on hardwood logging. See jammer.
log-lifter A log pirate. (P.N.F.)
log loaders The various types of machines and equipment for loading and unloading logs, both in the woods and at the mill.
log maker A bucker. (P.C.F.) LL
log mark See log brand.
log patrol An association of loggers and lumbermen that supports a protective patrol to fight log pirates. (P.N.F.)
log plate See bridle iron, rope plate and sled plate.
log pond A pond, slough, bay or river where the logs are dumped so that they may be economically floated up to the mill. Log ponds are disappearing as logs are now trucked to the mill and cold-decked. A loader brings a log to a conveyor that moves the log to the barker and then to the headrig..
log pusher A highball foreman of a logging crew. (P.C.F.)
log-rolling A sport where two men vied with one another to stay on a rolling log in the water. (N.F.) Also juggler. See birling. LL
log rule A volume table giving board foot contents of logs according to measurements of diameter and length. Log rules in common use are the Scribner, Spaulding and Humboldt. The earliest known rule was developed in Maine in 1825. Also log scale, scale rule, scale stick.
log scaler See scaler.
log slide 1. A fore-and-aft road made of poles in which logs are skidded with horses or cats. (N.W.) 2. A V-shaped trough built down a slope for the purpose of sliding logs to a landing at the bottom. (West Virginia) Also timber chute, timber slide. See pole chute.
log snipe The man who rounded off the head end of the logs so that they could skid easier in the bull team and ground lead days. Logs were also sniped for tailing with a locomotive. (P.C.F.)
log splitter A device used sometimes to split large logs before sending them up to the mill. It consisted of a float with a drag saw mounted overhead. The log was floated under the saw and was moved ahead by means of a line and hand winch.
log stacker A rubber-tired front end log loader. Also gobbler, stacker.
log stamper A branding hammer. LL
log tally The amount of board feet or cubic feet in a log. The logger is paid on log scale or tally for the timber delivered to the point of sale.

A **log slide** of F.H. Goodyear on Half Mile Run, near Gardeau, in McKean County, Pennsylvania, about 1892. These slides were also known as **pole chutes**, and in the West were often called **pole roads**. *L. Kilmer photo, courtesy W. Casler*

log tramway An overhead cable carrier that carried logs along a stationary cable. Also overhead roading.

An unidentified **log tramway**. *From Willamette Iron & Steel Company catalog, 1920, Glen Comstock collection*

log truck See logging truck.
log wagon A heavy wagon having from four to eight wheels designed for carrying logs. They were used chiefly in the Pine Country, and were pulled by animals or tractors. LL
log watch An expert river driver stationed at a point where a jam was feared during a river drive. (N.F.) LL See head driver.
log wrench 1. A peavey. 2. The forerunner of the peavey, a long-handled tool with a dog on the end which swung freely in all directions around the handle. Also swinging bitch.

The eight-wheel **log wagon** was popular in the dry pine areas of the West. This one was used by the Hallack & Howard Lumber Company, of La Madera, New Mexico, in the 1920s. *W.J. Duteau photo, courtesy J. Hereford, Jr.*

lokey man A man who worked on a locomotive that hauled logs from the woods to the mill. (L.S.) LL
lokie (locie, lockie, loky) A locomotive, particularly a logging locomotive. The term "lokie" or "locie" was used in British Columbia and the Pacific Northwest. In the East, the geared engines were stemwinders. Shays were either Limeys or sidewinders. In the Sierras they were sometimes called screwdrivers. In the East the term dinkey was used for any small engine, and in the West a dinkey was a small tank engine. In Mendocino County in Northern California, a locomotive was always called a train.
Lombard 1. A common brand of steam hauler. Alvin O. Lombard of Waterville, Maine, who first built the steam hauler, held the patent on the so-called caterpillar track. The later builders of crawler track machines lost a court battle and were forced to pay royalties for the use of this system. The Lombard put the horse out of business and opened the door to the development of heavy equipment. 2. Big and strong: "He's a Lombard" (lembed). (Maine)

The **Lombard** Steam Hauler was designed and built by Alvin O. Lombard, of Waterville, Maine. They first appeared in 1900, and incorporated the first practical crawler tread, which was patented by Lombard in 1901. These machines enjoyed a short period of popularity in the woods of New England before being replaced by the tractor and truck. *L. Easton photo, Labbe collection*

long ballast Poles, slabs or lumber used under the ties as ballast.
long-bodied Said of a tree, particularly a redwood with a long trunk and branches beginning high above the ground. (Red.)
long-butt 1. A cut taken from the butt log to remove a defect. 2. To cut a short section from the end of a log to remove a defect or to upgrade a log. See butt-off.
Long butt hook A heavy butt hook with a closing latch.
long car A long railroad car designed for handling extra length logs.
long corner A small area of timber that is left because of some hindrance to logging, to satisfy seed source requirements, or because it is out of reach of the yarder.
long crew 1. A complete crew for a job. 2. More men than are actually needed for a job. (P.C.F.)
long greens Bills in pay envelopes. (L.S.) LL
long handle underwear Heavy wool underwear with long sleeves and legs. LL
long haul The opposite of short haul and generally a nonprofitable haul. (L.S.) LL
long haul donkey See road engine.
long haul road engine A donkey with much line capacity, sometimes carrying as much as 8,000 feet of mainline.
long hog Pork side meat; sowbelly. (West Virginia) LL
long johns Long handle underwear.
long-line skidding See skyline and skyline logging.
long log 1. The standard log length in fir areas was 40 feet for logging, while the standard length for the mill was usually 20 feet or 22 feet or longer cut for a special order. In pine the length was 16 feet

A **long haul road engine**. *From Willamette Iron & Steel Works catalog, 1920, Glen Comstock collection*

Long logs were the hallmark of the Benson Timber Company of Clatskanie, Oregon. They were used in the construction of the big ocean-going **cigar rafts** that supplied the sawmill at San Diego, California. The longest log pictured here was 207 feet in length. *Labbe collection*

for both, although many pine logs came out in 32-foot lengths. 2. A tree-length log.

long log country The Redwood region of Northwestern California and the Douglas fir region of Southern British Columbia, Washington and Oregon west of the Cascades. (P.C.F.)

long log loader 1. An experienced loader who could handle long logs. 2. A loader designed to handle the extra logs brought out tree length for use in building ocean-going rafts. (P.N.F.)

long log scale The volume of a long log measured as a single log; no increase of volume due to taper results. (P.C.F.)

long log system The system of bucking, yarding and hauling logs over 40 feet long. (P.C.F.)

long neck A bottle of whiskey. (L.S.)

long oat The horse whip used by teamsters on tote roads and long haul roads. (L.S.) LL

long route Said of a lumberjack who remains in one camp for a long period of time. (L.S.)

long shingles See split shingles.

long span show (skidding) A logging operation where logs are yarded over very long distances, particularly in skidder shows.

long spar skidder 1. Another term for a tower skidder which came with 75-foot or 100-foot towers. 2. A Lidgerwood skidder. (P.N.F.)

long splice A long splice used on the mainlines and haulbacks; may be up to 20 feet long, carefully rolled into the lay of the other line.

long sweet'nin Black molasses to sweeten coffee. (L.S.) LL

look-after man A representative of a company on a contract job. (P.C.F.)

looking at the daisies from the root end Said of a man who has been killed or who has died. (P.C.F.)

looking for a stretcher ride Said of a man who takes chances or is careless on the job. (P.C.F.)

lookout 1. A watchman who warns fallers working on loose-bark snags. 2. A watchman who warns of blasting operations. 3. A man stationed on a high point to watch for fires on a logging operation. 4. The watchman left behind at the landing when a side has been shut down because of fire danger. 5. The danger where men are stationed on fire watch. (P.C.F.)

looper A small moth (hemlock looper) whose larval stage, a small green worm, is a serious killer of West Coast timber.

loose bark pole A pole cut during spring or early summer when the sap is up. As it dries, the bark loosens and is easily dislodged. (P.C.F.)

loose logs Marked or unmarked logs that had for some reason escaped being rafted. (N.F.) Also scrabble or prize logs. LL

loose-tongued sloop See swing dingle.

lop 1. To chop the branches off a log in order to scatter the slash. 2. To cut off branches or knots (even on the mill carriage).

lose the air 1. To lose control of a logging train or a truck because of failure of the air brakes. 2. To lose one's temper. (P.C.F.)

lot 1. A piece of ground other than the usual 1/16 of a section which is usually found along the north line and along water. 2. In British Columbia, a piece of ground of various sizes and shapes. 3. A piece of timber, small in area. (N.F.)

loud talk him To argue a man out of something. (P.C.F.)

louse cage 1. A hat. 2. A bunkhouse. (L.S.) LL

low bore Said of a poor logger. (P.C.F.)

lower a tree To lower a raised spar tree after logging has been completed in a specific area.

lowering car See block car.

lowering engine An engine used to control the lowering of log cars down an incline or to control logs running down a steep pole road. A big donkey with extra big brakes was often used. (P.C.F.) Also incline engine, snubber.

lowering rig Any type of engine used for lowering logs.

lower the boom 1. To lower the boom in loading logs. 2. To "throw the book" at a man. 3. To fire a man.

low lead, low line See ground lead.

low man on the totem pole Said of a man with the poorest job in camp who cannot pass the buck to anybody. First used by loggers in the Pacific Northwest where there were many totem poles.

low-stump falling 1. To cut timber without the use of rigging. 2. To cut the tree as near the ground as possible to leave the ground open for yarding.

L shackle A type of shackle used in the early days when the skyline tail hold was secured only by passing the line around a stump once and choking it with a shackle. The pin was L-shaped and had to be pounded out with a hammer. Now against the law because the cable will fly around. Also admiralty shackle.

lubber-lift To raise the end of a log by means of a pry, by weight instead of strength. (N.F.)

lubricator See ickter pooch.

lucifer A match, so-called because of the sulphur content and small tip (the devil and hell relationship). (L.S.)

luff A rigging setup using two blocks and two lines. The first line was carried from a tail hold through a block attached to the item to be moved. A second block was then hung on the free end of the line and the second line carried from a second tail hold, through the second block, to the power source. A third line and a third block would constitute a double luff, etc. (P.C.F.) Also block and whip, block hold.

lug 1. A projection of some kind. 2. The cross bar on a cat tread. 3. To pull hard at low speed. 4. To carry a heavy load.

Top of the Yosemite Lumber Company **incline** near El Portal, California. The **lowering engine** is housed in the structure at the top of the photo. *Courtesy Lyle Spears*

lug hooks A pair of tongs attached to the middle of a short bar and used by two men to carry small logs, timbers, or railroad ties. Also swiveled timber carrier, swede hooks, timber-carrier, timber grapple, timber hook and timber tongs, yokes. Yokes is the popular term on the West Coast.

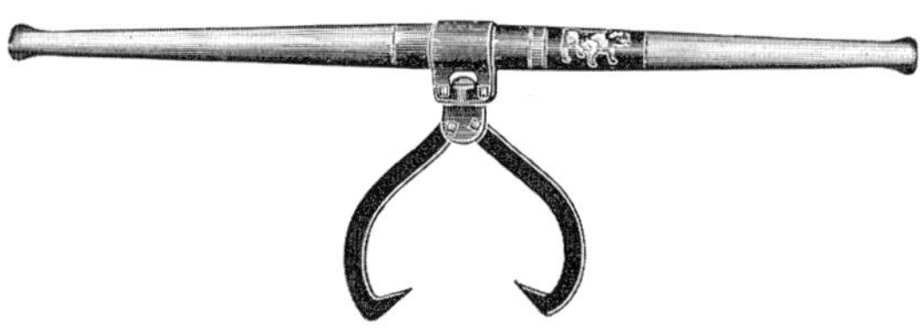

Lug hooks. *From Mill and Mine Supply Company catalog, 1916, Glen Comstock collection*

lug pin A heavy steel pin dropped through the lug on a donkey sled plate; anchored with a cotter key. It is used to hold blocks on line when moving or tieing down the donkey.
lulu 1. A thick growth of twigs or limbs on a tree. 2. Something spectacular: "It's a lulu." (P.C.F.)
lumber 1. In the early seventeenth century, there were no virgin forests in England, and by the end of the century there was such a serious timber shortage that the timber resources of New England attracted many investors and settlers. Before the nineteenth century, nearly every human activity depended on timber in New England, and the word "lumber" was used to mean "timber sawn into rough planks or otherwise roughly prepared for market" (recorded in the O.E.D. for 1662, Suffolk, Massachusetts). Today "lumber" is a manufactured product derived from a log in a sawmill, or in a sawmill and planing mill, which when rough shall have been sawed, edged and trimmed at least to the extent of showing saw marks in the wood on the four longitudinal surfaces of each piece for its overall length, and which has not been further manufactured than by cross-cutting, ripping, resawing, joining crosswise and/or endwise in a flat plane, surfacing with or without end-matching, and working. 2. To log or manufacture logs into lumber. (West Virginia)
lumber a tract To cut the trees on a certain specified area. (N.F.) LL
lumbered-up Said of a harbor congested with logs, timber and boards in the seventeenth century. (Maine)
lumberer An early nineteenth century term for a logger. (N.W.)
lumbering 1. The lumber business. 2. Working in the white pine in the 1840s and 1850s. (L.S.) 3. Often used in the West to cover the combination of logging and sawmilling.
lumberjack 1. A person who works in the logging operation in the Lake States or Canada. 2. Sometimes a derogatory term for a West Coast logger. "Lumberjack" is becoming popular in the West because reporters and new people are ignorant of the logging business.

With the changes that came to the lumber industry in the Lake States in the 1870s, the workers in the woods increased in number. The shanty-boy from Maine or other New England states were joined in ever-increasing numbers by "Ole" from Norway or "Hans" from Sweden, and by the Canadians, French and Scotch also. The crews were bigger; the shanty gave way to a camp; the whole organization became more complex. Then it was that the term "lumberjack" succeeded shanty-boy, pinery-boy, woodsman and lumberer. The term was used more or less in the eighteen seventies; in the eighties it was used frequently; in the nineties it was the only term used. (Agnes M. Larson, "History of the White Pine Industry in Minnesota.") Also rosin-belly, shanty-boy, wood head, wood hick.

lumberjack bird The Canadian jay bird. See camp robber and whiskey jack. LL

lumberjack small pox 1. The red ugly spots on a man after being caulked in a fight. 2. The scars that remain after being caulked. (N.F.)

lumberjack strawberries 1. Prunes. 2. Beans. (N.F.)

lumberman's thread Haywire. (L.S.)

lumberman's trinity Bread, pork and beans in the early 1800s. (N.F.) See bean hole.

lumber mill 1. Specifically the sawmill. 2. The lumber plant that includes the log pond, sawmill, sorting tables, planing mill, drying yards, etc.

lumber woods A specific area where there is logging. (N.F.)

lump it off To guess the amount of timber on a specific piece of ground. (P.N.F.)

lunch One-third of a day on a river drive, usually about 10 A.M. (L.S.) LL

lunch can A large round can flattened on one side, carried on the back of a lunch carrier or cookee, to carry lunch to the river drivers. It was divided into compartments and would hold lunch for fifty men or more. (L.S.) LL

lunch carriers Cook's helpers or cookees who carried hot lunch to the lumberjacks in the woods or to the river pigs ahead of the wanigan. (L.S.) LL

lunch grounds The area where a meal was served outdoors, usually with a bonfire. (N.F.) LL

lunch hook A lumberjack's hand. (L.S.) LL

lunch-in The noon meal served in the dining quarters of the logging camp. (L.S.)

lunch-out The noon meal taken to the men in the woods, either by themselves or by the cookee. (L.S.) LL

lunch sack A square sack about sixteen inches square usually made of oil cloth with straps to sling over the shoulders of the river pigs as they worked on the drive. (L.S.) Also river pig nose bag. LL

lung powder Snuff, Copenhagen snoose. (West Virginia) LL

lye 1. Strong or poor coffee. (P.C.F.) 2. Extremely strong tea. (L.S.)

M

M, MM 1. In the woods, "M" means one thousand board feet in logs or standing timber. 2. A thousand board feet of lumber. 3. "MM" refers to a million board feet.
macaroni Long curls of sawdust cut by a well-filed crosscut saw. (P.C.F.)
macaroni side A logging side that handles very small logs, usually prelogging or relogging. (P.C.F.)
MacFarlane system A system of slack skyline logging used over rough, steep country in which the chokers were attached to a hook on the carriage and no butt line was used; developed by Charles and Fred MacFarlane, prominent logging engineers in the State of Washington.
machine 1. Any piece of equipment used in the woods. 2. A donkey. (P.C.F.).
mackinaw 1. The standard overcoat of the lumberjack and logger, a short wool plaid coat. Also pea jacket, reefer, windbreaker. 2. The heavy woolen cloth used for loggers' shirts, jackets and pants, first made by the Hudson's Bay Company in 1812.
mackinaw boat A flat-bottomed boat with a pointed bow and square stem that was used with oars or sails to haul lumber, supplies and men. The boats were used on the Great Lakes and tributaries. (L.S.)
mad as a bear with a sore tail Said of a very angry man. (P.C.F.)
mad-ax See mattock.
Mae West A pair of large conks on a tree or log. (P.C.F.)
mail-order woman A wife obtained through a matrimonial bureau; a mail-order bride.
Maineites Transplanted lumbermen or settlers from Maine in the late 1800s. (L.S.)
main gut A mainline. (P.C.F.)
mainline 1. The main railroad on a logging show. 2. The hauling cable in donkey logging. See haul-in line.
mainline block The big block on a spar tree through which the mainline runs in yarding.
mainliner An engineer who likes to roar into camp with a lot of noise. (P.C.F.)
main push The head man or superintendent of an outfit. (P.C.F.)
main road A road cut from the bush and used by logging sleighs once the snow and ice had made it firm and smooth. (Canada)
main-say The woods boss, logging superintendent or boss on a logging crew. (L.S.) Also main-spring. LL
main squeeze The boss. (P.C.F.)
make a bag To start a job. (P.C.F.)
make a bed To level off the land in the direction that a redwood will fall so that it will not break. (Red.)
make a fist at it To make some effort to do a job. (P.C.F.)
make a wigwam In felling trees, to lodge several in such a way that they support each other. (N.F.)
make 'er out Said to the ink slinger when a logger wants to quit and wants his final paycheck. (P.C.F.)
make 'er pay 1. To do a good job. 2. To operate a successful logging show. (P.C.F.)
make logs 1. To work as a bucker. 2. The business of making trees into logs. (P.C.F.)
makins, makens Cigarette papers and tobacco. LL
makes a racket like an empty bed Said of an empty container in the cookhouse. (S.F.)
make out his time To fire a man.
make up a hat To take a collection for some worthy cause. (L.S.) LL
make-up yard The switch yard on the logging railroad where the loads were assembled for the final trip to the dump. It was used mainly on the larger operations which used swing locomotives from the woods spurs to the mainline and mainline engines between the woods and the dump. Where disconnected trucks were used, trains were made up with the more solid loads at the head end to avoid as much as possible the chance of loads being pulled apart on a pull.
Mallet A direct-connected, articulated, compound steam locomotive especially designed for use on track with sharp curves. The configuration consisted essentially of two sets of engines, each powering four or more drivers, and supplied with steam from one boiler. The rear driver set was fixed to the frame, while the front set was hinged. This articulation permitted the locomotive to negotiate sharper curves than that possible with a standard long-wheelbase engine. Technically, a true Mallet was also a compound engine, in that exhausted steam from the rear cylinders was piped to the front cylinders and reused. This required that the front cylinders be of a larger diameter than the rear, in order to achieve the same power with low-pressure exhaust steam. These engines were often referred to as "compound articulated." Engines of this type that did not have compound cylinders, but instead used the steam only once were referred to as "simple articulated."
Mallory butt hook A self-locking butt hook with two heavy iron flaps which closed around the open part of the hook when the butt rigging was pulled. F. B. Mallory & Company was a Portland, Oregon equipment dealer, supplying the logging industry. (P.C.F.)

The **MacFarlane system** of overhead logging. *From Willamette Iron & Steel Company catalog, 1925, Glen Comstock collection*

malones Wool pants; taken from the trade name of winter clothing popular in the late 1800s and early 1900s. (P.C.F.)

management license In British Columbia, a perpetual permit granted on public forest land which allowed a logger to carry on a sustained yield operation.

man catcher 1. An employment agent paid by the lumber companies to hire woodsmen on the slave market to get a crew for the woods. 2. A foreman or company man trying to hire a crew. Also man getter and man grabber.

man down Said of a logger hurt in a woods accident. (P.C.F.)

mandril A heavy iron cone used in the blacksmith shop in shaping rounds of all kinds. Also cone, pyramid. LL

manila fall block Usually a small, light fall block, used with manila rope.

man shark An employment agent. (P.C.F.)

manual release Hand-clearing of brush (as contrasted with the use of herbicides) to lessen the competition around desired trees.

mark 1. A letter or sign indicating ownership, which is stamped on the ends of logs. See brand. 2. To select and mark trees to be felled in a logging operation. 3. To brand logs.

Ostrander Railway & Timber Company No. 7 was a **Baldwin**-built 2-6-6-2T **Mallet**. *Philip Schnell collection*

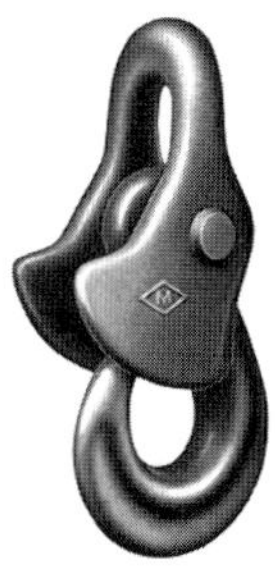

A **Mallory butt hook**. *From Mallory Catalog No. 12, 1925, Glen Comstock collection*

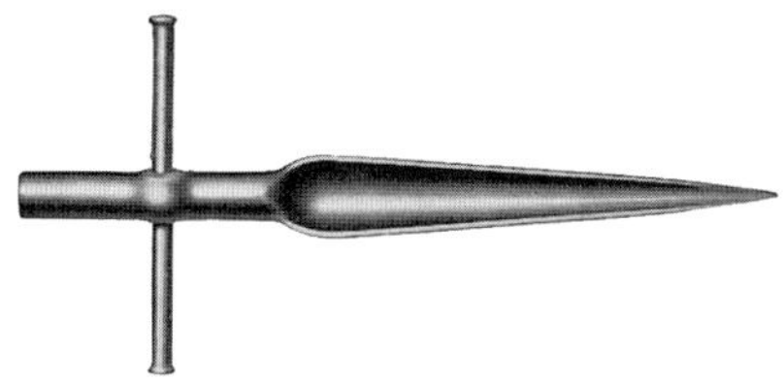

A **marlin spike**. *From Mallory Catalog No. 12, 1925, Glen Comstock collection*

mark caller The man who stands at the lower end of the sorting jack and calls the different marks so that the logs may be guided into the proper pockets. (L.S.) LL

marked tree 1. A tree marked by the foresters to be cut by the choppers. 2. A tree marked X to be saved for use as a spar tree, not to be cut. 3. A corner mark. Early surveyors often blazed the corner marks on convenient trees.

marker 1. A man who selects and marks the trees to cut. 2. The man who marks logs so that the buckers can buck for grade. 3. The one who puts the mark on the end or side of logs. (L.S.) 4. At a portable sawmill plant, one who takes lumber from behind the saw, scales, marks and tallies it. (N.F.) 5. One who grades lumber and places the grade symbol on each piece. (P.C.F.)

market A unit of measurement formerly used in New York: A log 19 inches in diameter and 13 feet long.

marking ax, marking hammer, marking iron A light ax used to stamp a brand on a tree or log. Also log stamper. See branding ax and branding hammer.

marking gun An oil squirt gun filled with paint to mark a big C for cut in the Pine Country. The modern logger prefers the spray can for marking.

marlin spike A spike that is used to splice wire rope. (P.C.F.) See bull prick, needle.

marmit A large cast-iron pot in which food was prepared. (L.S.) LL

marriage The act of uniting the strands of wire rope in splicing.

marsh feathers Marsh hay used to cushion the sleeping bunk in the logging camps. (L.S.) LL

Martin grip hook See cup hook.

mash the line To squeeze a line so hard that the hemp core breaks or squeezes out.

mast An older term for a spar tree. (P.N.F.)

master chopper The head man of the choppers. (Maine) See chopper.

master mechanic 1. The man in charge of power on a logging railroad. 2. The man in charge of all machinery, as in the mill and woods.

mat 1. Four or five logs cabled together to make a base for a raised spar tree. 2. A set of timbers bolted together to support a shovel when operating in the mud. 3. Small poles and brush matted together with old line to form a temporary truck road in swampy ground. 4. The bottom layer of logs in an ocean-going raft.

match, mate To place logs of similar size in a raft.

mattock A tool with a hoe blade on one side and an ax on the other. Also mad-ax, pick ax.

mature timber A stand of trees at a stage (age or size) which satisfies the purpose (primarily economical) for which it is managed.

maul 1. A large wooden or metal sledge bound with a metal ring and used for heavy stakes or posts. (L.S.) 2. A metal hammer used primarily in splitting wood or driving wedges in felling and bucking. (P.C.F.)

McCanse flyer The flyer, used successfully in steep canyons in Mendocino County, California, consisted of two Murray Brothers steam donkeys (technically a double slack highline flyer). A single drum donkey was placed on a hillside, and one end of a double slack line was anchored. On the opposite side of the canyon wall, the block was attached to a stump on the hillside. The lines contained a traveler with a spreader suspended below it; the spreader consisted of a 12- to 16-foot light log with about a 25-foot choker at each end. The spreader prevented the chokers from tangling before they were placed on the log. The lines could be lowered to any depth within the canyon, or on the side wall and with the use of a double pulley and the choke wires, the log could be quickly and easily transported to the waiting cars. Sam McCanse, while woods boss for the L. E. White Lumber Company, developed the McCanse Flyer in 1904. (Red.) Also aeroplane logging.

McClaren castings A ball and socket arrangement bolted to the sleigh beam on the runner to give the runner an up and down motion or rocking motion. (L.S.) LL

McGiffert (loader) Used both as a loader and yarder when logs were fairly close to the track. It did not require a regulation landing. Logs could be yarded to the railroad and picked up as the loader moved along. However, unlike the car-mounted loaders, the McGiffert traveled under its own power. When operating, the wheels were drawn up and the machine rested on the ends of the ties. Empty cars were drawn beneath the loader for loading. This was the design of J. R. McGiffert, who in 1907 assigned the patents to the Clyde Iron Works of Duluth. McGiffert was president of the company. See jammer, decker.

The **McGiffert** loader, built by the Clyde Iron Works, of Duluth, Minnesota, was a very popular machine in the Pine Country, or any place where short logs were handled. Commonly called a **jammer**, it was capable of traveling on its own wheels, which were raised out of the way when in use as a loader. Empty cars were fed from beneath the loader, making it a **tunnel loader**. Also, because of the supporting legs which rested on the ties, it was sometimes called a **bow leg loader**. This one was at work on the operations of the Watson Lands Lumber Company, at Mayburg, Pennsylvania. *George Deshner photo, courtesy W. Casler*

McLean boom loader A loading boom made of two parallel poles or timbers attached to a spar tree. Claude C. McLean, who was superintendent of the Cobbs-Mitchell Lumber Company at Valsetz, Oregon, first used a single timber (British Columbia in 1917), but many changes have occurred in the same pattern. Also double tong loader. See hayrack boom, heel boom.

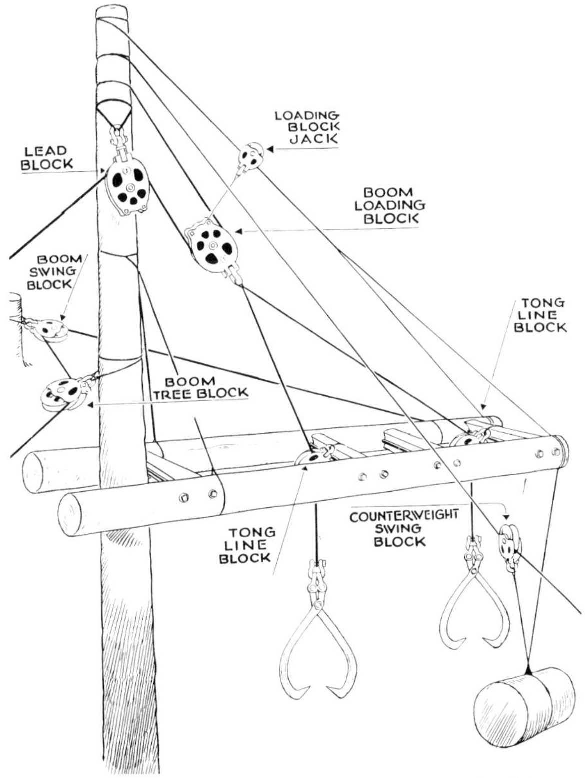

The **McLean boom loader**, also called **hayrack boom**. *From Young Iron Works catalog, Glen Comstock collection*

McLeod A fire tool that is part hoe and part rake; developed by Ranger Mel McLeod, Sierra National Forest.

measure The unorthodox loggers' way of measuring things: in the Northwest, bigness was compared to Erickson's bar in Portland, Oregon; length was measured in ax handles; width was compared to a fat schoolmarm; and smallness was measured by plugs of Star tobacco. (P.C.F.)

meat ball 1. A disparaging term for a logger. 2. A foreigner. (P.C.F.)

meat block 1. A large short block. 2. The butt log of a tree. (P.C.F.)

meat burner A cook. (P.N.F.)

meat head A stupid person (too much meat in the head). (P.C.F.)

meat run A special trip to the woods to bring out an injured or dead man after an accident. (P.C.F.)

mechanical lubricator See ickter pooch.

mechanical outhaul An early system in which the haulback line was used to return the mainline to the woods in ground lead logging. (P.C.F.)

medders Good work horses. (Maine)

medicine maker A trouble-maker. (P.N.F.)

memaloose Death (from the Chinook jargon). (P.N.F.)

men catchers Labor recruiters from the lumber companies who would recruit men in the cities and pay their railroad fare to the logging camp. (L.S.) LL

mensuration A science dealing with the measurement of volume, growth and development of individual trees and stands, and the determination of various products obtainable from them.

meow 1. A bend or snarl in a line. 2. A crook in a road. 3. Anything out of line. (P.C.F.) Also asshole, wahoo, wow.

merchantable 1. Timber that is big enough to sell. 2. A specific grade.

merchantable log A log that will make lumber of a quality and in sufficient amount to make it profitable to take it to a mill and have it sawed. LL

merchantable timber Usually interpreted to mean timber that can be manufactured and sold at not less than cost. The purpose for which the timber is to be used and local custom are factors which influence the degree of utilization. LL
merrygoround A railroad turntable. (P.C.F.)
messenger line An older term for a haulback line. (P.N.F.)
mess house The cookshack.
methodist An older term for a two-faced ax. (P.C.F.)
mexican powder Pepper. (L.S.)
Michigan ax First of the two-bitted axes. (L.S.) LL
Michigan jumper A lumberjack who worked only a few days in one place and then moved on to another camp. (L.S.) LL
mick An Irish lumberjack; often taken to be offensive. (L.S.)
middlers The troublesome flies in a bunkhouse that do not light. (P.C.F.)
midway The roofed-over space between the bunkhouse and the cookshack where barreled beef, pork, venison, lard and other supplies were stored. (L.S.)
mikado, mike A locomotive engine having a 2-8-2 wheel arrangement: 2 pony wheels, 8 drivers and 2 trailing wheels.

No. 117, a 2-8-2, or **mikado**, stands at the **dump** at Kerry, Oregon, on the Columbia and Nehalem River Railroad. *Labbe collection*

military bounty warrants The warrants that could be assigned to any purchase of pine land, making valuable pine land very cheap at such bargain prices. Government land was cheap, about $1.25 an acre. (L.S.) LL
mill 1. A sawmill. 2. The entire lumber plant that includes the log pond, sawmill, sorting tables, planing mill, drying yard, dry sorter, etc. 3. A beat-up old locomotive that still runs. 4. An old truck or truck engine.
mill up To make something. (P.C.F.)
mine 1. To use an ax clumsily. 2. To chop into the ground. (P.N.F.)
mine props 1. Small, short logs cut in interior camps for shoring up mine tunnels. 2. Timbers. 3. Round stock cut on just two faces.
minors Lumber suitable for temporary construction.
mired down 1. Said of a man or machine stuck on the job. 2. A man or machine stuck in the mud. 3. Said of a machine overloaded.
misery harp, misery whip See crosscut saw.
missing link The chain-repair link that comes in two halves. It is fastened together by burring its built-in rivets.
mix me up a hike An expression meaning "Give me my check; I'm quitting." Also "Mix me a mope," or "Mix me up a walk." (N.F.) (P.C.F.)
moccasins Special wide runners for sleds used on snow roads, consisting of a wide steel or wooden shoe installed between the wooden runner and its metal shoe. (N.F.)
modified North Bend system See South Bend system.
mogie See go-devil.
mogul 1. A locomotive engine having a 2-6-0 wheel arrangement. 2. Anything oversized or unusually strong and powerful. (P.C.F.)
mogul fairleader A fairlead designed to take the heavy strain of big yarders. (P.C.F.)
mogul yarder A big yarder with a large boiler, which operated at high pressure. It had a large drum capacity, high speed and was used for roading and occasionally for yarding long distances. (P.C.F.)

A **mogul yarder**. *From Willamette Iron & Steel Works catalog, 1913, Glen Comstock collection*

molasses In the absence of a cheap sugar, molasses, costing perhaps fifty cents a gallon, was universally used in the Maine woods. It was an exchange product given in return for lumber in the West India trade.

"Foresters float down timber that seamen may build ships and go to the saccharine islands of the South for molasses: for without molasses no lumberman could be happy in the unsweetened wilderness. Pork lubricated the joints; molasses gave tenacity to his muscles." (Richard G. Wood, A History of Lumbering in Maine, 1820-1861)
molle 1. Strands of wire rope woven into a circle and used as a temporary link between the eye splices of two lines, supposedly named for a "loose woman." 2. A ring of wire to replace a cotter key. 3. A rough splice. See Molly Hogan.
Molly (Molley) Hogan Another name for molle, an improvised eye on the end of a logging cable. A strawline is taken into the woods in short lengths of about 200 feet, with an eye splice at each end. These are tied together with a molle. When finished the molle or Molly Hogan is cut with a wire ax. See rolled eye, wild eye.
Monday Another name for the Mundy donkey. (P.N.F.)
Monday leg, Monday morning disease A disease of horses accustomed to work and exercise but left tied up in the barn for a couple of days. The disease made the horses' legs swell. (L.S.) Also black water. LL
money in the bank 1. The down and bucked timber. 2. Logs stored in rafts. 3. A cold deck bought and paid for.
money-maker See scale rule.
monitor A floating winch on a log pond.
monkey Burned sugar used by the cook to color gravy with a rich dark brown color. (L.S.) LL
monkey block See bunk block.
monkey meat Prepared sandwich meat. (P.N.F.)
monkey motion 1. A type of valve gear used on Willamette donkeys. 2. Said of any type of valve gear. (P.C.F.)
monkey shirt A heavy wool shirt with the tails hemmed up to hip length and worn outside the pants. (L.S.) LL
monkey wrencher 1. A mechanic. 2. A person who tinkers with machinery even if in good condition.

monkey-wrenching 1. Repairing a machine or piece of equipment. 2. Said of a donkey puncher when he pays too much useless attention to his machine rather than getting out the logs. 3. Tinkering with the wrong tools and too little knowledge. (P.C.F.)
monoculture The raising of a tree crop that consists of a single species.
Montana larch The Western larch or tamarack.
Montelius hook A latched butt hook with a swinging gate that closed the opening to keep the choker locked in; invented by Ira Montelius of the Trap Creek Logging Company of Nalpee, Washington.

A **Montelius hook**. *Mallory Catalog No. 12, 1925, Glen Comstock collection*

monthly insult A paycheck. (P.N.F.)
mooley A cant hook. (L.S.) LL
mooley hogan One who is an expert with a mooley. (L.S.) LL
moon bucket The kerosene lantern that was used by the loggers. The legend is that this name was given to the lantern by an Indian when he saw it for the first time. (Red.)
mooner A mythical creature in the logging woods. (N.F.) LL
moonshine Whiskey.
moose birds Canadian jays, the reincarnated souls of the dead lumberjacks. (L.S.)
moose cat 1. A log that goes up onto the deck or landing in the proper manner. 2. Anything unusually large. 3. A very good lumberjack. (L.S.) LL
moose milk A drink made of coffee, milk and rum or whiskey. (P.C.F.)
mop-up 1. To clean up the remaining small fires after a forest fire has been brought under control. 2. To clean the slash after a logging operation. 3. To clean up after any job.
mop up strays To bring in the lost or stranded logs after a drive.
more airs than a country studhorse Said of a very conceited man. (Maine)
more you stir it the more it stinks Similar to "let sleeping dogs lie." (P.N.F.)
morphodite (from "hermaphrodite") 1. Any vehicle that is different from a standard truck, such as a lift-truck, a straddle-truck, etc. 2. Also a combination of machines, such as the straddle trucks that incorporate lift forks on the front legs. 3. Any machine that was altered to improve its usefulness. 4. Any home-built copy of a standard machine. Loggers were very prone to free lance their machines, or to alter them to suit their particular needs. (P.C.F.)
morning glories Pancakes. (L.S.) LL
Morris skidding grab A patented adjustable chain with grab hooks, used in animal logging. It consisted of a chain with a large ring at each end. The grab hooks were attached to the chain by narrow-throated links which could be set at any point in order to make the distance between grabs conform to the size of the log. The draft power was attached to another narrow-throated ring which could be placed midway between the grabs and thus equalize the power.
moss See chink.
mossbacks People who come in to settle on the cutover lands. (L.S.)

A diesel **donkey** aboard a **moving car**, of Merril & Ring Lumber Company at Pysht, Washington, on the Straits of Juan de Fuca. A **bullseye fairleader** is mounted on the front of the donkey. *D. Kinsey photo, Labbe collection*

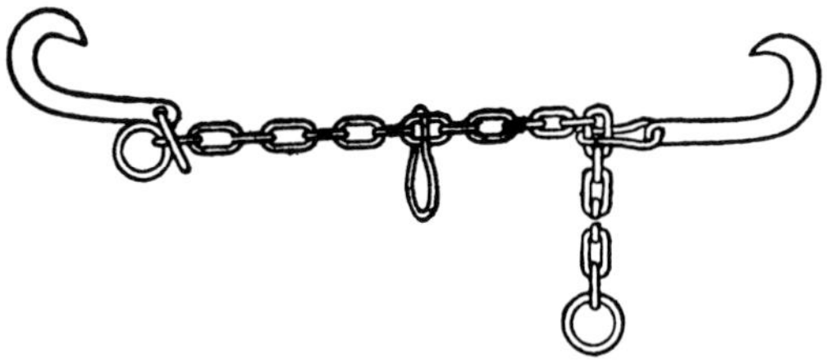

A **Morris skidding grab**. *From "Logging," Bryant, Wylie & Sons, 1923, Philip Schnell collection*

mother tree The female tree parent, the seed bearer; any tree capable of producing seed.
move To change a donkey from one setting to another. (P.C.F.)
move-road To remove the logging lines from one skidroad to the next area to be yarded or skidded. (P.C.F.)
moving block A heavy block used in moving donkeys.

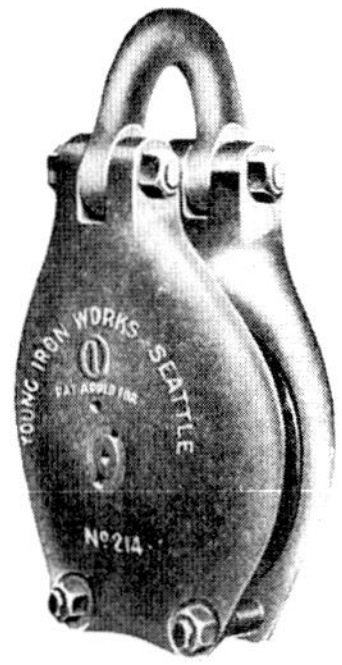

A **moving block**. *From Young Iron Works catalog, Glen Comstock collection*

moving car Moving cars fall into two categories: the steel moving car for heavy equipment, such as donkeys; and the cat moving car that is usually a car with a wooden deck for handling cats and trucks. See bull car, donkey mover.
mowich Venison (from the Chinook jargon). (P.N.F.)
Mrs. Murphy's chowder Unrecognizable food on the table. (P.N.F.)
mucker A man who works around a cat or shovel.
muck stick A shovel.
muckuks Bowls or containers made from birch bark. (L.S.) LL
mud 1. Coffee. 2. Slush in the winter. 3. To fill the crevices between logs in a cabin with soft clay, oakum or moss. Also daub. (N.W.) LL
mudboat 1. A flat sled that slid easily over a mud road; used like a pig to carry dogs and other gear. 2. A low sled with wide runners used for hauling equipment from camp to the operations. (L.S.) LL
mud cap 1. A pile of mud over a dynamite charge. 2. To put a big heap of mud over a dynamite charge on a rock to direct the force of the blast downward. Also mudshot.
mudder 1. A logging camp in a swamp or wet ground. 2. A logging camp on a dirt road after the fall rains have started. (P.C.F.)
mudding out logs Working in the mud to get the logs out. (P.C.F.)
mud'em in 1. To swing from a cold deck instead of rigging up a skyline for the swing because of the short distance. 2. To ground-log or cat-log in the mud. (P.C.F.)
mud hens The logging crews who work in low swampy places. (S.F.)
mud movers A derogatory term for a road-grading crew. (P.C.F.)
mudshot See mud cap.
mud sill 1. In skidder logging, a short heavy timber on which the hydraulic jacks rested to raise the skidder off the tracks so that empty logging cars could be run beneath it. 2. A log or timber buried in the mud to support a trestle pier, a sawmill frame, etc. 3. The bed piece or bottom timber of a dam which is placed across the stream, usually resting on ricks or in mud. (N.F.) Also bottom sill.

Mudding out logs on the Oregon Coast. This scene was at the operations of the Western Cooperage Company east of Astoria, Oregon. *C.T. Stanley photo, Labbe collection*

mug up To have coffee other than at meal times. (P.C.F.)
mukluks Soft uncaulked boots. (P.N.F.)
mule 1. A set of rope blocks, used to tighten guy lines on skidder tail trees. 2. An ornery man. 3. An ornery machine. 4. Rotgut whiskey. (L.S.) 5. A trade name for a brand of boom boat.
mule cart A four-wheeled vehicle for hauling logs which are suspended under the axle of the rear wheels. (S.F.)
mule tail The broomed-out end of a wire rope. (P.C.F.)
mule train A truck with a separate trailer.
mullet Any type of fish on the loggers' table. (Oregon)
mulligan 1. Another name for a caboose. In the early days of logging, the noon meals were taken out to the men in the woods in a special car called the mulligan. The term referred to the milligan stew which was served. Later the term was applied to the caboose or to any car that transported men from the woods to the cookhouse. (P.C.F.) See crummy. 2. Stew.
mulligan car A railroad car where the midday meal was served. (L.S.) LL See boarding car.
mulligan mixer A camp cook. (P.N.F.) (British Columbia)
multiple entry Entering a stand for commercial harvesting more than once in any one continuous rotation.
multiple use Management that encourages several distinct uses of the forest.
Mundy A early brand of hoisting machinery and donkey engines, made in Newark, New Jersey.
Murdy An old-time logging donkey.
murphys Potatoes. LL
mushing out Said of the movement of the lower logs in a pile at a rollway toward the water as the pile was broken out. (L.S.) LL
mush log The outside log on a load. LL
muskeg A lowland area with heavy organic or peat soil usually supporting heath.
mustard drive Said of the lumberjacks, who after the spring breakup and after celebrating, would go to the Dakotas and pull mustard out of the wheat fields. (L.S.) LL
mutt and jeff Two open faced hooks connected by a link. The upper hook goes on a guy wire wrap around the spar tree. The lower hook holds the strap of the high lead block when changing the lead.
muzzle loader 1. A bunk in the late 1800s and early 1900s that had to be crawled into from the foot because the bunks were too close together to get into from the side. Also shotgun bunk. See bunks. 2. Anything loaded or entered from the head end. 3. A camp car with the door in the end.
muzzler A foreman. (P.N.F.)
mycorrhizae The structure formed when a beneficial fungus invades a tree root, generally improving the roots' ability to function.

N

NG (no good) A poor cook. (L.S.) LL

narrow drum One stage in donkey engine development was the narrowing of the drums. This allowed fairleads to be placed on the donkey sled, which saved wear on lines and aided yarding and moving operations. The first donkeys to be built with drums cut down from the old wide-faced roaders were called narrow drums.

narrow drum yarder An early yarder with narrow drums mounted inside the donkey frame which was unsuccessful because the narrow drums were too hard on the lines.

narrow gauge A railroad less than 4 feet 8 inches wide. The most common gauge was 3 feet. See bastard gauge.

natural regeneration Renewal of the forest achieved either by natural seeding or from the vegetative reproduction of plants on the site.

near ox The left ox of a team of oxen. Off ox was the one on the right. Oxen were driven from the left side. LL

necktie A choker. (P.C.F.)

neck yoke A bar with a ring at both ends and in the middle, suspended from the harness collars to hold the tongue of the sleigh in hauling or backing up. (N.F.) LL

needle A steel tool used in splicing cable. Also bull prick. See marlin spike.

needle fire A light burn which only singes the needles off the slash. (P.C.F.)

needle gate In a logging dam sluiceway, narrow timbers or poles

Stoddard Lumber Company operated a 3-foot **narrow gauge** railroad in the Blue Mountains of Eastern Oregon. Here their No. 4, a 40-ton wood-burning **Heisler** built in 1922, handles a train of loaded **flats**. *Photo courtesy of Jim Brady*

with two or more squared faces which are placed in contact across the opening of the sluice to prevent the outflow of water. One or more "needles" may be removed without disturbing the remainder. LL Also bracket gate.

An upstream view of a **rafter dam**, somewhere in the Appalachian Mountains, having a **needle gate**. *From "Logging," R.C. Bryant, Wiley & Sons, 1923, Philip Schnell collection*

needle-scratching Said of the foresters who make a seedling count on cutover lands. (P.C.F.)
nester A settler who established claim to an area of timberland which he later sold to a lumberman or timber speculator. Many speculators sponsored these nesters, leading to wide-spread land fraud scandals. See public domain, shacker.
nest log A log that lies tight on the ground or fouled in some way so that the bucker can only saw it partly free from the next log. (P.C.F.)
Nestos snubber A method of lowering logs down from high ground, using a standing line and a snubbing line run through a carriage. Stops fixed on the standing line governed the position of the carriage at loading and unloading points. Probably named for Robert Nestos, prominent logger of the Bellingham area of Washington State.
net scale Merchantable volume of a log expressed in board feet, figured by subtracting visible defects from the gross scale.
never-slip An early choker hook. (P.N.F.)
neverstill 1. An eddy in a river. 2. A busy logging center. (P.N.F.)
never-sweat A lazy man. (P.C.F.)
new-era cableway A log tramway invented by W. F. Izett, Oak Harbor, Washington. (P.N.F.)
nick 1. An undercut. (S.F.) (L.S.) 2. To cut a notch in a tree. (S.F.) (L.S.)
nicknames An endless variety are found in camps, some of which can be printed, such as: one-eye, frog-face, the bear, step-n-a-half, pump handle, highpockets, mick, yellowjacket Joe, panicky Pete, hobo slim, bullbuck Andy, woodenshoes, cougar Bill, etc.
nickel sticks 2x4x4-foot culls which were sold very cheaply (about five cents each) and advertised as "nickel sticks." (Oregon)
nigger day Saturday night in the Southern woods when the black workers would hang their lanterns and gamble with dice. (S.F.)
nigger hair 1. A popular smoking and chewing tobacco in the 1800s. (L.S.) 2. A species of wild grass growing in swampy places.
nigger-head 1. The outside gypsy drum on the shaft of a donkey drum, used for light work around the landing. Also friction nigger, gypsy, gypsy head, overhung gypsy. 2. A rock in a stream bed which interferes with driving or splashing. 3. A boulder the size of a man's head or a clump of rocks uncovered on a grading job. 4. A log with one end floating and the other resting on the bottom. 5. A pile partially submerged.
nigger lip A simple choker hook with a broad lip that is turned outward. (P.C.F.)
night watch 1. The man who is left out in the woods to watch and to fire the donkey in the morning before the crew arrives. 2. The night watchmen who patrol the lumber mill.
nimrod A popular tobacco plug. (L.S.) LL
nipper A man working on the steel gang, usually the one who holds up ties to be spiked to the rail.
nips Pieces of old blankets folded around the feet inside the shoe packs for warmth. (L.S.)
no-bend choker A choker made of heavy cable which is difficult to bend around a log. (P.C.F.)
no-bill A non-union man. (P.N.F.)

This unidentified view shows early donkey engines with large **gypsys**, or **nigger-heads**, being moved on wooden **disconnected trucks**. *Rod Crossley collection*

no-blanket camp The first camps that supplied bedding for the loggers. At one time all loggers carried their own blankets. (P.N.F.)
no more teeth than an old hoss turned out to pasture Said of shoes without caulks. (P.N.F.)
non-fee foreign timber Timber owned by others, both public and private.
nooning Taking lunch in the woods so that no time was lost going to and from camp. A hot lunch was brought to the lumberjacks in the woods. (L.S.) LL
noose A sling for loading logs. (P.C.F.)
nootka pine See Douglas fir, Oregon pine.
no place to stood his foots Said of a log pond worker who has jumped from a log to what he thought was another log, but it was only a piece of bark. (P.N.F.)
North Bend system A method of yarding logs by which a carriage travels on a line anchored to stumps at both ends. The haulback is tied to a fall block, and takes it and the bight of the skidding line back to the woods; the end of the line is put through a fall block and fastened to the carriage that runs on the skyline. The system was developed by the North Bend Lumber Company, at Edgewick, Washington (P.C.F.) See South Bend system.
North Bend trolley A special trolley and fall block assembly designed to be used with the North Bend system of overhead logging.
North Carolina pine Pine lumber cut in the Coastal Plain region of Virginia, North Carolina and South Carolina. (S.F.)
North Sea piano A concertina played by a Scandinavian logger in the bunkhouse. (P.N.F.)
Northwest loggers The Pacific Northwest Loggers Association with headquarters in Seattle. (P.N.F.)
Norton jack A heavy screw jack used in rerailing railroad equipment and other logging purposes.
Norwegian snowstorm Said of the snow that was shoveled by hand when the snow road was getting bare. (L.S.)
nose 1. The front end of a donkey sled, boom, etc. 2. To snipe the end of a log. 3. To dig the front end of a log into the ground when being pulled along the skidroad.
nose auger An auger with a down-cutting bit edge for boring large holes in wood. (L.S.) LL

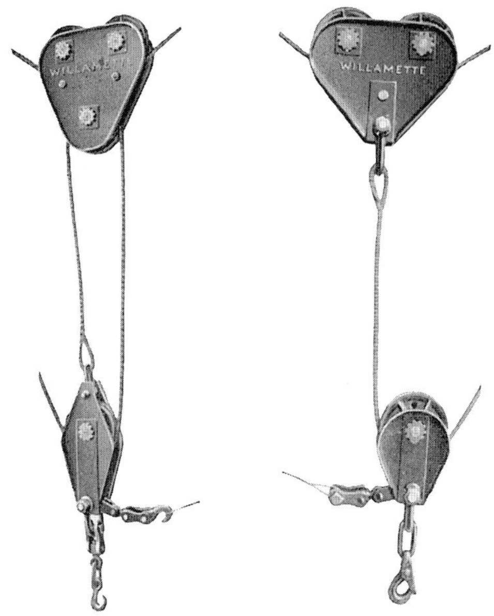

Two types of **North Bend trolleys**. *From Willamette Iron & Steel Co. catalog, 1920, Glen Comstock collection*

nosebag 1. A lunch pail. 2. A tin lunch bucket carried by a river driver, generally strapped to his back. (L.S.) 3. A grain bag hung under the nose of a horse when he is fed on the job. LL
nosebag show 1. A job which necessitates packing a lunch every day. 2. A job which will take all day away from camp. 3. A camp where the midday meal is taken to the woods in lunch buckets. (L.S.) LL
no-see-ums Midges, the tiny North Woods flies with a terrific bite.
nose guy A guy line holding a loading boom. See sail guy.

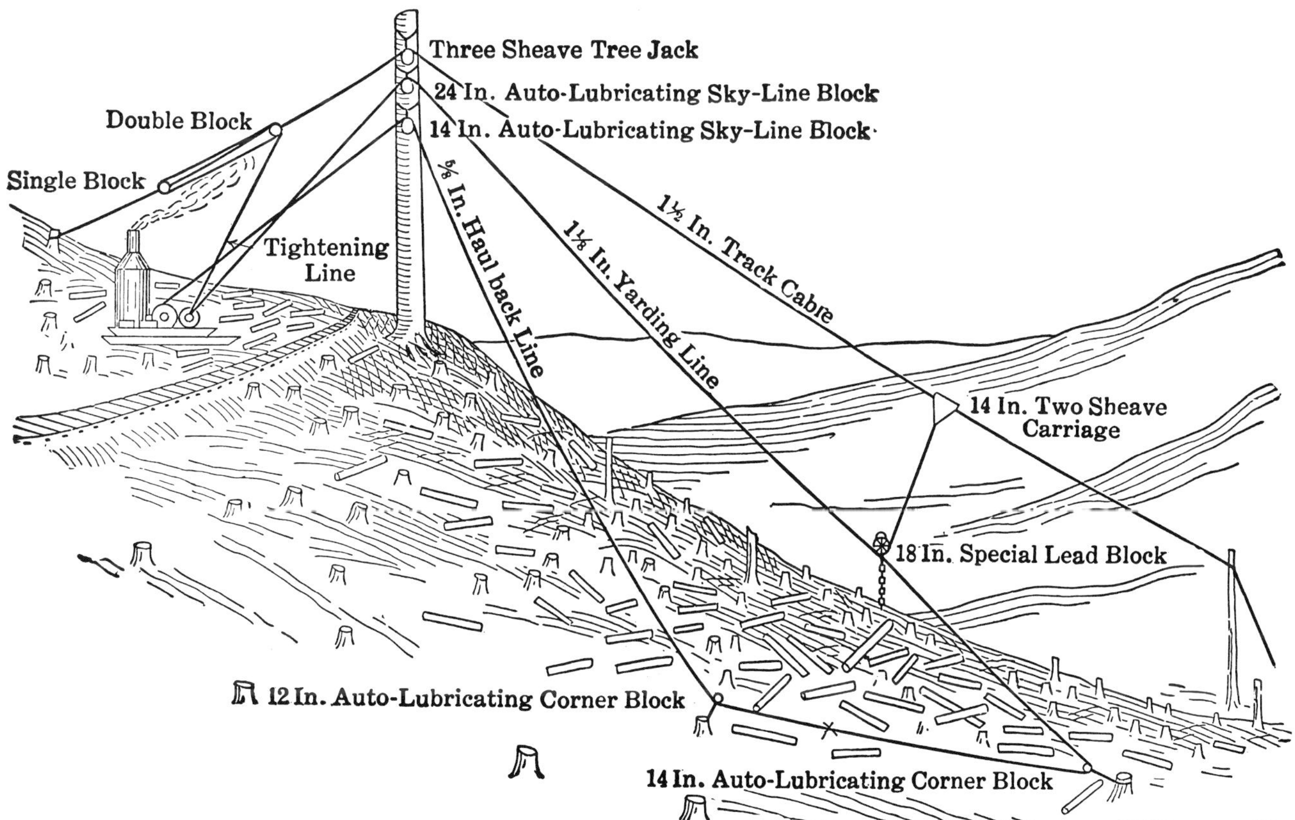

The **North Bend system**. *From "Logging," R.C. Bryant, Wiley & Sons, 1923, Philip Schnell collection*

A **Norton jack**. *From Clyde Equipment Co. catalog, Glen Comstock collection*

nosing 1. The front end of a sniped log. 2. The sniped part of a donkey sled. 3. The tendency of a log being yarded downhill to nose into obstructions or behind stumps as it tends to run up on the line.

notch 1. An older term for an undercut. 2. A groove cut into a stump for a guy line. 3. A groove cut around the end of a log in the ground lead days. 4. To make a slanting undercut with a power saw.

notcher 1. In the 1800s, the specialized logger or lumberjack who only put in the undercuts. 2. The man who cut the notches for the chokers in ground logging. Also knotcher.

notch'er up, notch up 1. To open up the throttle of a steam engine. 2. To notch back on the Johnson bar to bring it closer to the center notch. Also notch back.

now we're logging, now you're logging A popular expression that originated in the woods, meaning that things are back to normal, or that things are going very well.

nubbin A steel sleeve that is babbitted on the end of a choker wire to make a knob that is put into the slot of a choker hook. Also ferrule.

number one A top grade log, such as a number one peeler or a number one sawlog.

nursery An area set aside for the raising of young trees.

nursery stock Any seedling or transplant grown in a nursery.

nut-buster A machinist, usually one who works on locomotives.

This view shows Donovan-Dorkery Logging Company's No. 5, an 80-ton three-truck **Climax**, built in July of 1925. This engine represented the state of the art in Climax locomotive design, with cast steel trucks, piston valves and Walschaerts valve gear. The man with the large wrench at right in the photo might have been called a **nut-buster**. The man to the left of him holds a **brake hickey**. *Clark Kinsey photo, from the Martin E. Hansen collection*

O

O.B.U. The abbreviation for One Big Union. An attempt was made to organize all woods labor into one big union, setting a pattern later followed by other labor organizations. (P.N.F.)
O&C The timber lands of the Oregon and California railroad land grant, located in Western Oregon and administered by the Bureau of Land Management, Department of the Interior. These lands were taken back by the government when the railroad failed to carry out the terms of the grant.
oakum and tar The standard mixture for bateau and boat seam caulking. (N.F.) LL
oarsmen The men who powered the lumber rafts on the rivers in the early 1800s. These were usually 12 oarsmen on each raft, and the oars were sometimes 45 feet long. (L.S.)
O'Brien lock butt hook An early hook designed to prevent chokers from unhooking; it had a twin hook swiveling around a center pin and was kept shut by tension on the line which was attached to it. (P.C.F.)
ocean-going raft At one time there was quite a business of rafting logs from Oregon and Washington coastal camps to mills in California. There is still a lot of rafting between up-coast camps and city mills in British Columbia. An ocean raft (of squared timbers) was built on the Kennebec River in Maine in 1792 and set out by sail for Europe, but was abandoned off Labrador. Another venture from Quebec in 1825 was a raft so big for those days (300 feet long) that it was considered a ship and was named "The Baron of Renfrew." It made the Atlantic crossing safely but broke up on the French side of the English Channel. See Baker raft, Benson raft, Davis raft, Gibson raft, Robertson raft and Wheeler raft.
odd as Dick's hatband Said of a person who is peculiar. (Maine)
odd corner 1. A long corner of timber. 2. A difficult area to reach on a logging show. (P.C.F.)
off-highway truck An oversized truck with wide bunks built to haul heavy loads on private logging roads.
off his feed Said of a man who is sick. (P.C.F.)
off ox The right ox of a team of oxen. Near ox was the one on the left side. Oxen were driven from the left side. See dumber than Adam's off ox. LL
off the reservation Said of a man whose conduct is out of line. (P.C.F.)
O'Gorman system A tightline skyline yarding method, using a five-drum donkey and many lines, including a trip line to pull down the fall block. J. S. O'Gorman of Oak Point, Washington was the inventor. (P.C.F.)

One of the earliest big **off-highway trucks**, built by the Kenworth company in Seattle, Washington, hustles a big load of yellow pine logs down from the mountains to Lakeview, Oregon for the J.T. McDonald Logging Company in August of 1939. Operating on **private roads** and without licenses, the loads of these trucks were limited only by the capacity of their tires. *From "The Timberman" collection*

oil bottle A bottle with an iron hook that was filled with oil and fitted with a notched cork to sprinkle oil on a crosscut saw to cut pitch when sawing.

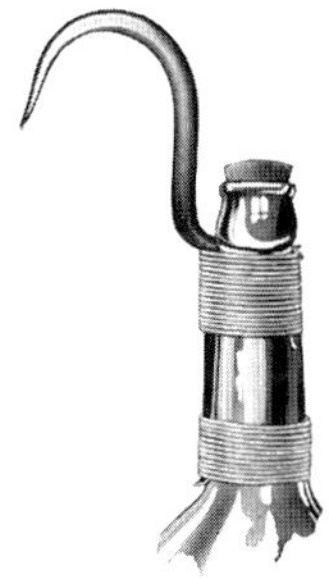

An **oil bottle**. *From Mallory Catalog No. 12, 1925, Glen Comstock collection*

okie 1. A native of Oklahoma. 2. Any worker from the South. 3. A term of disparagement. (P.C.F.)

Oklahoma credit card A tube that is used to siphon gas out of a parked vehicle into a five-gallon can. (P.C.F.)
Oklahoma hooktender A poor whistle punk. (P.C.F.)
old folks home The old Grays Harbor Commercial Company, a Pope & Talbot operation at Cosmopolis, Washington. The Grays Harbor mills were underselling the Pope & Talbot mill on Puget Sound, so Pope & Talbot decided to meet the competition with a mill of their own which operated on a very unusual principle. The company would provide a job and board and room for any and all comers, good times or bad. But they paid low wages, as low as $.75 a day. The men could come and go as they pleased. There was always a job, and no one said a word if the men walked off their jobs. The only requirement was that, if the company paid a worker's fare, he had to work the fare off. As a result, crews were always coming and going, and the company had ads for employment in the newspapers as far east as Minneapolis. But oddly enough, the workers were very loyal, and the unions were never able to penetrate. For years the company was one of the largest producers of all types of wood products on the West Coast. Also western penitentiary.
old growth Virgin timber that usually attained a size and fineness of texture not found in newer timber.
old head A veteran logger with much experience in his particular field. (P.C.F.)
old man 1. The number one boss. 2. A holding device used when pulling a pin out of a tight fit on a machine, or to buck up against when pressing a pin into a tight fit. (P.C.F.)
old pill A camp doctor. (L.S.) (S.F.) LL
old woman slept on the tail of my nightshirt An excuse for being late to work. (P.C.F.)
on a round drive Said of a person getting nowhere; from Paul Bunyan's famous round drive.
one-armed Pete A water pump. (L.S.) LL
one-choker show A tough logging show. (P.C.F.)
one-donkey show A small logging outfit. (P.C.F.)
one-lick method A way of making a fire line around a forest fire. Each man on a crew takes one or two licks with a tool before stepping ahead, leaving the following men to complete the work. (P.C.F.)
one-match burn Using only one match to set slash afire because of a hot, dry day. (P.N.F.)
one-spot Locomotive number one. Engine number two would be the two-spot. (P.C.F.)
one-way incline An incline railroad in which a loaded car was lowered without counterbalancing an empty car going back up the hill. Also single line system.
on his uppers Said of a man who is poverty-stricken. (Maine)
on the ball 1. Said of a worker doing a good job. 2. Said of an outfit doing a good job.
on the boat Said of a man traveling to a camp by steamer. (P.N.F.)
on the chokers To work as a choker setter: "I was on the chokers for a year." (P.C.F.)
on the dirt Said of hauling logs on a dirt road. (P.C.F.)
on the fly Working while logs are moving. (P.C.F.)
on the hog Said of anything broken down, haywire or out of order. (P.C.F.)
on the long route Said of one who works for a long time for one outfit. (P.N.F.)
on the prod 1. Said of one anxious to finish the work. 2. Said of an angry person. (P.C.F.)
on the prowl 1. Said of a man moving to another job. 2. Said of a man who is looking for trouble. (P.C.F.)
on the rods Said of one who drifts from one job to another.
on the skids Said of a man who is slipping, not doing a good job anymore (from the skidroad in town).
on the spot 1. A railroad expression to take it easy. 2. Said of a railroad crew that is waiting in a siding for another train to show up. (P.C.F.)
on the toboggan 1. Said of a poorly kept camp. 2. Said of any person on the downgrade. (L.S.) LL
on tic Said of buying credit at the company store. (S.F.)
open An opening in the timber. (P.C.F.)

This 47% **one-way incline** was operated by Wisconsin Logging & Timber Company, near Stella, Washington, on the Lower Columbia. Wisconsin bought out Simon Benson's holdings in the area in 1906. *E.A. Jones photo, Labbe collection*

open face 1. A block with a wide throat. See Tommy Moore. 2. A donkey engine placed on two side beams, the frame closed by a cross beam at the back end under the boiler, and open at the front and beyond the main drum. Usually the open face was a big road donkey. (P.C.F.)
open-faced Said of a boxing glove choker hook. (P.C.F.)
open-faced pie A pie without an upper crust. (S.F.) (L.S.)
open-fire kiln See Arkansas dry kiln.
open-grained A common classification for types of wood with large pores, such as oak, ash, chestnut and walnut.
open market Before World War II, logs were dumped in the Columbia River and Puget Sound by independent loggers for sale on the open market with much buying and selling and trading of logs. After the war, the independent loggers joined up with a mill and by 1956 the historic open market disappeared. (P.N.F.)
open side block A logging block that has one side plate hinged so that it falls open, permitting a line to be laid in the block without the trouble of threading it from the end. See Bouse block and snatch block.
open-side trolley An early skidder carriage with one side open, allowing it to pass through the hangars on a very long haul.
operation 1. A logging show. 2. The woods plant of a logging company.
operational cruise A timber inventory including estimation of timber volumes or other stand information on specific geographic areas for specific purposes as contrasted with more broadly based

estimates for forest wide planning.
operator 1. A man who owns and/or operates a logging show. 2. Any man who operates an engine or machine in the woods or the lumber mill.
operator's fire 1. A forest fire occurring on an active logging operation. 2. A forest fire on cutover lands not released from slash disposal requirements.
Opsal log loading boom A patented boom loading system similar to the McLean hayrack boom system, with so-called refinements and improvements. Invented by Andrew Opsal of Vancouver, British Columbia.
Opsal skidding system A skyline skidding system that resembled a trolley wire used for street cars. The skyline was suspended from guys attached at right angles to the standing line every 200 to 300 feet. The guys were run out from trees along the main travel line. A special carriage was used that passed over the guy line junctions. Invented by Andrew Opsal of Vancouver, British Columbia. (P.N.F.)
oral auction sale A sale at which interested parties bid openly for timber.
orchard show A show where logging is somewhat easy. See good ground.
Oregon block A stump used instead of a block to change the direction of a line. (P.C.F.) See siwash.
Oregon fir Douglas fir. (P.C.F.)
Oregon larch Noble fir, found at higher elevations in the Coast and Cascade ranges. (P.C.F.)
Oregon lead The use of a stump to change the direction of a line instead of hanging a block. The term is usually applied when moving a donkey around to change the lead. (P.C.F.) See Oregon block and siwash.
Oregon pine This name was given to Douglas fir by Dr. John McLoughlin as early as 1833 when shipping timber to the Sandwich Islands. Fir was an unknown word, and he thought that the cargo would sell better since the term "pine" was better known. Originally Western lumbermen found it easier to sell fir in the Eastern markets under this name, as Eastern buyers were used to Southern pine. See Douglas fir.
Oregon spruce Sitka spruce. (P.C.F.)
Oregon white cedar Port Orford white cedar. Also Lawson cypress. (P.C.F.)
Oregon wrench A cold chisel and hammer that the early steam engineers carried with them. (P.C.F.)
Osborne fire finder A device for sighting the location of forest fires that was developed by Bush Osborne of the U.S. Forest Service in the Pacific Northwest.
other side of the range Where the logging shows are always the best. (P.C.F.)
otter slide A saloon in the basement of a building. (L.S.) LL
outboard drum An extra, small drum mounted outside the frame of a donkey; used for various jobs. See straw drum.

A **Tacoma compound geared yarder** equipped with an **outboard drum**. *From Mallory Catalog No. 6, Glen Comstock collection*

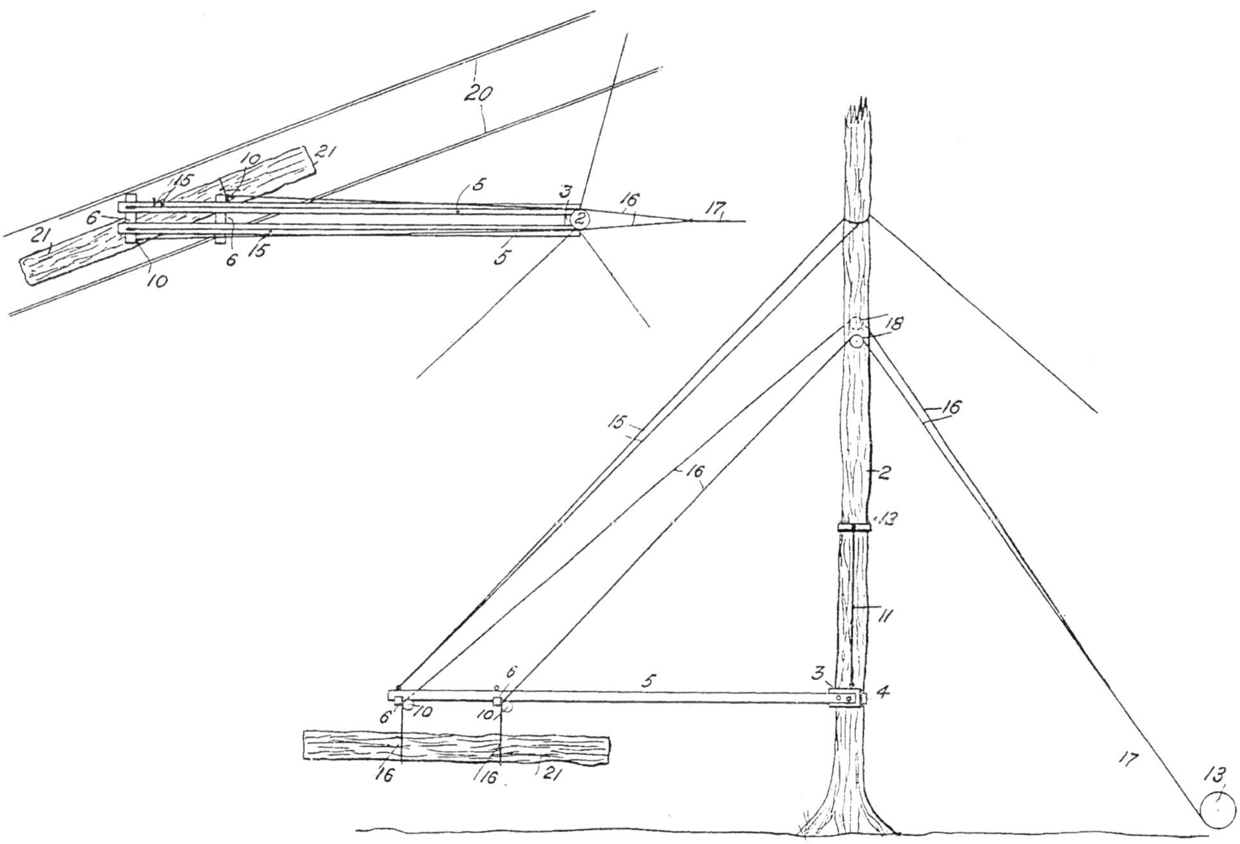

The **Opsal log loading boom**. *From the U.S. Patent Office*

outfit 1. The company for which a man works. 2. Personal gear. 3. Equipment and materials for a job.
outfit cars The traveling camp in the days of big-time railroad logging. (P.C.F.) See camp car.
outhaul To take the mainline back to the woods with a line horse in the days of ground lead logging. (P.C.F.)
outhaul block A tail block that is used at the back end of the haulback line.
outhaul line, rope In the late 1800s, a term for a haulback line. (P.C.F.)
outhaul system A development in logging by adding a second drum to a donkey. The haulback line was wound on the second drum, and as it was paid in, it outhauled the mainline to which it was attached. This system eliminated the line horse which used to take the line back to the woods. See high lead system.
outhouse A small outdoor building used as a toilet.
out in the tules 1. Said of a logger working in the brush or back country. 2. Said of a logger working in swampy ground. (P.C.F.)
outlaw 1. A worker with a record or bad reputation. See agitator. 2. A tree that is dangerous to fall. See renegade. 3. A machine that has killed a man. 4. A logging sleigh having bunks longer than 15 feet, outlawed by the state because of overloading of teams. (L.S.) LL
outlaw camp A temporary camp that houses a few men building a main camp ahead of a big crew. (P.C.F.)
out of lead Said of a block on a spar tree that is not in line with a drum on a donkey, causing wear on the wire rope.
out-of-lead block Any yarding block that is used to put a bight in a line, changing it out of its usual course.
out of round Said of a circular saw when its periphery is not a perfect circle.
out of the woods 1. Said by the loggers when they have almost finished an operation successfully. 2. Said by the loggers when things are back to normal after being in difficulty.
out of wind Said of a stalled donkey. (P.C.F.)
output See cut.
out-turning Said of an empty tractor going back to the woods. (P.C.F.)
overage A term often used to describe the overrun.
overalls and snoose A logger's pay. From the days when the men relied on the company store and seldom got ahead of the company. (P.C.F.)
oven-dry wood Wood completely dried, seasoned until it is without any moisture content.
overcut 1. A trim allowance of several inches left on the end of a log, the amount varying with log length. 2. The cutting of a quantity of timber in excess of the annual or periodic cutting budget, or in excess of the estimated volume for the area.
overhaul 1. To pull out a line against brake pressure on a donkey drum. 2. To pull in a light line attached to a mainline, such as in rigging up a skidder show. 3. To have a line paid out by its own weight pulling it off the drum. 4. To repair or rebuild a machine of any kind.
overhead loading The modern loading systems by which logs are lifted clear off the ground by a donkey or shovel loader.
overhead roading Hauling logs by tramway. See log tramway.
overhead wire system Any skyline method.
overhead yarder An early skidder. (P.N.F.)
overhead yarding An older term for bringing in the logs to the landing by a skyline. (P.N.F.)
overhung drum See outboard drum.
overhung gypsy See nigger-head.
overland trout Bacon. (P.N.F.)
overlay To spool line onto a drum from the top side.
overlength The trim allowance on a log. See trim.
overload Any load exceeding the legal limit. According to the law, too much weight on a truck. Overloads cause many serious accidents and ruin the roads.
overlooked his hand To describe the camp dispatcher's fatal lapse of memory who let a train get by the last passing track where it could meet an opposing train, thus causing a collision. (P.C.F.)
over-mature timber Trees or stands of trees showing a decline in vigor, health and soundness associated with aging.
overscale Said when the fallers and buckers get more scale in the woods than the company gets at the log dump; the difference between woods scale and water scale when the woods scale is greater.
overside delivery Delivery direct from a vessel to a scow or barge.
overstory That portion of the trees, in a forest or stand of more than one story, forming the upper or uppermost canopy layer.
overwind To put more line on a drum than it can hold, thus ruining the line.
overwound cable Said of line that is reeled on or paid off the top of a drum. Also overshot cable or drum.
ox, oxen The use of oxen in logging can be traced back directly to the period before the American Revolution when the English and French governments began to draw upon the forests of Maine and New Brunswick for white pine for masts and spars. During the early 1800s, the ox was the favorite draft animal in the Maine woods. Horses were not used in any great numbers until later. Although the oxen gradually disappeared in the woods after Dolbeer invented his steam donkey in 1881, they have appeared in the woods again to log the small logs.
ox bow 1. The loop of maple or other wood placed under the neck of the ox to keep the yoke from falling off the top of the neck. 2. A loop in a river or road.
ox bow pin A wooden pin to hold the ox bow in the yoke. LL
ox fork A short handled fork with two broad tines used in driving hay into sleigh ruts to keep a sleigh from going downhill too fast. (L.S.) LL
ox prone A frame used to hold the ox still while it was being shod.

An **ox** is held firmly in the **ox prone** while being fitted with new shoes. *Labbe collection*

ox team The ox teams moved slower than the horses, but were sure of foot and could outpull the horse team. Eight to twelve oxen were used in an ox team. Also bull team.
ox teamster 1. The man who drove the bull teams in the early days of logging. He had the care of the cattle and was the first to rise in the morning so that the animals might be fed, and he was the last man into camp for the same reason. Outside of Maine and New Brunswick, the whip was used in place of the goad in the Northern woods. (Isaac I. Stephenson, Recollections of a Long Life, 1829-1915) 2. The man who handled ox teams freighting supplies to camp.
ox tender The barn man who fed and watered the animals.
ox yoke A wooden block shaped to fit over the neck of oxen to push against when pulling. LL

A team of **oxen**, or **ox team**, dragging a string of logs along a **skid road** for the D.H. Kelly Logging Company, near Stringtown, in Clackamas County, Oregon. Note the heavy **ox yokes** on the necks of the oxen. *Courtesy of Mrs. McPherson.*

This unusual **skidroad** "action" view shows a **bull puncher** running to the aid of an **ox** that has lost its footing and tumbled down into a gully. Location is somewhere in the **Redwood Region** of Mendocino County, California. *From the Steven R. Gatke collection*

P

P.L.A. The Penobscot Lumbering Association, which was chartered by the Maine Legislature in 1854, undertook all new construction, paid the taxes, did its own rafting and paid a toll of 10 cents a thousand to the Penobscot Boom Corporation. (Richard G. Wood, "A History of Lumbering in Maine, 1820-1861")

P.L.D. The Penobscot Log Driving Association, which was chartered in 1846, owned all dams and booms and annually sold the drive at auction to the bidder contracting to take the logs down at the lowest rate per thousand. (Fannie Hardy Eckstorm, "The Penobscot Man")

pack a card To belong to a union.

packer A shortened form for water packer, the man or boy who packed water to the skidroads.

packin' a balloon, pack the balloon 1. To pack up personal gear and to leave camp. 2. To carry one's blankets from camp to camp. (L.S.) LL

pack rat 1. A woods rat which packed off everything it could from the camps. Also trader rat. 2. Usually a man who collects odds and ends.

packs Rubber-bottomed shoes with no caulks which were used in the winter.

packsack A canvas sack with shoulder straps in which the early loggers packed all their belongings. Also war bag.

pack the hook 1. To work as a hooktender. 2. To be able to handle the job of hooktender. (P.C.F.)

pack the riggin' 1. To carry the line out by hand. When rigging a side, the strawline had to be carried out before the haulback could be pulled out. 2. To carry a card to show that one has the authority to represent an organization. 3. An expression used by the wobblies to identify a man who was carrying I.W.W. literature. (L.S.)

pad 1. Another term for mat. 2. The sections of a Caterpillar track. 3. A bed. 4. To move equipment on pads.

paddle In turpentine orcharding, a small piece of wood that is placed over a cup during the chipping process in order to keep trash from falling into the receptacle. (S.F.)

pad man A helper around a shovel who attaches the lines to move the pads or mats on which the shovel works. See pit man.

paint To coat the ends of a log with mud so a defect will be hidden from the scaler. (N.W.)

paint a log To mark a log with paint to show it belongs to a certain company. LL

paint gun A squirt gun that is used to mark the trees to be cut.

paint mark Different colored paints were used in the 1800s in pine logging and river drives to mark logs for ownership; later changed to end and side marks made with an ax or stamping hammer. (N.F.)

A Bucyrus D-2 gas **shovel** at work building railroad **grade** for the Sound Timber Company, out of Darrington, Washington. The shovel rests on wood **mats,** or **pads,** to keep it from becoming mired down, and to help compact the earth. As it progressed, it could swing the pads forward from behind. The **pad man** is standing behind the shovel. *D. Kinsey photo, Labbe collection*

paired grab hooks See dogs.

paling A short board or piece of wood. (L.S.) LL

palooka 1. An Arab. 2. Another term for sucker. 3. A punch drunk fighter (from Joe Palooka of the comic strip).
palouser A makeshift lantern made by sticking a candle through a hole in a tin can. (P.N.F.)
pan See skidding pan.
pan bread Bread or biscuits made in a frying pan set on the coals of a camp fire.
pangoram 1. A dish made by a poor cook. 2. Any ingredient mixed with water. (P.N.F.)
pannikin A small pan or cup for tea or coffee. (L.S.)
pan road A road built for tractors with attached pans. In recent years tractors with attached pans were generally used in place of horses. One end of a log was placed on the pan for skidding logs out of the woods. See skidding pan. (L.S.) LL
pan skid 1. To skid logs with the front ends resting on a large steel pan turned up in front in skidding with horses. 2. To use the same procedure in skidding with cats. See skidding pan.
panther juice, panther piss A strong drink. (P.C.F.)
panther sweat A strong alcoholic drink. (P.C.F.)
pants rabbits Body lice. LL
paper Bankable or negotiable notes. (App.)
parbuckle 1. To roll a log sideways onto a railroad car. In the early days of railroad logging this was the method in which logs were loaded onto railroad cars. A cable was passed under and back over a log, and when the cable was pulled, the log rolled onto the car. Also crosshaul, gilhooley. 2. To free a log hung up behind a stump by rolling the log over the stump. 3. To use the same procedure in unloading cars or trucks at times.
pardner, partner Usually the other half of a set of fallers, or a working team.
paring chisel See slick.
park To collect cross-ties along a strip road, usually by hand. (R.M.F.)
partial cut Any harvesting method in which only part of the mature trees are removed, as opposed to a clearcut. (Idaho)
pass block A light block that is carried up by the rigger and hung at the top of a spar tree to haul up the rigging for a spar tree. Also climber's pass line block. See high climber block, rigger's block.

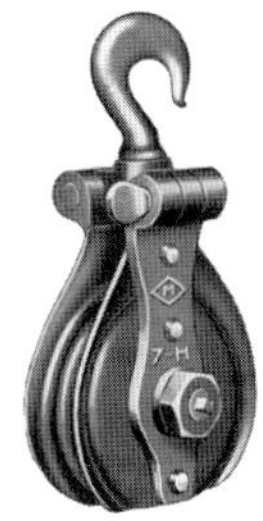

A **pass block**, used by **high riggers** for rigging **spar trees**. *From Mallory Catalog No. 12, 1925, Glen Comstock collection*

pass chain A chain that is used to make a chair for the climber so that he can be hauled up the tree. Also known as a grip chain when used on the end of a pass line in rigging the tree.

A **pass chain**, used as a chair by **high riggers** to be hoisted up a **spar tree**. *From Young Iron Works Catalog No. 52, 1952, Philip Schnell collection*

pass line A light wire rope used to haul gear up to the top of the spar tree for rigging. Also high climber's elevator, rigger's pass line. See jack line.
pass the 44s To pass the beans. Loggers graded the beans according to size as 22s, 30-30s, 44s, etc. (P.C.F.)
pass the hat To take up a collection, usually for somebody hurt or killed in an accident. LL
patch A tract or patch of timber. (P.C.F.)

Loading logs for the Chippewa Lumber Company near Rodney, Michigan. The logging was handled by O.P. Pillsbury & Company. The logs are being rolled up the **skids** by **ox team**, aided by loggers with **peaveys**. This system of loading was called a **cross haul**, or **parbuckle**. *Labbe collection*

patch logging See staggered settings, block cutting.
pat him on the lip To beat up a person. (L.S.) LL
Paul Bunyan The legendary folklore hero of the shanty boys of the Lake States. Historically the most likely account puts Paul Bunyan's origin in the Quebec country, where a powerful woodsman was famous over a century ago under the name of "Bunyan" or "Bon Jean." "The Round River Drive" is the oldest authentic bunkhouse tale of Paul Bunyan. The first printed Paul Bunyan story appeared in the Detroit News in 1910; the second was in a poem in the April issue of the American Lumberman in 1914. "The Round River Drive" was the main story in both selections. The story tells how in the wilderness Paul Bunyan took a log drive down a strange stream for months and months because the river was round. How did he ever get his logs out? The fun is in trying to solve his problem.
pavement pusher A logger looking for a job at the employment offices. (P.N.F.)
paving blocks Creosoted wood blocks used for surfacing city streets. (L.S.)
pay cheater A company timekeeper.
pay cut See back cut.
pay dirt A method that works out well.
pay hole The high gear on a donkey. (P.C.F.)
pay off To unreel a line from a drum.
pay out 1. To unreel a line from a drum. 2. To unreel wire rope. 3. To come out of a deal or job successfully.
pay pole A good log. (P.C.F.)
pea jacket See mackinaw.
peak In turpentine orcharding, the apex of the triangle above a box or cup at which the two faces meet. (S.F.)
peaker 1. A load of logs narrowing sharply toward the top, and shaped like an inverted "V." 2. The top log of a load. Also windsplitter. LL
peaker log The top log on a load of logs.
peanut post A small portable steel tower for quick rigging up in relogging or prelogging. (P.N.F.)
peanut roaster A small steam donkey or locomotive. (P.N.F.)
pea picker A farmer who worked part-time in a sawmill. (L.S.)
pearl diver A dishwasher.
pea-souper 1. A French-Canadian lumberjack. 2. A Frenchman or Quebecer working in the woods (from their fondness for pea soup). 3. A very heavy fog. LL
peaver See sacker.
peavey The famous woods tool, heavy handled, with a spike point and free-swinging hook, used in rolling logs all over North America. Said to have been invented in 1858 by J. H. Peavey, a blacksmith of Bangor, Maine. Also life preserver, peewee, Quebec choker.

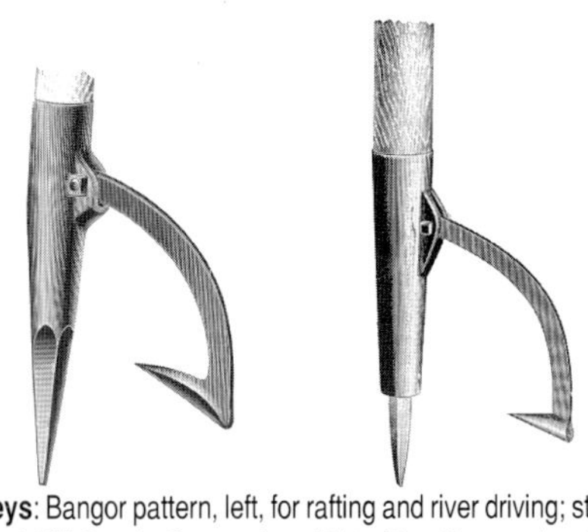
Two styles of **peaveys**: Bangor pattern, left, for rafting and river driving; standard pattern, right. *From Mine and Mill Supply Co. catalog, 1916, Glen Comstock collection*

peck Pockets or areas of disintegrated wood caused by advanced stages of decay in a living tree. Usually associated with cypress and incense cedar.
pecker pole A small log, pole or pile. (P.C.F.)
pecker pole show A gyppo outfit that handles small timber. (P.C.F.)
peckerwood outfit 1. A small gyppo outfit. 2. A small portable sawmill in the woods. (P.C.F.) (S.F.)
pecky 1. Partly decayed wood. Also peggy. 2. A term applied to an unsoundness most common in bald cypress. (S.F.)
pede The handcar or velocipede used on the railroad. (L.S.) LL
peel 1. To remove the bark on a redwood for easier sawing. (Red.) 2. To take the bark off piling or pulp timber. 3. To remove bark as in hemlock bark camps. (L.S.) LL
peeler 1. A log suitable for plywood. 2. A man who peels logs before they are sawed. (Red.) 3. A rotary veneer machine that cuts or peels a thin endless sheet of wood from a round log. 4. One who peels bark in gathering tanbark. (L.S.) LL
peeler log A log suitable for the manufacture of rotary-cut veneer, the top grade of log. See peeler.
peeler picker The man who grades logs to pick out the best peeler logs. (P.C.F.)
peeling bar, See barking bar, spud.
peeling chisel A chisel to remove bark. (L.S.) LL
peen hammer See ball hammer.
Peerless A brand name for an old favorite chewing and smoking tobacco which was popular in the logging camps and the mills. It was so strong that the name was applied to anything very strong or tough.
peewee 1. A small merchantable log, suitable for sawing in a gang mill. (P.C.F.) 2. A peavey. (L.S.)
peewee peeler A small log suitable for peeling in a gang mill. (P.C.F.)
peggy See pecky.
pegs Oats. (Maine)
pelican hook See finger link. (P.C.F.)
pencil pusher A bookkeeper or clerk.
penga, penger A Norwegian word for money. (L.S.) LL
pengi A Swedish word for pay. (L.S.) LL
Peninsula The Olympic Peninsula of Washington which is a famous logging area. (P.N.F.)
Pennsylvania salve Apple butter. (West Virginia)
penny dog An assistant foreman. (L.S.) LL
pension An easy job. (P.C.F.)
pensioner A lazy man. (P.C.F.)
percolate 1. To get something going. 2. To get to work. 3. Any piece of equipment running well is said to percolate. (P.C.F.)
perfume pie A pie made from packaged pie filling. (N.F.) LL
persuader Anything used as a lever. (P.C.F.) Also swede.
per-thousand The unit measure of timber value and of logging costs, so much per thousand board feet.
peter hook A large flat heavy hook that was used on the chokers in the late 1800s and early 1900s. So-called because it was manufactured by the Peters Machine Works of Kelso, Washington. (P.C.F.)
Peters choker hook A type of choker hook with a hinged joint between the hook and the sleeve.

A **choker** equipped with a **Peters choker hook**. *From Mallory Catalog No. 48, 1948, Philip Schnell collection*

A **hemlock special** behind a **Phoenix** steam hauler, built by the Phoenix Manufacturing Company of Eau Claire, Wisconsin. Similar to the **Lombard**, and equally popular, the Phoenix used the Lombard track system under license. However, the two differed in that the Phoenix carried water in a tank beneath the boiler for greater stability, and each track was driven by a small two-cylinder engine mounted alongside the **smokebox**. *Photo from M. Koch collection*

petticoat 1. A wide flared steel net put around a donkey stack to stop the sparks. 2. A metal baffle in the smoke box of a locomotive, used to control the draft.
petticoat pipe The pipe that extends from the smoke box up into the stack of a locomotive to direct the exhaust from the cylinders.
P.F. man The polite term for the pig fucker who took care of the skidroad pig, or boat, and kept the wire rope on the spools around the curves in the skidroad. Employment offices on the West Coast used to list "P.F. Man Wanted." See pig..
Phoenix A popular steam hauler made by the Phoenix Manufacturing Company of Eau Claire, Wisconsin. (L.S.) (Canada)
pick A heavy wooden-handled iron or steel tool pointed at one or both ends.
pickaroon A tool shaped like a short-handled ax, but with a pointed head.

A **pickaroon**. *From Mill and Mine Supply Co. catalog, 1916, Glen Comstock collection*

pick ax See mattock.
pickpole In river driving, a twelve-foot long slender ash handle with a metal tip twisted to a sharp point; used to pull logs and as a lever to push a log that was to be rafted up against the joint by sticking the pick in the bottom of the river and pulling on the handle. See pike pole. (N.F.)
pick the rear See sack the rear.
pick-up 1. The tightening of a line that lifts a logging block in the air. 2. To raise the rigging by tightening up on one or more of the lines.
pick-up block See loading block.
pick-up derrick An older term for a cherry picker or portable loading rig. Used to pick up logs spilled along the right-of-way. (P.C.F.)
pick-up man A logger who looks for stray logs that have slipped out of a boom. (P.N.F.)
piece cutter A lumberjack who cuts logs or pulpwood at so much a stick. (L.S.) LL See busheler.
piece system Bushel or contract work that is paid by the piece rather than by the hour. (P.C.F.)
pie fork. pie lifter, pie peel A fork-shaped long-handled gadget used in lifting hot dishes, especially pies, from the ovens of the shanty stove. (L.S.) LL
pie in the sky Anything that is good and usually unobtainable. (P.C.F.)
pier dam See wing dam.
piers The rock-filled cribwork structures out in the river from which the booms were strung or against which they were supported. (N.W.) See wing dam.
pig 1. A log that was hollowed out to take the dogs and other tools back to the woods or to the yarder donkey on a fore-and-aft road. The pig was dogged to the last log. When the logs were unfastened at the landing, the dogs were thrown into the pig that was returned to the woods. (P.C.F.) (Idaho) Also dog boat. See riggin' sled. 2. A pressure vessel used to force oil through an engine after the engine has been overhauled and before it is used.
pig boat Another term for the pig.
pig fucker The man who took care of the pig and kept the wire rope on the spools around the curves on the skidroad. The polite term is P.F. man. (P.C.F.) See P.F. man, pig, zoogler.
piggy back See humpback.

Roading logs with a **road donkey**, or **roader**, on a Hammond Lumber Company operation in Humboldt County, California. The **P.F. man** rides the **pig** behind the **turn** of **redwood** logs. *Hammond photo, Carranco collection*

pig iron Hardwood logs. (L.S.) LL
pigman See chaser.
pig's ear 1. A saloon. (West Virginia) 2. The first camp in the woods where the men lived who were building the main camp. (Maine)
pig's foot jammer hook See loading hook.
pig tail 1. A short piece of iron, twisted to make an open loop, and driven into a tree trunk to serve as a guide for a light line or rope. (P.C.F.) 2. A device to shorten the grip for short logs in crotchline loading. 3. A loop kinked into a line by careless handling.
pike See tote road.
pike lever See hand pike.
pike pole A long pole with a spike and hook on the end; used on logs floating in the water. (N.F.) (Canada) Also jam pole.

A **pike pole**. *From Mill and Mine Supply Co. catalog, 1916, Glen Comstock collection*

pike road See tote road.
pile 1. A pile of logs such as a cold deck. 2. A pile of lumber. 3. A long straight pole, used for foundations. Also piling.
pile and burn To dispose of slash in the Pine Country by piling limbs and tops to be burned after the fall rains or snows reduce the fire danger.
pile buck A pile driver engineer. (P.C.F.)
pile driver A rig that lifts and drops a heavy weight to drive piling into the ground for trestles, docks, log ponds, foundations for buildings, etc. Also spile driver.
pile man The man who unhooks logs at a log pile. (P.N.F.)
pilgrim 1. A visitor in camp. 2. A logger who works for a short time in a camp. (P.N.F.)
piling piece A place in the woods where tops and brush are piled. (Maine)
piling railroad A logging railroad on trestle work across swampy ground.
piling show 1. The operation of cutting tall, straight trees to make good piling. 2. A timber stand that contains tall, straight trees.
piling track In the late 1800s in railroad logging, a light spur built entirely on piling without benefit of trestle work, usually on the bottom of a canyon. (P.C.F.)
pill faker, pill roller 1. A first aid man. 2. A doctor. (P.C.F.)
pilot A man in charge of a raft who handled the front oar. (App.)
pilot car 1. Same as block car, a lead car used in some incline systems. 2. An empty flat car, or idler, used to hold the lowered spar of a steam tower skidder when moving the machine to a new setting.
pimp sticks Tailor-made cigarettes. (L.S.) LL
pinch bar 1. An iron bar used to pinch logs tightly together. (L.S.) 2. A bar used to move logging railroad cars short distances. LL
pinched 1. Said of a saw bound in a cut. LL 2. Said of someone stuck in a bad situation.
pin dote Small rotten spots on the ends of logs.
pine (pinus) Eastern white pine (pinus strobus) is an important eastern conifer. For about 250 years Eastern white pine was the leading commercial species. Because of its former abundance and the cheapness and varied usefulness of its lumber, it was important

A **pile driver** at work building a low trestle for Wisconsin Logging & Timber Company, out of Stella, Washington, on the Lower Columbia River. *From Knute Fahlstrom collection*

in the development of the states in which it grew, as well as of the regions far outside its natural range.

Eastern white pine grows naturally from Newfoundland to Lake Winnipeg in Canada, southward through the Lake States and New England. It also grows in the Southern Appalachians, as far south as Northern Georgia. The wood of Eastern pine is light, soft, durable and very easily worked. It is straight-grained and takes paint well. Practically all of the Eastern white pine taken from the forest today is converted into lumber.

Pine bull donkey Wire rope was introduced about 1884 when N. H. Pine of Eureka, California developed his bull donkey. The new donkey was equipped to haul logs over long distances, 1,000 to 2,000 feet. It was a portable machine with an upright boiler in back. The working parts of the donkey were two large drums. Two upright and two horizontal spools, which slid back and forth on a spiral-geared shaft to feed the line on and off the drums, were in front of each of these two drums. The main drum held over 4,000 feet of 3/4-inch steel line which hauled in the logs. The smaller drum held the 1/2-inch line which hauled the heavier line back to the starting point at the top of the skidway.

The early bull donkeys were comparatively small compared to the later bull donkeys which could carry up to 1-1/2 miles of 1-1/8-inch line and 2-1/2 miles of 3/4-inch haulback line. The two horizontal and the two upright spools were never too successful because the drums would get larger as the line was reeled in, causing the fairlead device to go out of timing. (P.C.F.) See bull donkey.

Pine Country 1. On the West Coast this term means the interior pine country: the plateau country in British Columbia, the east side of the Cascades in Washington and Oregon, both sides of the Cascades and the Sierra in California, and the white pine country in Northeastern Washington and Northern Idaho. 2. Any area with a heavy growth of white pine.

pine groves An early term in the Lake States for stands of white pine. (L.S.)

Piling railroad of Smith-Powers Logging Company in Coos County, Oregon, where the ground was too unstable for normal **grading**. *Courtesy of John and George Powers*

A **Pine bull donkey,** which replaced the **bull teams** on the **skid roads** of Humboldt County, California. *Carranco collection*

pine hog See timber baron.
pine logger 1. Any logger or lumberjack working in the Pine Country. 2. A Lidgerwood yarder and loader in one unit, used for pine operations. 3. A derogatory term for an Oregon logger who knows nothing about redwood logging. (Red.)
pinery, pineries The area in Northern Minnesota, Wisconsin and Michigan. The tree that gave character and distinction to the whole region, which was larger than all of the state of Maine, was the white pine. This tree dominated the lumber industry for at least five decades. The section which lay nearest the Mississippi and in the region of the St. Croix and its tributaries was most prominently the home of the white pine and its relative the red (Norway) pine and a remote relative, the jack pine, but the white pine was master, and that region was generally thought of as the real "pineries." (Agnes M. Larson, History of the White Pine Industry in Minnesota)
pinery-boy See lumberjack.
pinery road A road that leads to the pine woods in Northern Wisconsin. (L.S.) LL
pine sawyer A beetle of the genus Monohammus that attacks the sapwood of pine logs. (S.F.)
pine stroke In bucking with a crosscut saw, the method of bringing the arms along the side of the body and pushing forward as opposed to the redwood stroke of pushing from the chest forward.
pine top An alcoholic drink. (L.S.) (S.F.)
pinion Usually the small gear on a donkey that was fastened to the main shaft and drove the bull wheel. (P.C.F.)
pin knot A knot that is not over one-half inch in diameter. Usually found near the heart and does not show up on the surface of a log.
pink one In forest surveys, a close measurement used facetiously (pink cunt hair); for example when the transit man wants the rod moved to the left or right a very small distance. (P.C.F.)
pins One and one-half-inch square ash pins, nine inches long, driven into auger holes where the bow has been placed, to tighten and hold the bow to the lash pole, when making up rafts. (N.W.)
pin whackers The men who drove in pins of metal or wood attached to chains or driven over ropes in setting up log booms. (L.S.)
Pioneer 1. A certain make of logging donkey. 2. To rough out a new road. 3. To prepare a grade for later improvement.
pioneer road A rough road that is made before the actual construction of a road to bring in men and supplies.
pioneer work The first stage of a construction job. (P.C.F.)
pipe man, pipe monkey The man who handled the water and water pumps for steam engines on a logging show. (P.C.F.) Also pump man.
piperot Rotted wood that occurs usually in the center of the white pine. (N.F.)
pirate 1. A log thief. (P.N.F.) 2. A manufacturer who sells direct to a consumer at a place where retailers are in business.
PI's Men from Prince Edward Island. (Maine)
piss fir 1. White fir or lowland white fir; called thus because of its strong odor and because of its water content. (P.C.F.) 2. Larch found in Oregon and Washington. (P.N.F.) See larch.
piss fir willies The foresters. (Red.)
pistol grip 1. The handle of a chain saw. 2. A tree with a sharp crook near the butt.
pit 1. A skidway elevated so that its base is level with the logging car bunks. (App.) 2. A cleared space at the tail of a portable mill which is used for storing lumber. 3. A hole in which the pit sawyer stands when whip-sawing lumber.
pitch 1. An accumulation of resin which occurs in separations in the wood or in the wood cells. 2. A slope. 3. On wire rope, the straight line distance along the rope required for one wire to make a complete turn around the rope.
pitch butt The bottom log in a tree which is soaked with pitch.

pitchers Small hills. (N.F.)
pitch pocket A lump of pitch inside a log.
pitch ring A circle or part of a circle of pitch following around the annual ring in a log.
pitch seam The pitch in a crack in the wood.
pitch spike A spear of wood that is soaked with pitch and that is found in an old rotted stump, usually an old knot.
pitch streak In coniferous woods, a well-defined accumulation of pitch at one point. LL
pitch up To seal anything with any kind of sticky substance.
pith The small soft core occurring in the center of a log.
pith knot A sound knot with a pith hole in the center in cut lumber which is not more than one-fourth of an inch in diameter.
pit man The man who works with a power shovel or a bulldozer when moving dirt. (P.C.F.)
pitoune A bolt. (Quebec)
place See turpentine orchard.
plank road 1. A wooden truck road, varying from makeshift planks thrown on the ground to expensive roads built almost to railroad standards. 2. Any road built of sawn planks.
plug a boom To jam a wooden plug into the hole in a boomstick to hold the toggle of the boom chain in place. LL
plug and knock down A device for fastening boom sticks together when there are no chains. It consists of a withe secured by wooden plugs in holes bored in the boom sticks. LL
plug boom A boom of logs connected by ropes passed through holes bored in the ends of the logs. Wooden plugs driven into the holes wedged the ropes that pulled small bunches of logs to the main boom. (N.F.) LL
plug for pitch To tap an old fir to get the valuable pitch in the late 1800s. (P.N.F.)
plug tobacco Chewing tobacco sold in long strands and blocks and was clipped off with a tobacco cutter.
plumb-bob squirrel A counterweight for a loading boom that runs up and down a tree or guy line. Also called squirrel, squirrel chunk. See squirrel.
plunder The logger's personal gear which he brings to camp. (P.C.F.)
plystran core A material used for cores (center layer) in plywood. It is composed of oriented fibers (flakes) mixed with phenolic resin and a small amount of wax which are then pressed under heat.

A **motor truck** on a **plank road**, hauling three **sitka spruce** logs. This was one of the operations of the **Spruce Production Division** of the U.S. Army Signal Corps, which logged and milled sitka spruce for building airplanes during W.W.I. *From "Pictorial Review of World's War Activities Spruce Production Division, U.S. Army, Oregon & Washington," 1918, A.M. Prentiss photo, Philip Schnell collection*

plank road man A man who keeps the plank road in repair and watches for loose spikes and possible jill-pokes.
planting hoe A short-handled long-bladed narrow hoe used to dig holes in tree planting.
play hell with To make a bad mess of anything. LL
plot cruise To estimate timber by tallying trees on square plots laid out at intervals along a compass line.
plug 1. Logs piled up in a river drive. (L.S.) 2. A steel pin about 2 inches in diameter and 18 inches long. Two of the plugs are joined together by chains which are attached to a large ring. They are used on pullboat operations in a cypress swamp in place of skidding tongs. (S.F.) 3. To put on the air brakes on a locomotive in an emergency. 4. To drill a shallow hole in a boulder to take a small charge of dynamite in order to shatter the rock.
Pocatello yardmaster On the railroads, a boomer who brags of his exploits and switching prowess on other railroads. (P.C.F.)
pocket 1. An area of timber that is left uncut because of rough ground or other reasons. 2. A well-defined opening between the rings of annual growth which develops during the growth of the tree, and usually contains pitch or bark. See pitch pocket. 3. An iron stake pocket on the side of a railroad car into which stakes are driven to hold logs or piling on the car. 4. An area into which logs are kept before making up a boom.
pocket boom A boom in which logs are held after they have been sorted. (N.F.) LL
pocket rot Advanced decay which appears in the form of a hole, pocket or area of soft rot usually surrounded by apparently sound wood.

pocket undercutter A small undercutter that could be clamped to an ax handle. The ax, embedded into the log to be bucked, supported the undercutter. See undercutter.

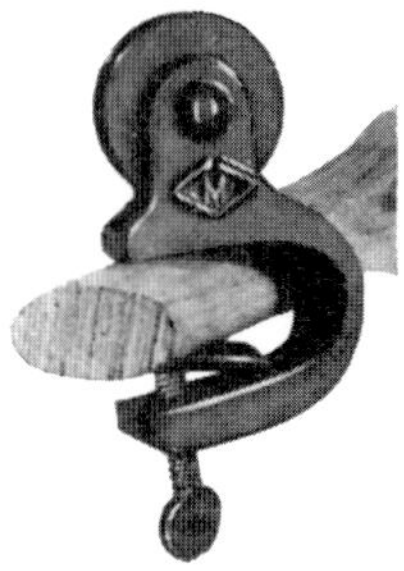

A **pocket undercutter**. *From Mallory Catalog No. 48, 1948, Philip Schnell collection*

pod A derail to prevent cars from running out of a siding onto the mainline. (P.C.F.)
point In hoop making, to bevel both sides of the lapped end of the hoop.
pointer 1. A light, sharp-ended rowboat used on river drives. (Canada) 2. See tripper.
pokelogan A bay or pocket into which logs may float during a drive. (N.F.)
polack A woodsman or mill worker from Eastern Europe; this term is often used disparagingly. See bohunk.
pole 1. A spar tree. (Red.) 2. A round timber of any required size or length and usually used with the larger end in the ground. 3. A young tree at least 4 inches and less than 8-1/2 inches in diameter, 4.5 feet above ground.
pole ax A single-bitted ax used for chipping and driving wedges and for releasing a crosscut saw from binding when trees were being felled. (N.F.) (Canada)
pole camp A camp which only cuts poles. Also piling camp.
pole cats The men who make ties for the railroads. (N.F.)
pole chute A light fore-and-aft road made of poles which could be on a steep pitch, where logs could be moved by gravity, or flat ground where horses or cats provide the power. See log slide.
pole cutter A man who fells, limbs, tops and peels the poles, getting them ready for market. Also pole maker.
pole road 1. A V-shaped trough made of small logs used for hauling logs from the woods. 2. A crudely-made trough of small poles to ease the skidding on poor ground. 3. A railroad using small poles for rails. The rolling stock had concave wheels. Also wooden railroad. See tram. 4. To haul logs by road donkey or otherwise, on a pole road or fore-and-aft road.
pole road hauler Any type of power on a pole road.
pole road man A man who took care of the pole road.
polers Men who direct pulpwood and logs from rafts in the water to its proper location for loading on cars. (N.F.) LL
pole size Trees from 5 inches to 11 inches in diameter, 4.5 feet above ground.

A **pole road** somewhere in the South. Typical of such inexpensive installations, this shows a Class A **Climax** locomotive and cars operating on a railroad built of poles laid on top of the ground. *Courtesy of W. Casler*

pole stand Young timber, suitable for poles rather than sawlogs. (P.N.F.)
pole stock Lumber used in the manufacture of poles for vehicles.
pole team The team nearest to the load. (L.S.) See butt team and wheelers. LL
pole thicket A dense stand of young trees that is difficult to walk through. (P.N.F.)
pole tie A tie having the bark still on and sometimes hewn, or sawn to size on one or two sides.
pole timber Small trees suitable for poles, and not big enough for sawlogs; generally trees 12 inches to 18 inches in diameter, 4.5 feet above ground. Also piling timber.
pole tram road A logging road, the rails of which are round poles. (S.F.) See pole road.
poling the rear In river driving, said of the man who rode herd on loose logs by shoving them ahead of him to clear the river faster. (N.W.)
pond A body of water into which logs are dumped for sorting, booming, reloading or sawing at the mill.
pond boss The man in charge of a log pond.
pond-buck To cut long logs into various lengths in the log pond.
ponderosa pine (pinus ponderosa) Formerly known as Western yellow pine, it is one of the most widely distributed and important of western timber trees. It is found in every state of the Great Plains, as well as in Southern British Columbia and Northern Mexico. It is the most valuable species of the Southwest, where it forms great open forests at 3,500 to 8,500 feet above sea level and is to that region what eastern pine has been to the Northeast or longleaf pine to the Southeast. This pine has a wide range of uses from highly finished products to the roughest construction.
pond lily 1. Same as lily pad. 2. The flat piece cut off the end of a peeler block to bring it to exact size for placing in the chucks before peeling.
pond man 1. Usually a boom man who builds rafts of logs to go from camp to the mill. 2. Also the man who leads logs to the jack ladder in a mill pond. Also pond monkey. Called jacker in the North

A **stand** of **yellow**, or **ponderosa pine**, in Eastern Oregon. *Labbe collection*

woods and jackerman in the Southern forests.
pondosa pine Another term for ponderosa pine. (P.C.F.)
pontoon A raft on the log pond which is used to recover sinkers.
pony express A speeder that brings mail to camp. (P.N.F.)
pool drive The association of lumber companies that pooled their logs into one drive. (N.F.) LL
poontang Sexual intercourse. The word is a heavily nasalized Creole pronunciation of the French word "putain," whore. The word probably spread from French-speaking Louisiana. (Harold Wentworth and Stuart Flexner, Dictionary of American Slang)
poontanger A logger's penis. (P.C.F.)
poor box A tobacco box near the door containing free tobacco. Men who broke rules were punished by fines paid in tobacco for the free use by the men. (L.S.) LL
poor steamer 1. Usually a leaky steam engine, but some engines were just naturally poor steamers, for no apparent reason. 2. A lazy man.
popcorn crew A loading crew that made up a poor load of logs on the sleigh that might fall off while moving. (L.S.) LL
popple 1. An aspen. (R.M.F.) 2. A poplar tree. (L.S.)
porcupine Frayed wires on the unravelled end of a wire rope. (P.C.F.)
porcupine boarding house A tree with an excessive hollow in it. (L.S.) LL
porcupine stew Bad food. (P.N.F.)
porcupine tracks Porcupine paths led everywhere through the white pine regions of the Lake States. The newcomer to camp was told these were the tracks of some fearful animal, and that, if its path was disturbed in any way, some awful calamity would befall the camp. The newcomer making swamping roads was cautiously advised to be sure they did not cross porcupine tracks. He took the advice seriously and would cut a road far out of the way to avoid disturbing the path of this fierce animal. The cat came out of the bag when the foreman called him to task, but the lumberjacks had a good laugh at his expense. (Agnes M. Larson, History of the White Pine Industry in Minnesota)
portable pine logger An early high lead system using an A-frame instead of a tree, and a horse instead of a haulback. (P.N.F.)
portage road See tote road.
portage trail A trail used to carry in supplies in a packsack. (N.F.)
Port Blakely snipe A crude sniping job in the Pacific Northwest. (P.N.F.)
Porter A make of rod engine built primarily for industrial use but was popular with the loggers; built by H. K. Porter & Company of Pittsburgh, Pennsylvania.
Port Gamble snipe A snipe in which the cut was carried farther back on the log than in the usual bevel; originated at Port Gamble, Washington. (P.N.F.)
Portland An early steam donkey that was used for yarding and loading; built by the Portland (Oregon) Iron Works. (P.N.F.)
Port Lawson's high flyer An improvement over the McCanse flyer used in Mendocino County, California. Lawson placed all of his drums on one steel frame, which eliminated two firemen and two engineers. Four drums were placed on a sled about 100 feet long. An upright boiler powered the engines driving the drums. A brow line was placed around an entire canyon by selecting large trees on the opposite side. On the opposite sides, spars were made and anchored and a brow line was placed on these spar trees. To this brow line, a cable was attached, and the logs were moved as desired. The flyer could stay in one spot until the area was logged out. The flyer could also pull the standing trees over, thus eliminating the chopping crews! (P.C.F.) See Lawson system.
Port Orford snipe The Oregon version of the Port Gamble snipe. (P.N.F.)
pot 1. A common term for a donkey. 2. A locomotive.
potato histers The farmers who worked in the woods in the wintertime. (L.S.) LL
potato in the pocket A rheumatism remedy in which a logger carried a potato in his pocket until it became hard and dry. The potato was supposed to draw the rheumatism out of the body. (P.C.F.)
potential yield The maximum sustained yield cutting level (stated for a period of time such as a year or decade) attainable with intensive forestry considering the productivity of the land, conventional logging technology, standard cultural treatments and inter-relationship with other resource uses and the environment.

A **Porter** 2-8-2 locomotive built for the Carlton & Coast Railroad, a logging line out of Carlton, Oregon, in 1924. After service for the Port of Grays Harbor the **mike** was acquired by Shafer Brothers Logging Company and was later put on display at their tree farm near Satsop, Washington. The engine is now in railfan excursion service for the Mt. Rainier Scenic Railway, at Elbe, Washington. *Swanson photo, courtesy of Guy Dunscomb*

pothole 1. A small pond. 2. A small depression, like a pothole on a paved road. LL
pot hook 1. An S-shaped wire anchored over a camp fire, on which pails hung from the bottom loops. 2. A river with reverse bends in its course. (P.C.F.)
pothooks Hands. (P.C.F.)
potlatch 1. A gift. 2. A celebration; a native festival common to all Indian tribes of the Pacific Northwest which featured the distribution of gifts. The most noted chief was the one who held the largest potlatch and gave away the most valuable and largest number of presents (from the Chinook jargon). (P.N.F.)
potlatcher A generous man. (P.N.F.)
Potlatch tram The Willamette Iron & Steel Company of Portland, Oregon developed the Willamette Aerial Tram with the Potlatch Lumber Company of Idaho. This was a method of hauling logs by aerial tramway. Two stationary main lines were used, one for loads and the other for the returning empties. The hooks which held the logs were attached to an endless cable. (P.N.F.) See log tramway.
potter A round stick, 3 or 4 inches in diameter and 2-1/2 or 3 feet long, around the center of which is fitted an iron clasp to which is fastened a short piece of chain with a hook on the free end. It is used when loading sleds to prevent logs from rolling off the far side of the load until binding chains are placed in position. (N.W.)
pot walloper A cook. (L.S.) LL
pouch A French term applied derisively by loggers to the woods workers who drift from camp to camp. (N.W.)
pouched out (pooched out) 1. Said of a logger with a big gob of snoose in his mouth. 2. Said of any unusual bulge. (P.C.F.)
pound sand down a rat hole 1. To do some useless job. 2. To do nothing.
pound the screen To beat a spark arrestor with a bar to jar the carbon loose to let the engine draw better.
pound the ties 1. To leave camp, walking down the railroad in the early days of logging. 2. To quit. LL
powder 1. Black powder used in blasting. 2. Sticks of dynamite. LL
powder headache A severe headache, caused by handling dynamite.
powder house 1. A small storage shanty (later of cement) for explosives. 2. A privy.
powder man A man in charge of blasting around a logging operation.
powder monkey 1. A man who uses either dynamite or black powder. 2. The man in charge of blasting operations, usually when breaking a log jam in a stream. (L.S.) LL
powder packer The man who helps the powder monkey. (P.C.F.)
power All the motive power owned by a company. The expression is heard: "The company has lots of power."
power haul The ground lead system of logging, using steam engines instead of animals. (P.C.F.)
power horse A small tractor with rubber tires; used for yarding young timber. (P.N.F.)
power saw A motor-operated saw for felling and bucking.
power saw gyppoer A logger who owns his own chain saw and works by contract for different outfits. (P.C.F.)
power saw jockey A power saw mechanic. (P.C.F.)
pregnant women Dried apples. (L.S.) LL
preload 1. To circle several logs with binders so that the entire unit can be handled as one log. 2. To load trailers ready for the trucks to back under the load.
prelog 1. To remove the windfalls ahead of felling to prevent breakage, and to make yarding easier. 2. To remove small trees from a stand ahead of the main logging to prevent breakage. 3. To log one species, such as cedar, in advance of the main logging show.
presumascope A guessometer used by logging engineers in making estimates. (P.N.F.)
prime log A log that is clean and free from defects. (N.W.) LL
primer 1. A stick of dynamite with cap and fuse to set off a blast. 2. A man who starts trouble. (P.N.F.)
private road A logging road built by a company for its own trucks. This exempts the company from taxes and load restrictions. See off-highway truck.

Power saws have gone through a long process of evolution, but it was not until a lightweight portable power source was developed that such saws became practical. This print first appeared in 1879, and shows the Ransome falling and bucking machines, which had to rely on a mobile steam boiler for their power. *"The Timberman" collection*

A **preload**, in which a load is prepared for the truck while the truck is busy hauling a previously-prepared load. In this particular setup, the legs are folded up after the truck is in place and remain with the bunk. Sometimes a frame is used that remains on the site for the next load, and this support is called a **jack**. The same arrangement is used in a **transfer**, where the loads are stored at the transfer point to be hauled to the dump at a later time. *"The Timberman" collection*

prize log 1. A log with no brand, floating free of any boom, thus a prize for the first man who claimed it. Also scrabble log, stray log. 2. A wife or sweetheart. (L.S.)
prize up To pry up with some kind of lever; used by the Southerners who migrated west. (P.C.F.)
processor A modern machine used to delimb, top and cut trees into desired lengths for transport.
prod 1. Another name for the goad used in the ox team days. 2. A foreman.
prohibition shoe A well-oiled shoe which stays dry inside. (P.N.F.)
prospector A person looking for timber for speculation. (P.C.F.)
protection The preparations taken to protect the forest lands against fire, disease and insects. (P.C.F.)
protective association The people involved in forming an association to combat fires.
protestant cross A mattock. (L.S.) LL
prune burner A blacksmith. (L.S.) LL
prune picker 1. A native of California. (Red.) 2. A man who is a part-time logger and part-time farmer. (P.C.F.) 3. A family man in a stag camp. (P.C.F.) 4. A Willamette Valley (Oregon) logger. (P.N.F.)
public domain Land that was publicly owned. In the Redwood Country and the Northwest, public lands were stolen by fraud. The first

big fraud under the Timber and Stone Act of June 3, 1897, that ever occurred on the Pacific Coast, took place in Humboldt County, California. During this time a dense area of redwood timber in Northern Humboldt County was surveyed and opened to the public. Immediately, the California Redwood Company, with offices in Eureka, began to hire men to file on the entire tract. A man had only to appear at the land office to file and to exhibit his first papers to show that he was either a citizen of the United States or had declared his intention to become one. If he was a citizen, he would take an oath and his entry was allowed. Under these conditions, the company ran men into the office by the hundreds, and soon the whole tract was secured. By illegal means the company secured about 64,000 acres of redwood worth $6,400,000. The General Land Office in Washington, D.C. became aware that there was an abnormal rush for the issuance of patents, and suspended the unpatented claims, and special agents were sent out to investigate. The first three agents were bribed by the company in Eureka, but the fourth could not be bought off. After his report, between 150 and 200 entries were later cancelled. Many of the principal agents of the California Redwood Company were indicted by a Federal Grand Jury. (S. A. D. Puter and Horace Stevens, Looters of the Public Domain)

puckerbrush A common term for underbrush.

Puget Sound pine Douglas fir. (P.N.F.)

Puget Sound snipe One type of nose hewed on the end of a log for easier skidding. (P.N.F.)

Pulaski A light single-bitted ax with a straight handle, having a narrow adze-like trenching blade attached to its head and used for fire fighting. The tool was named after Ranger Pulaski, U.S. Forest Service; now used for pulling out the blocks of wood made in forming the undercut with a power saw, when felling timber.

pull 1. An older term for a turn of logs. 2. To reef on a wire rope when logging. 3. To redirect the fall of a tree by leaving more wood on one side of the undercut than on the other. 4. An older term for yarding with a donkey.

pull a job To work at a job: "He pulled a job as foreman at Camp 5." (P.C.F.)

pull back cable An older term for a haulback. The drum on which it was wound was called the pull back drum. (P.C.F.)

pull back line A haulback on a skyline carriage.

pull boat A flat boat, carrying a steam skidder or a donkey and used in logging cypress. (S.F.)

pull cable The mainline in early yarding operations. (P.C.F.)

puller 1. A good strong fast donkey. (P.C.F.) 2. In turpentine orcharding, a strong knife with a curved edge, fastened to a long handle. It is used to chip timber during the third and following years, when the face has advanced so far up the tree that it cannot be chipped with a hack. (S.F.)

pull freight To leave camp.

pulling In turpentine orcharding, a term applied to a cup or box that is hacked with a puller. (S.F.)

pulling boom The boom through which skidding lines were run in the early-day skidding machines.

pulling face In turpentine orcharding, one that is hacked with a puller. (S.F.)

pulling in line Another name for the mainline on a cable logging system.

pullin' second Said of the second rigger on a skidder. (P.C.F.)

pulling the oar Said of a man at the oar (sweep) on a lumber raft who would raise the stem (handle) of the oar so as to push the blade of the oar as deep as possible into the water, and then would take five steps as he pushed the oar. He would then drop the oar by letting go of the stem and swing back to his original position. On returning, it was "pushing the oar." LL

pull out a lung To pull the drawbar out of a railroad car, usually resulting in spilled logs. (P.C.F.)

pull riggin', pullin' riggin' 1. To work on the rigging crew. 2. To work on any job that includes handling cables or rigging. (P.C.F.)

pull rod A trip on a railroad car with built-on stakes. The trip releases the stakes from the opposite side of where the logs will roll off.

pull slack To pull extra line when needed in the woods. The line was pulled by hand by the crew.

pull the briar To handle a crosscut saw.

pull the pin 1. To quit: "He pulled the pin yesterday." 2. To pull the pin to release a car on the railroad. (P.C.F.)

pull the tit 1. To pull the throttle of a locomotive all the way out, thus making a fast run: "Pull the tit, let's go." 2. To shut a mill down: "Wish they would pull the tit." Also "Pull her tit." (P.C.F.)

pulp Mechanically-ground or chemically-digested wood used in the manufacture of paper and allied products.

pulp hook A curved steel hook with a wooden cross handle; used as a one-hand tool in lifting pulpwood bolts. Also birch hook, bolt hook. (N.F.) (Canada)

pulp log Usually a log that does not meet the 1/3 merchantability standard for a sawlog, but contains a minimum of 50 percent sound wood fiber by volume.

pulp rack Sides fastened to a truck to haul 4-foot pulpwood. (N.W.)

pulp sticks Individual pieces of pulpwood. (N.F.) LL

pulpwood 1. Small timber cut in 4-foot lengths to be used by a pulp mill. In the North Woods, 4-foot or 52-inch bolts; in the Lake States, usually 100-inch lengths for truck hauling. 2. Pulp species, such as hemlock, white fir or spruce.

pulpwood stands Standing trees to be harvested for pulpwood. (N.F.)

pumie, pummy The pumice soil that flies up in clouds of dust on many logging roads in the Pine Country.

pump can 1. A five-gallon water can with an inserted pump to put out small fires. Usually designed to be carried on a man's back. 2. An oil can with a built-in pump.

pumped water out of him A descriptive expression for a resuscitated log driver. (App.)

pump jack A hydraulic jack.

pump man A pump tender who looks after the water supply on a steam logging show. Also pipe man, pipe monkey.

pump up To fill engine boilers with water after a day's work.

An unidentified **pump man** tending the steam pump and pipes used to supply water to one or more donkey engines. *From Steven R. Gatke collection*

punch bar One of the arms on a log kicker. (P.C.F.)
punch donkey To work as a donkey engineer: "He punched donkey for Carson's." (P.C.F.)
puncheon Poles laid crossways to make a road across a swampy piece of ground. (P.C.F.) See corduroy.
puncher 1. The engineer on a donkey, a yarder, a loader, a swing donkey or a roader. 2. An ox team driver. See bull puncher.
punch hammer A short-handled, long-nosed hammer used by a blacksmith to enlarge holes in hot iron. LL
punching 1. The dirt, brush or slabs punched in between the skids on a skidroad to make a road or footing for the oxen or horses. The skids were a few inches above the road. (P.C.F.) 2. To work as an engineer. See puncher.
punchings Puncheons or logs split straight with the top surface adzed off to make a smooth floor for a shanty. (L.S.)
punch logs To work on the log pond. (P.C.F.)
punch the flues To clean out the flues on a donkey or locomotive. (P.C.F.)
pung 1. A logger's purse. (P.C.F.) 2. A crude oblong box on poles that served both as the shafts and sled runners. Originally used by Indians and pioneers in hauling. (L.S.) LL
punk 1. The whistle punk who passes signals from the choker setters to the donkey puncher. 2. Any green youngster in the woods. 3. Rotten wood. 4. Bread. 5. Molasses cake or white cake. (N.W.) LL
punk and skidgrease Bread and butter. (P.C.F.)
punk boy See chore boy.
punkies Gnats. (West Virginia)
punkin' whistle To work as a whistle punk. (P.C.F.)
punky knot A rotten knot with soft, partially decomposed wood.
pup, pup hook A small hook at the end of a loading line or chain. (N.F.) LL See end hook.
purchase 1. To get leverage in prying or lifting. 2. To get a hold on something.
purchase block A block used to obtain a greater pull on a line in rigging.
purring sack A bed. (P.N.F.)
push A camp boss, assistant superintendent or man in charge of a side in the woods.
push camp To run a logging camp. (P.N.F.)
pusher 1. A truck driver. (Red.) 2. The foreman or boss of the logging camp. (L.S.)
pusher cat A tractor used to push over snags in the Pine Country.
pushing hands Said of men working under an oar on a raft. (App.)
push side To be in charge of a logging side. (P.C.F.)
pussy pusher A cat skinner. (P.C.F.)
put another hole in the barrel After the passing of the old camboose with the smoke hole in the roof, round holes were cut into the bunkhouse roof and barrels, with the bottoms still in and holes for ventilation, were placed upside down over these holes. If the bunkhouse got too hot or too ripe from the odors of drying wet socks and other clothes, one of the jacks would call out, "Put another hole in the barrel," meaning that more ventilation was needed. LL

This unidentified **cat skinner** may have been referred to as a **pussy pusher**. *From Steven R. Gatke collection*

put a patch on hell To ruin a good situation. (P.N.F.)
put a roll on it To put rigging on a log to make it roll when the mainline is pulled, thus clearing a hang-up. (P.C.F.)
put binders on it 1. To set the brakes on a railroad car. 2. The expression is sometimes used with trucks; also "set the binders." 3. To wrap the binders around a load. (P.C.F.)
put'er (donkey) in bed, put the donkey to bed 1. To bank the fires to get the steam donkey ready for the night. Usually the job of the night watchman who also got up steam in the morning. 2. Also applied to any machinery, locomotive, truck or even a mill. (P.C.F.)
put'er in Canada To place a log on the gin-pole side when loading a car. (P.N.F.)
put'er in irons To chain up a truck or carload of logs after loading a car. (P.C.F.)
put'er in lead To set up a donkey so that it is directly in line with the pull. (P.C.F.)
put his brains in a blue jay and it will fly backwards Said of a dumb logger. (Red.)
put in 1. To deliver logs at a landing. LL 2. To join, as "I'll put in with you." (P.C.F.)
put lead in his (your) pencil 1. To encourage a worker to produce more work. 2. To be capable of and eager for sexual intercourse. (P.C.F.)
put on another horse To shift gears for more power. (P.C.F.)
put on the feed bag To eat. (L.S.) LL
put on the nosebag To start eating. (P.C.F.)
put the caulks to him To get a man down in a fight and to jump on him with caulked shoes. See logger's smallpox.
put the steel to it To drive in a wedge when sawing.
put your gloves on and stand close to the line Full steam ahead and watch out. (P.N.F.)
pyramid See mandril.

Q

Q A Chinese logger; from the long queue worn by the Chinese when they first came to the Western woods. At one time there were many Chinese fallers and buckers in the logging camps of British Columbia. The Chinese were employed in the Redwood Country as swampers and fire tenders, and as early as 1875 they came to Humboldt County, California to construct logging railroads.
quality cruise A cruise that tallies the grade of logs as well as the total volume.
quarter 1. That part of a setting being logged at any one time. 2. A quarter section of land.
quarter corner In surveying, the corner of a quarter section.
quartering to the tree To yard slantwise across a hill to the landing.
quarter section One-fourth of a land section, 160 acres.
quarter tie A tie made from a stick of timber yielding four or more ties. (S.F.)
Quebec choker See peavey.
Quebec standard 1. In logging, a white pine log 22 inches in diameter, inside the bark, at the small end and 12 feet long. A spruce or balsam log 14 inches in diameter inside the bar at the small end and 12 feet long. (Eastern Canada) 2. In lumber, formerly a unit for the measurement of deals in transactions between producer and shipper. Now confined to the measurement of Ottawa pine deals. It is based on a piece 3 inches by 11 inches in cross section by 10 feet in length and contains 27-1/2 feet board measure. One hundred (100) Quebec standards are known as a Quebec standard hundred. (Eastern Canada)
queen A hasher. (P.C.F.)
queen's palace Living quarters for women cookshack employees. (P.N.F.)
quick-change A two-speed yarder with a quick change from high to low and back again. (P.N.F.) See two-speed.
quick eye An opening pried apart in a line for temporary use. (P.C.F.)
quick lick A man in too much of a hurry. (P.N.F.)
quick like a cow Said of an awkward person who is always falling off logs. (P.N.F.)
quick like a squirrel Said of a worker who is quick on his feet. (P.C.F.)
quick-moving skidder A ready-rigged steel spar skidder mounted on a railroad car. (P.C.F.) See Lidgerwood, tower skidder.
quick on the throttle Said of a hot-tempered man. (P.C.F.)
quick return Said of a haulback on a highball show. (P.C.F.)
quick slip Said of a shoe without caulks. (P.N.F.)
quick step Said of a man who has dysentery. (P.N.F.)
quick water That part of a stream which has fall enough to create a decided current. Also rapid water.
quick water country A mountainous region where streams rise very quickly after a heavy rain. (P.N.F.)
quinine jimmy A camp doctor. (L.S.) LL

R

rack A sled body for hauling short bolts, such as pulpwood. (N.F.)
rack a load A teamster's expression for breaking loose frozen runners by making the team go right and left. (N.F.) LL
rack car A railroad car equipped with stakes or racks to handle wood.
rackheap 1. A piled-up mess of logs and trees in a river. 2. Logs piled up and ready to be splashed down a river. (P.N.F.)
racking line A line attached to the trolley on a tight line loading system.
rack the pole Moving the tongue back and forth by geeing the team to the right and by hawing them to the left in order to break the runners loose. By this process, the cross chains broke the rear runners loose. (N.F.)
Radley-Hunter A type of locomotive spark arrestor.

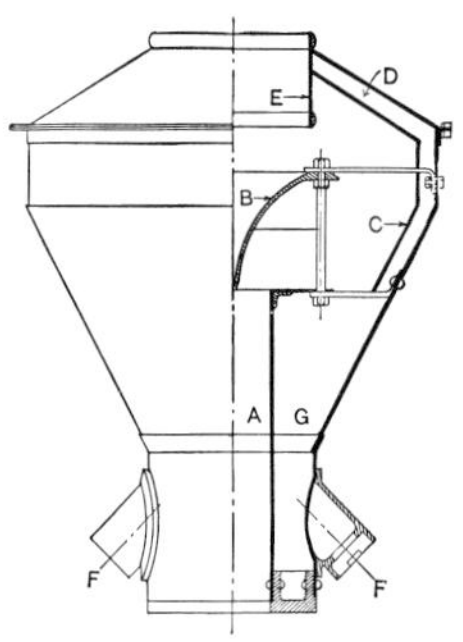

A **Radley-Hunter** spark arrestor. *From "Logging," Bryant, Wylie & Sons, 1923, Philip Schnell collection*

raft 1. A float made of logs or timbers. 2. Logs gathered together for transportation. 3. To make a raft out of logs. 4. To transport logs by raft.
raft bundle Logs bound together into a circular unit for towing. (S.F.)
rafted-out logs Said when there were no more logs or when the river drive was completed. (L.S.) LL
rafter dam A dam in which long timbers are set on the upstream side at an angle of from 20 to 40 degrees to the water surface. The pressure of the water against the timbers holds the dam solidly against the streambed. Also slant dam. LL
rafters River pigs, usually those who rafted lumber rather than logs. (L.S.) LL

Log **rafts** in the Clarion River at Cooksburg, Pennsylvania, are being prepared for the trip down the river. Here the logs have been **slabbed** to provide a more compact raft, and some lumber loaded on top as a deck load. The rafts are held together with **swifters**, and **sweeps** are attached to the **head sticks** to provide some control during the trip. The **ark** will accompany the rafts to provide for the **rafters**. *Courtesy of W. Casler*

A **rafter dam** used in driving **pulpwood** in Eastern Canada. *From "Logging, Principles and Practices," Brown, Wylie & Sons, 1934, Philip Schnell collection*

rafting auger A very long auger with an offset handle for drilling holes in raft logs. See boom auger. (L.S.) LL
rafting chance The area along a river usually with a high bank and deep water where logs are delivered and made into rafts. (L.S.) LL
rafting dog See boom dog.
rafting grounds 1. The area where logs are dumped into the water to be made up into rafts for towing to the mills. 2. The area where rafts are made up for ocean towing, which is not necessarily at a dump. Sometimes logs were rafted down to the rafting grounds.

Rafting grounds at Kerry, Oregon, on the Lower Columbia River. Here the Columbia & Nehalem River Railroad, commonly known as the "Kerry Line," dumped logs from as many as twelve different camps. The logs were sorted and made into **rafts** to be towed to one of the many sawmills located on the river. *Labbe collection*

rafting in Building a raft. (App.)
rafting picket A lane lined with catwalks leading out of a sorting picket in a log pond. Logs of the same ownership were steered into rafting pickets to be assembled into log rafts. (L.S.) LL
rafting pin A round or wedge-shaped wooden pin used to wedge cable in the rafting pin holes on a raft.
rafting pocket An area leading out of the sorting pocket at a log dump. Logs of the same kind are poled into the rafting pocket to be made up into rafts for towing.
rafting rigging Rope used for many purposes on a drive. (N.F.) LL
rafting sled Sleds on which lumber raft cribs were made. (L.S.) LL
rafting works See booming grounds.
raft of (a) A large accumulation of something. This logging term has become very popular in common usage.
raft shackles See boom dog.
raftsman 1. A raft builder. 2. A shantyman who crewed square timber rafts down the Ottawa and the St. Lawrence Rivers to the market at Quebec City in the 1800s. (Canada)
raft stick A long slender log to be made into a boom stick.
rag a wedge To roughen a wooden wedge with an ax, or a steel wedge with a cold chisel to make it hold better, especially in frozen timber. (N.W.)
rail The slot in the bar, which guides the chain on a chain saw.
railhead The end of a railroad line in the woods.
rail kinker A railroad brakeman. (L.S.) LL
railroad See logging railroad.
railroad camp A camp made up primarily of camp cars mounted on railroad trucks. Sometimes denotes portable camps in which the buildings are designed to be moved by rail.
railroad logging The system where logs were transferred directly to railroad cars from the yarder. In railroad logging, the spurs were constantly moved, whereas on a modern logging railroad both terminals are fixed.
rainbow To spring a peavey handle by lifting, i.e., evidence of energy applied. (App.)
raise a tree To put up a spar tree where none is available in the proper location. See dummy tree, lift a tree.
raised on the river Said of a man skilled in handling logs. (P.N.F.)
raise tree See dummy tree.
rake To clear away all grass and other inflammable material from the base of a bled tree in turpentine orcharding. (S.F.)
rakers The longer chisel-pointed teeth on a crosscut saw alternating with the side-sharpened teeth. Rakers break the shaving loose and curl it up into the gullet where it will be carried out of the cut, thus preventing binding.

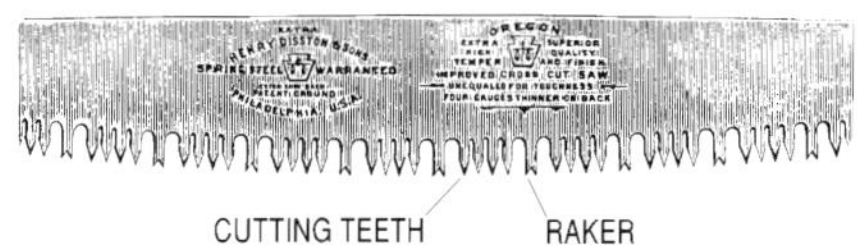

A **crosscut saw** showing the cutting teeth and **rakers**. *From "Henry Disston & Sons Handbook for Lumbermen with a Treatise on the Construction of Saws and How to Keep Them in Order"*

ramp See dock.
ram pasture 1. A bunkhouse. 2. A large hotel room in which the loggers or lumberjacks slept on cots or on the floor.
ram pike 1. A tall dead tree. 2. A tree broken off by wind or ice, with a splintered end on the portion left standing. 3. A snag.
ramrod A boss or foreman.
ramrod the job To hurry a piece of work through to a finish.
ranch hand A derogatory term for the chore boy who carried kitchen scraps to the pigpens in the early days. (P.N.F.)
rank To haul and pile bark or cord wood regularly. (N.F.)
ranking bar See handbarrow.
ranking jumper A wood-shod sled that is used for hauling tanbark. (L.S.) LL See bark dray.

Along the track of the Babcock Lumber Company, near Ashtola, Pennsylvania, **ranked** bark and **peeled** logs await the coming of the train crew and loaders. *Courtesy of W. Casler*

This view shows Red Star Camp of the Whitney Company's logging operation out of Idaville, Oregon. In this **railroad camp**, the buildings are not permanently mounted on railroad trucks, although they are built on skids to be easily moved by railroad cars. This was an up-to-date camp for its time. Note the **turntable** at the extreme left of the photo—a rarity on logging railroads. *Courtesy of J. Hess*

rapid loader A slide log loading machine manufactured by Clyde Iron Works, used in the pine regions. See slide loader.

A **rapid loader** loading short logs in Arkansas. *From Clyde "Self-Propelling Logging Machinery" catalog, 1912, Larry Nyce collection*

rapids pilot A log drive specialist who guided log and lumber rafts through treacherous rapids. (L.S.) LL
rapid unloader See Donovan unloader.
rapid water The current in a stream that is fast enough to make white foam. (L.S.) See quick water. LL
r'ar of snoose A gob of snoose. (P.C.F.)
r'ar on 'er 1. To put slack on the mainline and then to jerk it to free a hang-up in donkey yarding. 2. To open the throttle. (P.C.F.)
rate of growth The rate at which a tree has laid on wood, measured radially in the trunk or in lumber cut from the trunk. The unit of measure is the number of annual growth rings per inch.
ratline 1. A short section of small line used in rigging. See squirrel line. 2. A rope through which at intervals small pins are driven into the logs which are to compose a raft joint. Its purpose is to hold the logs together until the boom poles can be adjusted. (Eastern Canada)
rat's nest 1. A bushy branch growth. 2. A dirty cabin or camp. See boar's nest.
rattlers See disconnected trucks.
rattling line See ratline.
rave A piece of iron or wood that secures the beam to the runners of a logging sleigh. (N.F.) LL
rawhide 1. A loop of wire to repair a break in rigging. Rawhide was used before baling wire was introduced. 2. Any slipshod job. 3. To force men to work harder. 4. To carry on one's back; usually applied to carrying tanbark. (App.) 5. To hurry. (L.S.)
rawhider 1. An unfair boss. 2. A man who drives himself and his men to get the job done.
Raymond loader A jammer on sleighs with runners turned up on both ends to enable it to travel in both directions and used to load on sleighs. (L.S.) LL
razor front dog hook A sharp-nosed heavy hook used in dogging logs. (P.C.F.)
reach 1. The wooden or steel shaft that connects a logging truck and the trailer. 2. The distance spanned by a skyline. 3. A steel shod wooden pole for fastening sleighs together, then hauled by steam haulers or tractors. (L.S.) 4. See rooster.
ream To enlarge a hole in a log with a chain saw.
reamer A large spiral bit with sharp corners used to cut ice out of water holes in a sprinkler. (N.F.) Also ream awl. LL
rear 1. The upstream end of a drive; the logs may be either stranded or floating. "Floating rear" comprises those logs which may be floated back into the current; "dry rear," those which must be dragged or rolled back. 2. To roll stranded logs into the river behind a drive.

rear boat A heavy boat used in rearing a drive. When there were no roads the camp gear followed behind the crew in the rear boat.
rear foreman The head man of the cleanup crew at the back end of a drive.
rear it To figure the distance in river driving, e.g., "To rear it three miles today." The progress of a drive was measured by the advance of the rear, since the drive was not finished until all the logs were brought in.
reburn Another fire on an already burned area.
receding line The line which takes the carriage back to the woods on a skidder or other skyline show. See haulback.
receiver An older term for a compressed air tank mounted on a donkey, which was used to provide air to operate frictions, brakes, or a variety of tools. (P.C.F.)
receiving boom See storage boom.
red card A membership card in the I.W.W.
red cypress Cypress lumber which has a deep color. (S.F.)
red eye Whiskey. LL
Redfield An early air-operated drag saw. (P.C.F.)
red fir Douglas fir. The quality of Douglas fir differs greatly according to the conditions under which it grows. Fast growth usually produces the red fir which is a hard and durable wood. Slow growth usually produces the yellow fir, which is a fine grained, soft-textured wood, ideal for finishing material.

Red fir on the operations of the Oregon-American Lumber Company near Vernonia, Oregon. Lighter wood under the bark is the **sapwood**. *Harold M. Brown photo, Labbe collection*

redheart Rot which appears on the butt end of the log in the heartwood; appears red when wet and caused by fomes pini in Douglas fir.
red horse Salted beef or corned beef. (L.S.) LL
red-ink log A cull log, one which will not pay its way. (P.N.F.)
red lead Tomato ketchup. LL
red-light woman A whore. (Alaska)
red one In forest surveys, a facetious measurement (red cunt hair) between a pink one and a black one.
redring Rotted wood which appears as a moon-shaped reddish area on the ends of Douglas fir logs.
Red River car A four bunk connected truck logging car 36 feet long, built especially to haul 32-foot pine logs. Designed and built by Seattle Car & Foundry Company for the Red River Lumber Company of Westwood, California.

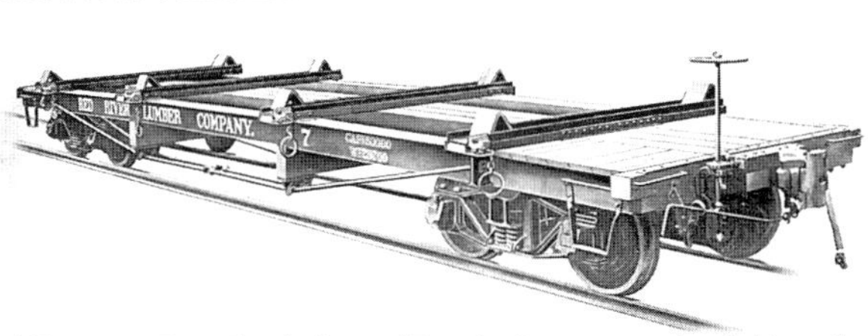

A **Red River car**. *From Seattle Car and Foundry Company catalog, Labbe collection*

red rot A heart rot fungus in pine tree trunks causing wood to become dry and flaky. (N.F.)
red top 1. The dying crown of a tree, turned a reddish-brown when girdled by porcupines. 2. A tree infested with beetles which turns reddish-brown. (P.N.F.)
redwood California redwood lumber of commerce is produced from the Coast Redwood (Sequoia sempervirens) which grows in the cool fog-bathed Northern California coastal area. In laboratory tests and in actual use, California redwood ranks high and is rated by the U.S. Forest Products Laboratory as one of the most durable of woods. It also rates high in workability, strength, dimensional stability or freedom from warping and shrinkage, and is particularly noted for its ability to hold paint coatings. Redwood has excellent nail-holding power and glues up well with a great variety of adhesives.

An enormous fallen coastal **redwood** (**Sequoia sempervirens**) from the Richardson Grove, along the Redwood Highway on the coast of California, 1936. *Courtesy of Don Smith*

Redwood Country, Redwood Region Northwestern California, specifically Humboldt County, the center of the redwood industry.
Redwood Inspection Service An organization for inspection and grading of redwood lumber.
Redwood Region Conservation Council An organization which sponsors programs of conservation of natural resources in the redwood region.
redwood special A saw chain with two tie straps between the cutting teeth instead of one to give more chip-carrying capacity when making cuts in big logs. (P.C.F.)
reef 1. To pull at anything. 2. To pull hard with the donkey throttle wide open.
reefer A jacket or short coat. (Michigan) See mackinaw.
reef his main line 1. To stop a man or action quickly (usually the operator of a machine). 2. To pull a man off a job that he is botching. (P.N.F.)
reefing 'er Pushing a boat with a pole. (L.S.)
reefing line 1. The tag line or extension to the skyline on a skidder. 2. In some British Columbia camps, a line which runs through the fall block and is spliced to the mainline of a North Bend system.
reel 1. A small spool on which the whistle punk winds the signal wire. 2. A drum on a donkey or cat. 3. The wooden spool on which wire rope is delivered.
reel engine A compound geared engine used for changing lines, to tighten the skyline on a skidder, and to store extra lengths of line.
reeve To run a line through a block or a trolley.
reflector In baking biscuits and bread, a box-like utensil of copper so constructed as to catch the heat when placed near a fire. (L.S.)
reforestation The natural or artificial restocking of an area with forest trees.
refuse That portion of a tree which cannot be removed profitably from the forest or utilized profitably at the manufacturing plant.
regeneration 1. The renewal of a tree crop, whether by natural or artificial means. 2. The young tree crop itself.
rehaul 1. A haulback. 2. To road logs (older term).
rehaul engine A donkey used to pull the mainline back to the woods, used in the period between the use of the line horse and the development of the two-drum yarding engine.
relay 1. The middle donkey on a three-donkey show. The middle donkey and the third donkey are also called swing donkeys. 2. In fire fighting, a water tank between the bottom and top of a hill, where the water is relayed or repumped to the top of a hill. 3. A rail that has been used and relaid in a new location.

The **reload** of Rayonier, Incorporated, at Crane Creek, north of Hoquiam, Washington. Trucks brought the logs down out of the mountains to the company's railroad where they were transferred to the cars for the long trip to the **dump** at Hoquiam. *Vernon Goe photo, Labbe collection*

relay rail A second-hand rail, usually used on logging railroads.
relay skidding A method of yarding from a tail tree to an intermediate tree, decking logs at this point, and then swinging to the landing.
released slash An area in which slash has been sufficiently burned or decayed to release the logger from the responsibility should a fire start in it, other than from his own actions.
reload A yard or area where the logs are transferred from one type of transportation to another, such as from truck to railroad or from the woods truck to the highway truck. See transfer.
reload pond A pond that was used to store logs at a point between the woods and the mill, for sorting and shipping, and sometimes in place of a cold deck.
relog 1. To salvage small timber after the area has been logged. 2. To salvage logs with defects that increased log prices make profitable to log.
renegade A tree that is dangerous to fall. (P.C.F.) Also outlaw.
renter In turpentine orcharding, one who subcontracts for or rents an area of cups or boxes and supplies all labor for chipping and dipping. Usually he is paid on the basis of the number of barrels of dip and scrape gathered. (S.F.)
reprod An abbreviation for reproduction; young trees growing on an area following fire or logging.
reproduction 1. Young seedlings which have seeded naturally. 2. Regeneration; the renewal of a tree crop, whether by natural or artificial means. 3. The young crop.
rerailer, rerailing frog A heavy steel device, shaped like a turtle; temporarily spiked to the railroad ties, and used to guide car wheels back onto the track after a derailment. Also camel back, car frog, frog, humpback, wrecking frog.

Stimson Lumber Company used this **reload pond** in their operation out of Forest Grove, Oregon. The **donkey**, and **spar tree** in the middle of the pond were used to fish logs from the pond and load them onto railroad cars for the trip to their mill in Scoggins Valley. *Courtesy of Ted Blair*

residuals 1. In tree crops or stands, those trees remaining after an intermediate or partial cutting. 2. In general, residuals are by-products of some operation. Examples are chips from lumber production and hog fuel from any wood processing.
residual stand In crops or stands of trees, that portion of the growing stock retained after an intermediate cutting (like a thinning) or a partial cutting.
resin Pitch. See crude turpentine.
rest powder Snuff. (L.S.) LL

return block The last block at the end of a skidroad to which the haulback line was pulled by the bull donkey. Also corner block, tail block.
return line See haulback. From this use, the return line drum meant the haulback drum on a donkey, and the return line head block was the haulback block at the spar tree.
return road 1. A road by which empty trucks return to the woods when the outgoing road is too steep to go back. 2. Roads used by cats in returning to the woods.
reversing engine A steam donkey engine used in loading logs, where the tong line was live and could be raised or lowered by the engine; did not depend on gravity. See duplex loader.
reyard To swing logs from a yarder tree to the landing.
rib log 1. A log in the middle of a load, below the peaker and on top of the bunk logs. 2. A sheer log used to change the direction of a turn. (P.C.F.)
rick 1. A pile of wood cut in lengths, usually 16 inches, piled 4 feet high and 8 feet long; three ricks to a cord. In 12-inch lengths, 4 feet high and 8 feet long; four ricks to a cord. 2. A pile of cordwood stave bolts or other material split from short logs; a cord 8 feet long, 4 feet high, and of a width equal to the length of one stick. (L.S.) (C.H.F.) 3. To pile wood into ricks. (L.S.) (C.H.F.) LL
rick up 1. To pile wood in ricks. 2. To pile up wood in any quantity or shape. 3. Said of the end of a floating log in a flume when it climbs and skids along the top board.
ride 1. The beveled length of a log so that it would "ride" down the skidroad without turning over: "Chop a ride on the log." 2. That side of a log on which it rests when being dragged. 3. Said of a floating log when it skids along on the top board of a flume. (P.N.F.)
ride a log 1. To stand on a floating log. 2. To stand on a log while it is being ground-logged or roaded.
ride of a log The side of a log that tends to remain in contact with the ground when yarded.
ride polly To draw half-pay as the result of an accident. (L.S.) LL
rider An extra hole, on a boom stick, bored about 12 feet from the usual hole at the end of the stick. At the front of the boom, the head stick is hooked onto this second hole; at the back of the boom the tail stick is hooked on.
ride side The side of the log which turns under when it is being pulled on the ground. The sniper bevels this side when preparing logs for the skidroad.
ride the donkey sleigh To quit the woods job and take a ride back to town with the tote teamster. (L.S.) LL
ride the hook To steal a ride on the tongs from the railroad car or truck being loaded by placing one's foot in the tongs and holding onto the line that is heading back to the log pile; against company regulations and Industrial Accident Commission rules.
ride (riding) the saw 1. To push down on a crosscut saw when cutting in the early days of logging. 2. To pull to one side to make a crooked cut. 3. Said of a lazy man who is accused of climbing on the saw to ride and not doing his share of the work.
ride the smoky end To hire out as a brakeman and ride the engine. (P.C.F.)
ridge runner A farmer who is an intermittent woodsman. (App.)
riding The pressure put on a crosscut saw in cutting down a tree. (L.S.) LL
ridin' 'er out Said of a river driver caught in a log jam that gives way so that he has to stick to a moving log and then navigate the breakup of tossing timbers in rushing waters. (L.S.)
ridin' shanks ponies To walk. (S.F.) (West Virginia)
riff board A rough-edged untrimmed board. (Maine)
riffles A term used sometimes for rapids, but usually water not as swift as rapids or white water. (L.S.) LL
rift crack A wide crack across the center of a log.
rig 1. Any outfit, apparatus or machine. 2. To hang the rigging up on a yarding or loading tree. 3. To get a machine in working order. 4. To set up any kind of rigging, such as rigging a skid road.
rig down To take the rigging down from a spar tree after yarding or loading has finished.
rigged-up 1. Said of a spar tree that is all set for logging, guyed and

Two large **Douglas fir** logs ride on wooden **disconnected trucks** on the operation of C.C. Masten, out of Svensen, Oregon, on the Lower Columbia River. The small 30-ton **Climax locomotive** was construction No. 248, turned out for Masten in March of 1901. The log nearest the engine shows evidence of having had a **ride** chopped into it. Both logs have been **sniped** to aid in **ground-lead yarding** and **roading**. *Labbe collection*

A large fir log on a White truck and trailer of the Woods Logging Company at Miller River, Washington. The log shows a **rift crack** and a **pitch ring**. The trailer's **reach** can be seen connected to the underside of the truck and extending through the trailer. Also visible is a **fid hook** securing the **bunk chain** on the rear **bunk**. *Pickett photo, courtesy of M. Koch*

hung with blocks and lines. 2. Said of a tree ready to be felled. 3. Said of a logger all ready for town.
rigger 1. Another term for rigging man. Also rig tender. 2. The head man on a skidder show.
rigger's block Also called rig-up block. A small block used by the high climber to lift the heavy blocks and rigging up into a spar tree. It had a wide throat to pass splices and fittings needed in rigging. The very small pass block was first taken up by the climber, and used to haul up the rigger's block. See pass block.

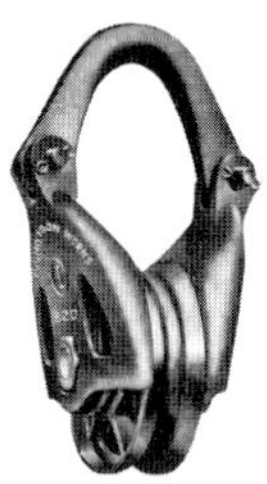

A **rigger's block.** *From Young Iron Works catalog, Glen Comstock collection*

rigger's drum A small drum on a donkey or skidder that handles the straw line or other light line used in rigging up. Also haywire drum, straw drum and strawline drum.
rigger's pass line See pass line.
riggin' 1. The cables, blocks, lines, hooks, chokers, etc. used in cable logging. 2. The equipment used to fell a tree. 3. The doubletrees, singletrees, swivel hook with hand ring, and chain or tongs for skidding logs with horses. (L.S.) LL
riggin' cap A steel cover, used at times to go on the top of a spar tree; it had lugs for attaching guy lines and high lead blocks. Also tree cap. See Geary high lead cap.
riggin' chain In the late 1800s, a small length of chain used as a strap to anchor the blocks on a head or tail tree.
riggin' crew (gang) 1. The chokermen who handle the rigging in a yarding operation. 2. The crew that rigs the spar tree ahead of logging. 3. The men who handle the rigging for construction work in a lumber mill. 4. The men who laid the logs in place on the skidway or loaded the sleds. (L.S.)
riggin' cut A cut made by a bucker to remove the long end of a tree so that it will not interfere with yarding. (P.C.F.)
riggin' hustler A rigging repairman who makes splices, puts in eyes, and maintains the rope and blocks. (P.C.F.)
riggin' man Usually any man working on the rigging. Also ape.
riggin' plate A heavy steel plate to which the mainline, haulback and butt rigging were attached in some high lead systems. See section plate.

A **riggin' plate. Shackles** were used to connect the lines to the plate. *From Mallory Catalog No. 12, 1925, Glen Comstock collection.*

riggin' puller, riggin' rustler See riggin' slinger.
riggin' shop The headquarters for the riggers and their equipment.
riggin'-shy Said of a man who is afraid to work on the rigging. (P.C.F.)

riggin' sled In the early days of animal logging, a flat sled or hollow log, similar to a pig, was used to haul dogs and other gear back to the woods. See pig, P.F. man, pig fucker.
riggin' slinger 1. A general term for a man working the rigging. (P.C.F.) 2. The head rigger, next in line to the hooker; he directs the choker setters. (P.C.F.) (British Columbia) 3. A member of a yarding crew, whose chief duty is to attach chokers which have been placed around logs to the pulling line. (P.C.F.) 4. One who attaches the rigging to spar or tail trees in steam skidding. (S.F.)
right on the pin Said of a load of logs that is balanced on a truck. (P.C.F.)
rig tender A rigger. (P.N.F.)
rig up 1. To put rigging in place on a spar tree or a skidder. See hang the riggin'. 2. To dress up for town.
rig up a tree 1. To prepare a redwood for felling by putting drivers into the driver holes and placing the staging boards on the drivers for the chopper to stand on. See board up a tree. (Red.) 2. To hang rigging on a spar tree.
rig-up block See rigger's block.
rig-up crew A crew which rigs spar trees ahead of logging. See riggin' crew.
rig-up donkey Usually a light, small machine, easily moved, and used to rig trees. (P.C.F.)
rig-up goat A small donkey mounted on a truck and used to rig up trees.
rig-up man See riggin' man.
rim 1. The edge of a canyon or break in the ground. 2. A curled-up chunk of tanbark, usually 4 feet long. 3. One of the two pieces which comprise the curved rim of a wheel, between the spokes and the tire. It is a straight piece of lumber which is bent in the shape of a semicircle.
rimrack To ruin or tear apart: "They sure rimracked that outfit."
rimrack a tree To fell a tree over some obstacle, such as a log, snag or hill, breaking the tree so that no merchantable lumber is left. (Red.)
rimracker One who tears or ruins something. (Red.)
ring 1. The ring of annual growth, that is the growth layers put on in a growth year. 2. A large ring that was slipped over the mainline, and chokers were attached to it in the early days of ground lead yarding. 3. A ring that was cut around through the thick redwood bark on the opposite side of an undercut. (Red.) 4. A ring that used to be cut around a redwood through the bark to help the peeler. (Red.) 5. A section of tanbark, usually 4 feet long. (N.F.) LL
ring boom Small boom used to gather logs or pulpwood that washed over the booms. (N.F.) LL
ring dog In rafting, a dog with a large ring through the eye. A cable passed through the ring had a certain amount of give as the raft rose and fell, and prevented the dog from pulling out.

A ring dog. *From "Logging and Lumbering," C.A. Schenk, Biltmore Forest School, Glen Comstock collection*

ringer A logger who used to cut rings around a felled redwood through the bark at intervals along the log with a razor-sharp ax to help the peelers in their work. Signs used to hang in employment offices: "Ringers wanted." (Red.)
ring rot Circular rot in a log; any rot localized in the springwood of the growth rings, giving a concentric pattern of decayed wood in the cross section of a tree or log. LL
rippers The heavy teeth welded to the bottom of a bulldozer blade for use in roadwork, land clearing or brush piling. Also buck teeth.
rip track 1. The specific tracks where railroad stock is repaired. 2. The track where discarded equipment is shunted. 3. The hospital where an injured logger or mill worker is taken. (P.C.F.)
rise 1. The crest of a freshet in a river. 2. The difference in diameter, or taper between two points in a log. LL
riser log A floating log used to help raise a sinker; the sunken log is attached to the rise log so it can be brought to shore. (P.C.F.)
ritchie The heavy woolen clothes worn by the lumbermen. (West Virginia)
rive 1. To split timber lengthwise. 2. To split shingles or shakes from bolts with a froe. (N.W.)

U.S. Army loggers of the W.W.I. **Spruce Production Division** using large logging jacks to **rive** spruce logs into **cants**, for easier ground transportation. The cants often weighed between two and three tons each. After the army installed railroads, this laborious and wasteful practice was abandoned. *From "Pictorial Review of World's War Activities Spruce Production Division, U.S. Army, Oregon & Washington," 1918, A.M. Prentiss photo, Philip Schnell collection*

river boss 1. The foreman in charge of a log drive. (N.F.) LL 2. The man in charge of river works in splash dam operations. (P.C.F.)
river dog A special dog with two right angles attached by a link to the chain on the doubletree. When a log was pulled into water deep enough to float it, the up end of the dog was pounded, and it pried the down end out of the log, freeing the stick.
river drive The movement of logs by high water in a river or creek; an important operation in the early days of logging.
river drivers See drivers.
river hog A logger or lumberjack working on a drive.
river jack A river driver. (S.F.) (West Virginia)
river pig See river hog.
river pig nose bag A tin lunch bucket carried by a river driver, generally strapped to his back. (N.F.) LL See lunch can, lunch sack.
river pirate The person who steals logs, saws off the ends with the log brands and remarks them with his own brand, or saws them into lumber. (West Virginia)
river rat A log driver working on river drives, as contrasted with a laker. (L.S.)
river scale The tally of logs scaled at a dump on the bank, or in the boom at the river; used as the basis for paying the loggers and lumberjacks.
road 1. The abbreviation for skidroad in high lead yarding where logs are hauled to a landing. 2. The dirt road in ground lead yarding on which logs were roaded from the yarder to a landing. 3. To haul logs down a fore-and-aft road by donkey on a logging show in the early days. 4. To skid logs from a yarder tree to the landing by cat. 5. To trail logs when a locomotive was used. See trail.
road donkey A swing donkey with a large drum capacity to haul logs long distances in the late 1800s to the early 1900s; it moved logs from the point of origin in the woods to the point where they were loaded out. Its primary purpose was to road logs on a skidroad.

A **road donkey** set up to haul logs on a **skidroad,** or **road**. This is Yeon & Pelton's operation above Rainier, Oregon. Note the large **roller** used to keep the **haulback** line from fouling the **mainline**. *J.F. Ford photo, courtesy of Jim Brady*

road engine 1. A road donkey. Also ground haul engine, long haul donkey. 2. A mainline locomotive on a logging railroad.
road engine line The mainline used in roading logs.
roader See road donkey.
road gang The men of a logging camp who cut roads and kept them in repair. (N.W.) (West Virginia) LL
road hook A hook used in hauling logs down a pole road. It was stuck in the end of the last log to the turn, which then shoved the others ahead of it. (P.N.F.)
roading rigs Various types of spools, blocks, rollers, etc. used to guide cable in roading operations.
road locomotive 1. Same as road engine. 2. In the Pine Country, a steam traction engine used to pull log carts. Also bug picker, dry land lokie, steam freighter.

A **road locomotive** or **steam hauler** of the Shasta Land & Timber Company in the Siskiyou Mountains of Northern California with a loaded **log wagon**. *From "The Timberman" collection*

road-making engine Before the days of bulldozers, a donkey used to pull a scraper in grading up a railroad line.
road monkey 1. A man employed in road maintenance. 2. The man who keeps a sleigh road or truck road in good shape. (N.F.) LL Also blue jay, chickadee, sandman.
road roller In roading operations, a roller to support cable, to prevent the line from rubbing on the ground. The ends of the roller were flanged, to keep the cable on the roller. (P.C.F.)

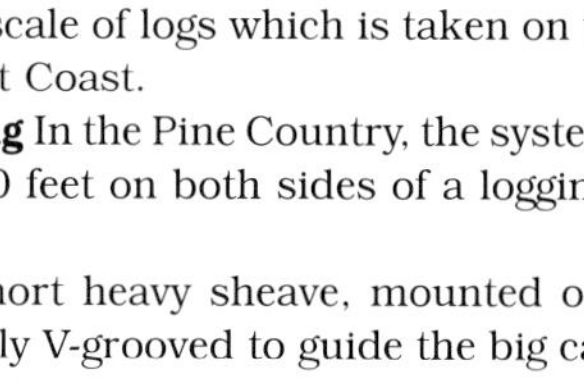

A **road roller**. *From Mallory Catalog No. 12, 1925, Glen Comstock collection*

road scale The scale of logs which is taken on the landing, formerly used on the West Coast.
roadside burning In the Pine Country, the system of piling and burning slash for 100 feet on both sides of a logging road to make firebreaks.
road spool A short heavy sheave, mounted on a heavy base. The sheave was deeply V-grooved to guide the big cable used in roading. (P.C.F.) Also spool.

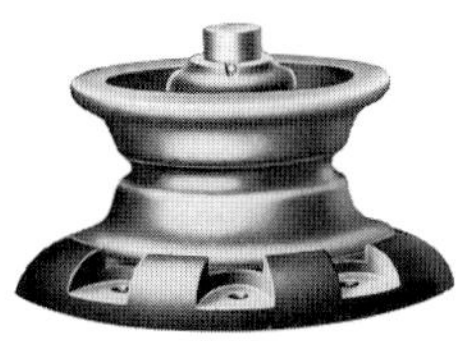

A **road spool**. *From Mallory Catalog No. 12, 1925, Glen Comstock collection*

road stake Money enough to help a man, looking for a new job, down the road from one camp to another.
rob 1. To steal logs that belong to another side, usually by going over the line to get the bigger logs. 2. To take parts from a machine to repair another. Also cannibalize.
robber's cane A long hardwood ruler used in estimating the board feet in logs or standing timber. (L.S.) See cheat stick. LL
robber's stick A scale rule used by a man underscaling logs: a dishonest scaler. (App.)
Robertson log yarder In the Pacific Northwest, a patented overhead cableway logging method using a standing line; the general rigup resembled the modern North Bend system. A raised tree was used. The first Robertson system was like the present high lead. In the early 1900s, Robertson tried to collect damages from every Northwest logger who used a high lead, but the entire industry united to fight the case, and won.
Robertson raft A cigar-shaped ocean-going raft assembled in a cradle to cut down on transportation costs between the stump and the lumberyard. As early as 1791, attempts were made to utilize this method off the East Coast, and in the 1880s a full scale business of rafting took place from Nova Scotia to New York. The leader in this venture was Captain Hugh R. Robertson of Joggins, Nova Scotia. Between 1883 and 1892, about a dozen of such rafts were completed, and although not too successful, a good deal was learned about the method or technique of rafting.
Robertson launched a redwood raft in July, 1889 at Noyo, Mendocino County, California which broke up in the Pacific Ocean. By 1895, after unsuccessful trials and many improvements, the rafts became popular on the Pacific Coast. (David C. Smith, "The Logging Frontier," Journal of Forest History, Vol. 18, No. 4). See Benson raft, cigar raft, ocean-going raft.
Robinson hydraulic wheels The evolutionary link between the big wheels and the arch; steel big wheels with a hydraulic jack to lift the load. Used in the Klamath pine regions of Oregon and California.

Robinson hydraulic wheels. *From "The Timberman" collection*

robissary, roberssary Camp commissary. See van.
rob the road To take logs from one logging road that should be yarded from another.
rob the show To steal logs from a donkey show with a cat. (P.C.F.)
rock A low grade coal for locomotive use. (P.C.F.)
rock a road To spread rocks over a logging road. (P.C.F.)
rock blade A bulldozer blade with teeth for combing out rocks.
rock crusher 1. A machine or rig that makes a lot of noise. 2. An early coffee grinder.
rocker A squared timber with a central swivel to enable front runners of a sleigh to turn. (N.F.) LL
rockin' chair money 1. Unemployment compensation. 2. Retirement.
rock powder The strong blasting powder used in blasting rocks.
rock road A gravel road.
rock work 1. Cutting through rock in making a road. 2. Preparing rock at the crusher for ballasting a road.
rocky mountain huckleberries Prunes. (West Virginia) LL
rod engine A steam locomotive in which the drive wheels were directly connected to the pistons by the main rods. Used on the mainline in the woods, or in areas where the grades seldom exceeded six percent. Also direct-connected engine.

Dolbeer & Carson Lumber Company's 2-6-2 **rod engine** No. 2 handles a log train in front of the Elk River, California shops. *Courtesy of Miller-Freeman Publications*

rodeur An itinerant lumberjack who continuously moves from one camp to another. (N.F.) (Eastern Canada)
rod locomotive See rod engine.
Roddis mutton Venison served in some of the early Roddis Co. camps in Wisconsin.
roll 1. Bedding. 2. The crossbar of a logging sled into which the tongue is set. (N.F.) 3. A short steel-shod bar used to hold the front end of sled runners in position. (N.W.) LL 4. To turn a log in the direction desired by putting a cable around a log and pulling. 5. To deck logs up on a rollway. 6. To roll a log away from a hang-up by wrapping the choker line around it so that the next pull will roll it to one side or the other before the log is moved forward.
roll a log 1. To attach a choker to a log in a certain way that when power is applied the log will roll sideways. (P.C.F.) 2. See roll.
roll bark Hemlock tanbark which is inferior because it has not been carefully dried. (N.F.) LL
roll dam A small dam without a gate. By building a dam without gates, a sufficient head of water could be obtained in the swell on the river's surface. The logs floated up an inclined plane and down the other side, thus causing the roll, after which the dam took its name. (Richard G. Wood, A History of Lumbering in Maine, 1820-1861)
rolled eye A rough eye in a wire rope made by splitting and rolling the two halves onto each other to make the eye. See Molly Hogan and wild eye.
roller 1. A roller placed at bends in the skidroad to reduce friction on the mainline in ground lead logging. See road spool. 2. A roller mounted above the mainline drum of a road donkey to keep the haulback line from fouling the mainline. 3. A short billet of wood or steel used to move a log lengthwise, as in a bucking chute. (N.W.)
roll in To go to bed. (L.S.) LL
rolling bind A choker fastened on a log so as to make it roll over; used to slide a log or to get free of a hang-up. (P.C.F.)
rolling chain See loading chain.
rolling crew The peavey men on a log drive. (P.N.F.)
rolling dam Logs implanted in the bank to narrow up the stream to deepen the water to get logs through a rough spot in the river. (L.S.) (Canada) LL
rolling hitch Two wraps of a wire rope around a log; when pressure is applied to the line, the log is twisted into position.
rolling place A place to spread out a bed roll and sleep. (P.N.F.)
rolling stock 1. All the trucks owned by a company. 2. On the railroad, it includes anything with wheels for use on the track, but usually confined to cars and locomotives. 3. Doughnuts.

In the late 1890s Yeon & Pelton, near Rainier, Oregon, were loading logs from a **rollway**, which was common practice in the West at that time. The logs were **yarded** in over the **skidroad** and stored on the rollway until the train was ready to load. The logs were then rolled onto the cars with the aid of **jacks** and **peaveys**. *J.F. Ford photo, Labbe collection*

rolling team A team of horses used to make up loads for big wheel hauling.
rollins Tobacco for a cigarette. (S.F.)
roll landing 1. A landing built to handle long logs and which had big brow logs down both sides of the track level with the car bunks. 2. An older term for a landing where logs were rolled on cars. Also rollway.
roll logs To turn over the logs on a landing so that the bark marks can be inspected by the scaler. (N.F.)
roll-out A log that rolls out of a chute.
roll out To roll out of bed; the cook's predawn cry. (N.F.) LL
roll out hook A heavy toothed hook mounted on the front of a bulldozer frame to roll over pine logs so that the knot bumper could get at the limbs on the underside; invented by W. K. Dyche. (P.N.F.)
roll out or roll up To go to work or roll up your personal gear and get out of camp. (P.N.F.)
roll splice 1. An eye splice. 2. A long splice rolled in a line with a marlin spike.
roll the boom To roll a boom of logs along the shore of a lake by the use of a cable operated by a steamboat. (N.F.)
roll the guff To talk. (L.S.) LL
roll up See bank up.
rollway 1. An older term for a landing where logs were piled to await loading. 2. An older term for a log dump. 3. Any single tier of decked logs. (L.S.) 4. Logs piled along a river to be rolled into the water. 5. A natural or prepared slope for rolling logs into a stream. (L.S.) See banking ground, skidway. LL
roll your blanket To be fired. (P.C.F.)
rooms On the skidroads on the West Coast, whore houses were labeled Rooms; for example, Alpine Rooms, Mission Rooms, Ruby Rooms and Tuxedo Rooms, and old-timers tell many humorous stories of trying to obtain lodging at these places when they first arrived in town.
rooster 1. An iron bar, usually a length of old rail, used for coupling disconnected trucks to form a car. (P.C.F.) 2. Also used as a spacer where overhang prevented couplers from joining. 3. A bar used to couple two logging sleighs. Also gooseneck, reach, slipper. (N.F.) 4. A river raftsman. (L.S.)

Noyes-Holland Logging Company, near Rainier, Oregon, loading **long logs. Roosters** were used between sets of **disconnected trucks** to allow the logs to overhang the ends. *C. Kinsey photo, Labbe collection*

rooster comb A small device used in the hook end of a cant hook to send up logs when loading with a single chain. (L.S.)
rooster tail The spray of water that went high in the air when the logs hit the log pond or river. (P.C.F.)
root cellar A cave-like depression, usually dug into the side of a small hill or knoll near the camp, shored up and roofed over with log rafters and earth in which perishable foodstuffs, such as onions, potatoes, etc. were kept from freezing in the winter. (N.F.) LL
rooter A log that digs into the ground as it is being yarded. (P.C.F.)
root hook A multi-pronged hook used in road building and land clearing.

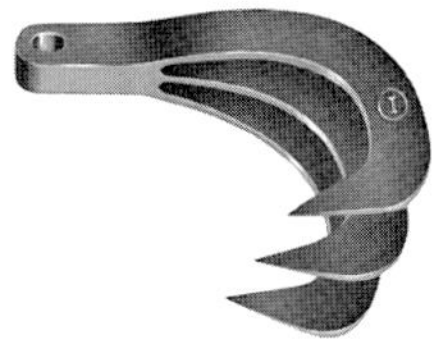

A **root hook**. *From Young Iron Works catalog, Glen Comstock collection*

root wad A mess of roots and dirt at the end of a tree that has blown down. (P.C.F.)
rope plate See bridle iron.
rope print The tracks that a wire rope makes when cutting into a worn sheave or drum.
rope work Any work associated with the moving of logs by wire rope.
rosin Pitch. In the distillation of crude turpentine, the residue remaining after the volatile oils have been driven off. (S.F.)
rosin-belly A lumberjack or logger. (S.F.)
ross 1. The loose, outside bark of a tree. (Maine) 2. To trim a log so that it will drag easily. 3. To hew the end of a log which slides on the ground when logs are to be hauled on a travoy so that it will slide easily. (N.F.) LL
ross a log To debark a log. (L.S.) LL
rossed wood Debarked pulpwood, free from bark. (N.W.)
rosser 1. A man who takes the bark off a log. 2. A rig that takes the bark off a log. 3. The man who barks and smoothes the ride of a log so that it may slide more easily. (L.S.) LL
rossy Said of shaggy-barked trees, chiefly of hardwood trees like swampy maple, but sometimes also of scurfy or scaly barked softwoods. (Maine)
rot See decay.
rotation The period of years required to establish and grow timber crops to a specified condition of maturity.
rot gut Cheap whiskey.
rotten knot A knot which is not as hard as the surrounding wood.
rough Said of low quality, as a rough tree which is full of branches.
rough and tumble landing A landing where no attempt is made to pile the logs in any order. (S.F.) (West Virginia) LL
rough as a card Said of a rough surface: "His hands are as rough as a card." The card is the wire brush used in carding horses. (Maine)
rough ground Country that is difficult to log and usually requires a skidder or a donkey show.
rough lock 1. A brake made by wrapping rope or chain around a wagon wheel to make it skid. 2. A brake made by putting a rope or chain around a log to prevent it from running into the animals in the ox team days.
round A turn of wire rope on a drum.
round boom A limber boom used to impound logs during towing. (L.S.) See bag boom.
round forty See rubber forty.
round head A Scandinavian logger or lumberjack. LL
round hook A half-round hook through which a chain would slide like a slip noose in a rope. (L.S.) LL
round stuff Logs. (P.C.F.)
round timber 1. Timber which has not been bled for crude turpentine. (S.F.) 2. Poles and piling. (P.C.F.)
round turn The loop at the end of a sleigh road. (N.F.) LL

A **Barnhart** loader works a **rough and tumble landing** of the Goodyear Lumber Company. Location is in Potter County, Pennsylvania. *From M. Koch collection*

roundwood Logs, poles, piling and cordwood of any size with or without bark.
roustabout 1. See drifter. 2. A handyman around camp. (P.C.F.)
route 1. The length of time a logger stays in one camp. 2. The length of time an operation stays in one place.
roving forty Said of a man who had a small acreage of land but who cut timber all over the country with no regard for ownership. (L.S.)
royalty A flat fee paid, in addition to stumpage, by a man who logs publicly-owned timber in British Columbia.
rub A metal bracket which is used to guide cables in roading.
rubber forty A timber steal in the early days when timber was plentiful and accurate surveys were scarce. A crooked buyer would buy 40 acres of land, and then stretch it by logging as far as he could in all directions until caught, or until he ran out of timber. In the late 1800s, it was said that a logger logged to the line, and as far beyond as he could throw an ax. (P.C.F.) See he logs on section 37.
rubber man Any gadget that the early fallers used to bring the saw back when working alone, e.g., a bent sapling or an inner tube. (P.C.F.) See spring pole falling.
rub irons Metal shoes used on swinging bunks and bolsters, located near the outer end. These shoes contact similar shoes fixed to the frame or car trucks, preventing a load from leaning too far off center and reducing the bind.
rub tree A tree used to change direction of a line around an angle in high lead logging. See siwashing.
rudder boom See fin boom.
run 1. A trip to the log dump and return. 2. A string of logs. 3. A skidroad. 4. The distance traveled by a fire at one time. 5. A line running downhill off a drum. 6. A narrow trail, cleared of brush and stumps, down which logs are pulled by a power skidder. (S.F.) 7. To make a try on a job. 8. Said of a log that moves ahead of the pull, when carried along by gravity or by its momentum. See running. 9. To float. (App.)
run a camp Said of a man who has full responsibility of a logging camp.
run a line To survey anything, e.g., a section line, a cruise strip, etc.

A **landing** of C.B. Howard & Company, of Emporium, Pennsylvania. Logs here are handled largely by gravity for loading aboard **Russell log cars**. This type of landing was known variously as a **drop landing**, **roll landing**, or **lightning landing**. *Courtesy of W. Casler*

run-around That portion of the stream where the river had changed its course, often because of a log jam. (L.S.) LL

run a sandy To play a trick on a man. (P.N.F.)

run-camp The man in charge of a logging camp. (P.C.F.)

run cutter 1. A logger who swamps out roads on a skidder show. 2. The man who clears narrow trails which radiate from a pullboat or from a head spar tree, down which logs are hauled by a power skidder. (S.F.)

run dam Tiers of hewn logs built into the bank at the curve of a driving stream. The logs hit the dam and run past instead of jamming. (L.S.) LL

run easy 1. Said of oil or grease, especially when applied externally to the running parts of donkeys, locomotives and other machinery. 2. Said of the oil carried in bottles by fallers and buckers to dissolve pitch and gum to let the saws cut freely. (P.C.F.)

run like a scairt deer Said of a very fast getaway. (P.C.F.)

runner 1. A locomotive engineer. 2. A donkey puncher. 3. An employment agent who recruits loggers. 4. A messenger sent downstream to notify the drivers that a sufficient head of water had accumulated at a dam to drive logs downstream. (L.S.) LL

runners 1. In making a crib to haul lumber by water, two pieces (8" x 10" x 16') laid end to end on each side of what was to be the crib. Two similar pieces were placed in the center, which lay parallel to the planks on either side. These boards were called runners. At right angles to the runners were laid planks 2 inches by 8 inches by 16 feet, which were known as binders. Three sets of binders were placed in parallel arrangement just as there were three parallel runners. (N.F.) 2. The large longitudinal timbers which make up a donkey sled.

runner chain A chain bound loosely around the front runners of a logging sled to act as a brake. (N.F.) LL

runner dog A curved iron attached to a runner at the hind sled of a logging sleigh, which holds the loaded sled on steep hills by being forced into the bed of the road by any backward movement. (N.F.)

running 1. Logs coasting ahead of the butt rigging on a downhill haul. 2. Logs running ahead of the pull. (P.C.F.)

running block See traveling block.

running foot A foot measured in a straight line, as a linear foot. (L.S.) LL

running guy A line attached to the end of a loading boom to swing the boom back and forth.

running hook A choker hook made to run easily on the choker wire.

running line A moving cable in logging operations.

running logs Logs drifting downstream on a drive. (L.S.) LL

running slide A slide on which logs run by gravity. (App.)

runout The amount of slack in a cable which permits it to be carried a short distance to one side of the line of travel. (P.C.F.)

run on a hill A sleigh load of logs that pushed the horses more than they could hold back with the harness due to improper sanding of a hill. Sometimes horses and drivers were killed on a run. (L.S.) LL

runs around like a blind dog in a meat house Said of a very confused man. (Red.)

runs by accident and Alaska Junk Said of a miserable show using second-hand or poor equipment and operating haphazardly. The Alaska Junk Company was the major buyer of used equipment in the Pacific Northwest, and a source of supply to many smaller operators. (P.N.F.)

A 2-pole **running slide** under construction near Corbett, Pennsylvania. *From W. Casler collection*

run train Usually the head brakeman on a logging railroad that had no conductor. (P.C.F.)
run-way See drag road.
run your face To ask for credit at the company store. (P.C.F.)

Russell log car A short logging railroad car used at one time in the Pine Country. Built by Russell Wheel & Foundry Company, Detroit, Michigan, one of the earliest suppliers of rolling stock to logging railroads.
Russell engine A steam traction engine, once used to haul log carts or wagons from the woods to the mill pond or landing. (P.C.F.)
Russell truck A make of disconnected railroad truck. See Russell log car.
Russian coupling See bucker's coupling.
Russian jump splice A splice halfway between a long splice and a short splice. (P.C.F.)
rustle To bring, e.g., "Rustle me up some grub." (P.C.F.)
rustle wood To bring wood. (L.S.)
rut cutter A form of plow for cutting ruts in a winter logging road for the runners of the sleds. (N.F.) Also rutter, gouger, groover and swamp angel. LL See ice a road.

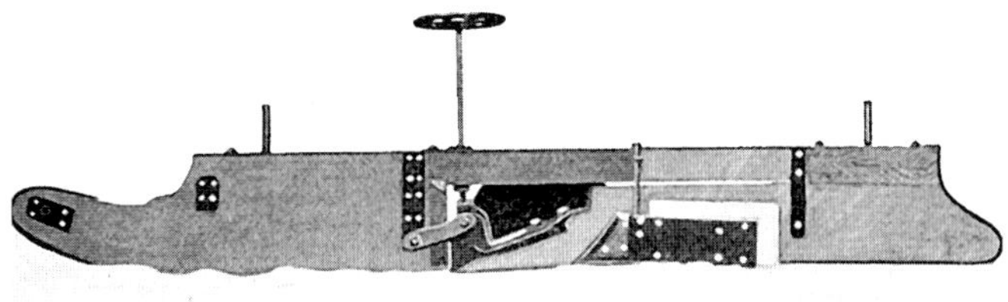

A Badger **rut cutter**. *From "Logging," Bryant, Wylie & Sons, 1923, Philip Schnell collection*

S

S&W A donkey engine made by the Smith & Watson Iron Works of Portland, Oregon. (P.N.F.)

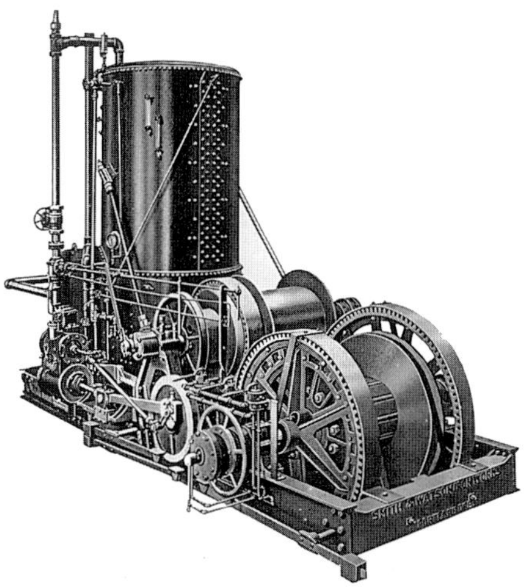

An **S&W** (Smith & Watson) 13x13 two-speed express **yarder**. *From Smith & Watson Iron Works Logging Equipment Catalog No. 9, Merv Johnson collection*

S.I. 1. An abbreviation for scenery inspector or "sivil ingineer." 2. A surveyor. (P.C.F.)
SOS An abbreviation for shit on a shingle or creamed chipped beef on toast; World War II use.
sack 1. To draw a circle of boomsticks together to close up a bag boom ready for towing. 2. To chase stray logs behind the drive. 3. To fire a man.
sacker 1. The man on the tail end of a river drive. 2. A log driver who rolled the logs from the bank or obstruction into the river with a peavey during a log drive after the main drive has gone down the river. (L.S.) LL
sacking 1. To roll logs down into the water that have been thrown on the shore during a log drive. 2. To roll or haul by manpower any logs from shallow to deep water. (N.F.) LL
sacking crew On a river drive, the men who looked for logs that might have become lodged in the inlets, swamps and sloughs along the way. (L.S.)
sacking dog In river driving, an S-shaped hook by means of which the hauling chain could be quickly attached to the log. The sharp end was flattened like a wedge, which made it easier to drive into the log. (N.F.)
sack the banks To hunt for logs hung up along the river banks following a drive.
sack the rear To follow a drive and roll in logs which have lodged or grounded. (N.F.) Also pick the rear, sweep the rear.
sack the river To add to the rear of the drive any logs hung up along the shoals.
sack the slide To replace logs which have jumped out of a chute or slide. (N.F.)
saddle 1. A pass between two higher points. 2. The trough made in the middle of the skids on a skidroad to keep the logs from rolling off. 3. The top part of a downed tree, called thus in peeling. 4. The depression between two parallel logs wherein a third log might be laid. 5. To put logs on or between other logs when loading a truck to prevent them from rolling off.
saddle a skid To chop a saddle in a skid or timber. (N.F.)
saddle bag A float of lumber that has become doubled around a tree or a sand bar or island when going down the river, or balanced sideways on a rock or bar in a river. (L.S.) LL
saddle blanket A tough flapjack.
saddle block A two-sheave, Y-shaped block which acts as a tree jack on a skidder show.
saddle strap A strap that secures a block to the tail tree on a duplex aerial yarding system. The strap is laid into a groove, or saddle, in the base of the tree. the block is connected to the strap by shackles and a tail end triangle.
saddle tank A rod locomotive with the water tank over the boiler to increase the weight on the driving wheels, to increase the pulling power and to make the locomotive more maneuverable. The maintenance required on the tender was eliminated. See side tank, tank.
safe burn Said of an area that has been burned to fireproof an area for a long time. The thinking now is that it helps but does not fireproof.
safe side 1. The side away from a dump in unloading cars. 2. The safe side of any loading or unloading rig.
safety first A camp welfare man. (L.S.) LL
safety guy A guy line that is rigged under the high lead block and the loading block; if the holding straps break, the guy line will carry the two blocks down to the ground.
safety shoes Shoes that have an iron reinforced toe for protection.
safety strap A strap fastened to the bull block and around a guy for safety purposes.

Wind River Lumber Company's small **saddle tank** locomotive heads a log train at a camp near Stevenson, Washington. *Philip Schnell collection*

sag 1. The slack in a cable. 2. A low place in a track or on the ground. 3. A saddle or pass in the mountains.
saginaw To retard or hold back the larger or butt end of a log in loading it up to a car. (N.F.) The opposite of Saint Croix. LL
saginaw stag Pants stagged (cut off very high above the tops of high boots). (P.C.F.)
sags A term used by the lumberjacks in the Upper Peninsula of Michigan for the lumberjacks who first came in from the famous valley of the Saginaw River in Lower Michigan. (L.S.) LL
sail block A small block suspended from the sail guy on a spar tree rigged for boom loading; used to support the loading block, through which the tong line(s) pass. Also loading block jack. See McLean boom loader, hayrack boom, heel boom.
sailer A flying piece of a tree that is sprung loose when a tree is felled. See widow maker.
sail(ing) guy A guy from the nose of the loading boom; usually runs through a tree jack and out to a stump. Also nose guy. See bracket guy line, McLean boom loader, hayrack boom, heel boom.
Saint Croix To retard or hold back the small end of a log in yarding. (N.F.) The opposite of Saginaw. LL
Saldern's logging college A method of cutting logging costs in the Pacific Northwest. Louis Saldern, a logger on the Lower Columbia, ran his logging camp as a "school," teaching young men how to log while getting out his logs and paying them little money. (P.N.F.)
sale A timber sale.
Salisbury hook An end hook for log loading.
saltchuck The ocean. (From the Chinook jargon.) (P.C.F.)
salucifer A match. (L.S.) LL
salvage logging 1. An operation with a small crew and light equipment to clean up after the main logging operation, bringing in merchantable trees that are too small to be handled with big equipment. 2. A logging operation after a fire.
salve Butter, lard or oleo. (L.S.) LL
sampson 1. In the bull-team days, a pry against the skidding line to change direction of pull to eliminate a hang-up or get around a sharp corner. 2. In horse-skidding, extra leverage applied to get a load started. 3. A system of setting chokers on a log to make it roll out of a hang up. 4. The use of a prop to give a lifting motion. 5. A pry used to help push a tree in felling. 6. A blocking rig-up to get a greater purchase. 7. A prop used under a log in bucking. 8. A heavy timber and a chain with a heavy swamp hook used for loosening or starting logs by horse power.
sampson a tree To direct the fall of a tree with a lever. (P.C.F.)
sand Sugar; at one time the brown crystal sugar used in the lumber camps.
sand hill Any hill that had to be sanded to hold back a sleigh and prevent sluicing. (N.F.) (Canada) LL
sandman See road monkey.
sand the flues To introduce sand into the firebox of a locomotive while under maximum draft, thus drawing the sand through the flues to remove the accumulation of carbon.
sand train Sugar. (West Virginia)
sanitary inspector The man who took care of the latrines. (P.C.F.)
sanitation cutting The removal of dead, damaged or susceptible trees, essentially to prevent the spread of pests or pathogens and so promote forest hygiene.
sap The sapwood.
sap fir Wet heavy timber, such as white fir. Also larch, noble fir or piss fir.
sap is down Said of the condition of a tree in the winter when it does not grow rapidly and the bark is tight. (P.C.F.)
sap is stopped Said of a tree in winter dormancy when the bark is tight and hard to remove. (L.S.)
sap is up Said of the condition of a tree in the spring or early sum-

mer when it grows rapidly and the bark is easy to peel. Also sap is running.
sapling 1. Any small tree that is unmerchantable. 2. A young tree trunk, diameter no more than four inches. The minimum diameter of a sapling is usually, though not invariably, placed at 2 inches.
sapling chopper The logger who cuts saplings ahead of the yarding crew as a safety measure. (P.C.F.)
sap peeled wood Wood with bark removed manually during the growth season. (R.M.F.)
sap pine An inferior pine which is heavy, knotty and pitchy. (Eastern Canada)
sap-stained Said of logs left in the woods or cold deck too long so that the sapwood will change color, causing a degrade. Also happens to green lumber when piled too tightly and not allowed to dry properly.
sapwood The layers of cream-colored wood next to the bark.
satchel stick A stick carried on the shoulder and used by a woodsman to support his turkey. (App.) LL
saucer hook A type of end hook used to load logs. The cup was to prevent the hook from pulling into the wood, and thereby pulling free. See cup hook.
sauna The Scandinavians' steam bath which was always found in camps where they worked.
saving himself work Said of a logger or mill worker who marries a widow with many youngsters. (P.C.F.)
savvy Knowledge or practical know-how: "He has plenty of savvy."
saw 1. The old reliable crosscut saw that was used in felling and bucking. 2. A chain or power saw. 3. A general term for any saw in the sawmill. 4. To rub one line against another. 5. To rub a line against or over a log. 6. To shunt or to work one train past another in the opposite direction when the siding is too small to hold all of either train. 7. In high lead logging, to pull the mainline back and forth, trying to free a hang-up.
saw alive To make all cuts on the log parallel.
saw bar The steel frame around which the chain runs on a power or chain saw.
saw bones A company or camp doctor.
saw boss A foreman of the felling and bucking crews. (S.F.) See bull buck(er).
saw box A long narrow box in which crosscut saws are carried to the woods in the crummy.
saw buck A saw horse; a short log with four legs extending through the log to form an X.
saw crew Fallers in the Pine Country.
sawdust city Eau Claire, Wisconsin. (L.S.)
sawdust savage A chopper. (Red.)
sawdust with legs Lice. (West Virginia)
sawed by a cross-eyed man Said usually of a butt log with a very uneven stump end, caused by one high and one low cut. (P.C.F.)
saw filer, saw fitter See filer.
saw guide A block nailed to a tree to start a level cut in the early days of the crosscut saw.
saw horse See saw buck.
sawing foreman A bullbuck. (P.C.F.)
saw oil Kerosene or light oil stored in a whisley bottle with an iron hook and fitted with a notched cork. The oil was sprinkled on a crosscut saw to cut pitch when sawing. See oil bottle.
sawing woods The logging woods. (S.F.)
saw kerf That part of a cut that is removed by the saw blade.
sawlog A log suitable for the production of lumber.
sawmill log A sawlog.
saw teeth Usually refers to the cutting teeth and not the raker or clearing teeth. See rakers.
sawtimber The trees that are big enough to be sawn into lumber.
saw to length To cut to a specified length.
saw vise A wooden or metal vise that held saws while being sharpened or set. See filer. LL
sawyer 1. In the late 1800s, the logger who determined the lengths that a redwood log should be bucked. (Red.) 2. A bucker. 3. A faller. (N.F.) (West Virginia) (S.F.)
scale 1. The board feet in a log. 2. A rig for weighing log trucks. 3. To tally board feet measure in logs.
scale bill The amount of board feet in a load, in a day's work, etc.
scale book 1. A book especially designed for recording the contents of scaled logs. LL 2. A book with a chart showing the footage for each measurement of logs already computed.
scaler The man who measures the scale of logs.
scale rule, scale stick 1. A long, heavy hardwood ruler marked in the scale for each inch; used for scaling logs. 2. A rule used in estimating the lumber in standing timber. (L.S.) Also cheat stick, moneymaker, robber-cane, thieving rod and thief-stick. LL
scaling The method or act of determining the volume of a log. All logs are scaled by measuring the length and average diameter of the small end, and determining its volume from a log rule.
scalper 1. A scaler. 2. One who sells lumber on commission. (L.S.)
scalping Removing small plants and duff or ashes from around the spot where a tree seedling will be planted; usually done by hand.
Scandihoovian A logger from Scandinavia. (P.C.F.) Also Skandihoovian, round head.
Scandinavian dynamite Snoose, chewing tobacco.
scare In the late 1800s, a term for the staging boards on which the choppers stand. (Red.)
scarf 1. An undercut. 2. The surface of an undercut. (N.F.) 3. To cut a deep groove around the end of a large log to hold the butt chain, to prevent it from slipping in the late 1800s. (P.C.F.)
scenery inspector A logging engineer or forester. (P.C.F.)
schoodic chain bind A method of binding logs to the bunk of a sled. A chain is passed around the log and then around the sled bunk in the opposite direction and held in place with a fid hook. (N.W.)

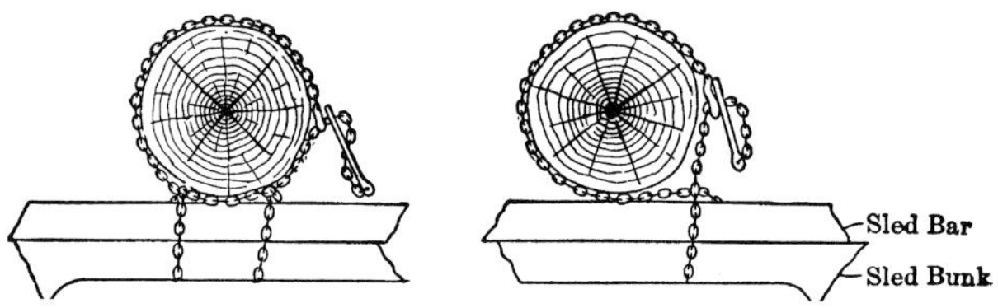

Two types of **schoodic chain binds**. *From "Logging," Bryant, Wylie & Sons, 1923, Philip Schnell collection*

school bell has rung Said of the forestry students and others when they go back to college in September, leaving the fire crews shorthanded. (P.C.F.)
schoolmarm A crotch, usually at the base of a tree trunk, commonly caused by two tree trunks growing together. The swell at the crotch is not good for lumber and has to be cut out. In the Lake States used for training river drivers in practicing the art of balancing. LL
school's out Said when something has gone wrong, and everyone is going to catch hell. (P.C.F.)
scissor bill 1. A non-union man. (P.C.F.) 2. A green worker. (P.N.F.) 3. A worker indifferent to the interest of the laboring classes. (L.S.) LL See hoosier.
scone A rich baking powder biscuit containing currants or raisins. (L.S.) LL
scoop 1. The bucket of a power shovel. 2. A scoop bucket that was used to move dirt by horses in the early days. Also Fresno or Fresno scraper.
scoop roof A roof made out of logs hollowed out and fitted to shed rain. (L.S.) LL
scoop scraper A very heavy dirt-moving scraper which was used with a donkey engine. See Bagley grader.
scoot 1. Another term for dray. 2. A crude log transport. (Maine)
score 1. The marks chopped into a log to indicate the location of cuts by the barker, marked by the bull buck. (P.C.F.) 2. To notch a timber with an ax before cutting it.
scorer A square timber maker who notched out rough dimensions of a log so that the hewer could square it. (Canada)
scoring ax An ax used to score timbers before squaring them with a broadax.

scorper A dried toasted bun. (P.N.F.)
scrabble log See prize log.
scratch To remove a car out of a train and to put it on a siding.
scratch a tell To explain to a man how to get somewhere by scratching a map in the dirt with a stick. (P.C.F.)
scratcher 1. A tool for cutting marks in trees; used in surveying. 2. A timber cruiser or forest engineer who marks trees for reference. (P.C.F.)
scratch grade A logging railroad grade on which only light work has been done. (P.C.F.) (N.F.)
scratch grading Said of a road that is made by hardly scratching up the surface of the ground. (P.C.F.)
screech cat A high stump with splinters that whistle or sing in the wind. LL See sliver cat.
screwdriver A geared locomotive, usually a Shay. Also corkscrew, sidewinder, stemwinder.
screw jack A jack that is turned by a bar that passes through the screw and used to "screw" it up or down. Also jack-screw.
screw the pops down To screw down the pop valves on a steam engine so that it will carry a higher head of steam than it is supposed to, to get the last measure of effort from a machine.
Scribner A log rule giving the scale of logs according to the diameter and length; now slightly changed and known as the Scribner decimal C. It is used by the U.S. Forest Service and the Pacific Northwest log scaling bureaus; published by J. M. Scribner at Rochester, New York in 1846.
scrip Money issued by logging companies which could only be used at the company store. LL
scroll knob Specifically the butt end of an ax handle. (L.S.) LL
scrub 1. The rough unmerchantable brush growing at timberline and on the edge of the desert. 2. Any kind of timber other than the giant redwoods or good-sized fir trees. (Red.)
seale rope A wire rope with a large center wire in each strand, surrounded by smaller wires, with an outside layer of larger wires.
seam A longitudinal crevice in a log or tree usually filled with bark.
seam squirrel Body lice. Spelled "seams-squirrels" in the Lake States. Also blue jackets, crumbs, sawdust with legs.
sea mule A boat in the log pond that is used to push logs. (Red.) See boom boat.
season (the) 1. The logging season, usually from September 1 through April 1. (L.S.) 2. The fire season.
season's cut The season's cut of logs.
Seattle An early steam donkey made by the Seattle Iron Works. (P.N.F.)
Seattle bunk A bunk on a railroad car that has patented locking cheese blocks. (P.C.F.)
Seattle car A product of the Seattle Car & Foundry Company, later part of Pacific Car & Foundry Company, which produced both disconnected trucks and connected trucks, as well as other logging railroad equipment. (P.C.F.)
second chain tender An older term for a second rigger.
second cook The meat cook in a large camp. (N.F.)
second faller The logger who follows the lead of the first faller who makes all the decisions. (P.C.F.) Also helper.
second growth 1. The trees of small diameter. 2. The trees that have grown since the virgin stand of timber was removed. 3. In Douglas fir stands, all stages of growth between saplings and red fir. (P.C.F.) Also young growth, third growth.
second hook A latched hook used with the head hook to carry additional chokers on a skidder show.
second hooker 1. The second in charge of the choker crew on a skidder. 2. The lead man on a second crew of choker setters. Also second hooktender.
second loader The first loader picked the loads and sent them up, while the second loader handled the loading and released the tongs. The second loader was the one up on the car. Also tong shaker.
second rigger 1. An assistant in rigging trees, specifically rigging tail trees on a skidder show. 2. A logger who fights hang-ups in yarding with high lead. (P.C.F.)
second riggin' rustler See second rigger.
second second rigger On a big skidder show, the second rigger's chief helper (not third rigger). (P.C.F.)
second train loader The top loader's helper in the days of railroad logging, who was needed when two tongs were used. (P.C.F.)
section 1. 640 acres (a square mile); there are 36 sections in a township. 2. One unit of a flat log raft, usually 66 x 66 feet or 70 x 70 feet.
section crew The men who build and repair railroad tracks.
section hand See gandy dancer.
section plate A heavy rigging plate with rings on each side. Extra sections could be added to the mainline through the section plate, allowing more chokers to be carried. See riggin' plate.
section 37 1. A township that exists in Ontario, Canada. 2. Something unusual; not supposed to exist since the usual township in the United States contains 36 sections. 3. A perfect logging area where all good lumberjacks go when they die. (L.S.)
sections Short pieces of line.
sector Part of the fire line, under the charge of a foreman who is called the sector boss. (P.C.F.)
seedbed An area prepared to receive seed, such as an area cleared of plants and duff, so that natural seedfall can establish a new forest.
seed block The uncut blocks of trees which are left between and around small clearcut blocks to provide seed for natural regeneration.
seedling 1. A young tree grown from seed, from the time of germination until it reaches sapling size. 2. In the nursery practices, a young tree that has not been transplanted.
seed trees Standing trees left behind by loggers, usually because of poor quality. (P.C.F.)
seed tree law The usual logger's term for state forest conservation laws. (P.C.F.)
see the dentist An excuse to get away from camp.
segment ground Said of a crosscut saw tapered in thickness from the teeth to the back, making a cut wider than the saw back, thus preventing binding. (P.C.F.)
seize To tie down the strands of wire rope before cutting it.
seizing iron A tool carrying a reel of wire on an iron handle, used in wrapping loose ends of wire rope.
selective chopping, selective cutting, selective logging 1. The operation of cutting only the selected trees, leaving the rest to grow. 2. Any kind of cutting which stops short of a clear cut. Usually applied to trees marked by a forester for removal under a forest management program; not used in early logging and later only by a few lumber companies.
self loader A logging truck equipped with a small boom and power winch by which it can load itself.
self-loading dam A dam in which long timbers were set on the upstream side at an angle to the water surface. (L.S.) Also slant dam. LL See rafter dam.
self-oiling Said of a logging block with a built-in reservoir which feeds oil slowly to the bearings.
semi-siwash 1. A camp between a gyppo and a big camp. 2. A camp partly haywire. (P.C.F.)
sender One of the loading crew who guided logs up the skidway pile. (Canada) Also ground loader.
send him down the road To fire a man.
send up In loading, to raise logs up skids with cant hooks or by steam or horse power.
send-up man 1. A logger working down in a canyon setting chokers on logs sent up from a cold deck. 2. A loading crew man who guided logs up the skids when loading from the ground onto flat cars. (P.C.F.) Also ground loader.
sent out on the tote road Said of one who was ordered to leave camp. (L.S.)
sequoiadendron gigantea The Sierra redwood or Big Tree found in the higher altitudes of Central Eastern California.
sequoia sempervirens The Latin name given to the coast redwoods that produce redwood lumber. The literal meaning is "always living."

A **stand** of **Sitka spruce** in the **fog belt** of Western Oregon which has been **selectively logged**. *From "Logging, Principles and Practices," Brown, Wylie & Sons, 1934, Philip Schnell collection*

Sequoia spark arrestor A make of steam donkey spark arrestor.

Sessoms system An incline system in which a line from a hoisting engine was passed through a huge block consisting of two large pieces of plate cut in a triangular form with smaller sheaves at each "corner" on a hoisting or block car and then to an anchor on the opposite side of the track from the lowering engine. Railroad cars loaded with logs were coupled to the huge block car which was then lowered down a steep hill. Designed by H. W. Sessoms of the Ebey Logging Company at Arlington, Washington and built by Willamette Iron & Steel Works, and used widely in the Pacific Northwest. (P.C.F.) Also two-line system.

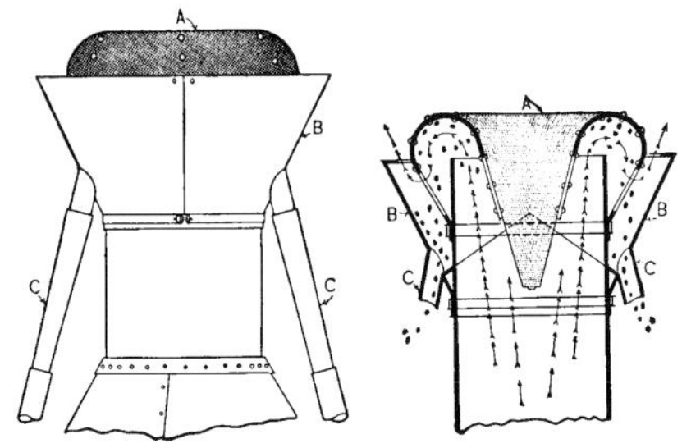

The **Sequoia spark arrestor**. *From "Logging," Bryant, Wylie & Sons, 1923, Philip Schnell collection*

set 1. A pair (2) of fallers or buckers. 2. The degree to which the teeth are flared in a saw. 3. A device in which a saw was set to fix the correct set of the teeth or the correct flare to the teeth. 4. The location for a skidder. (S.F.) 5. To go to work on a log job. (App.)

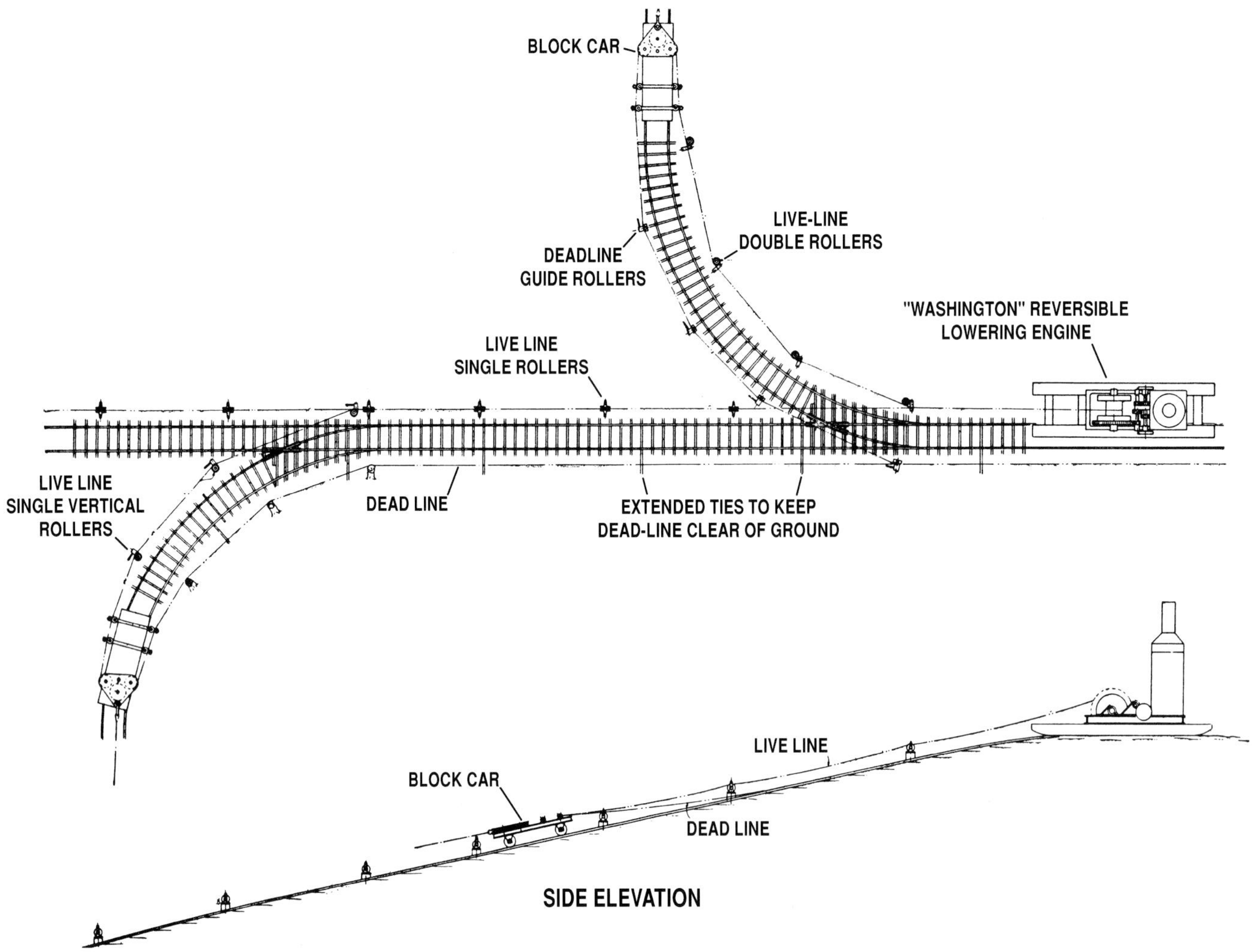

The **Sessoms system** of lowering cars on an **incline**. *From Washington Iron Works catalog, Labbe collection*

set a samp To rig up a sampson of any kind to get greater leverage. (P.C.F.)
set beads To set chokers on a logging show. (P.C.F.)
set block A steel block used in setting the teeth of a crosscut saw. The set block is used to back up he tooth while it is hit with a hammer. Also setting anvil, setting block.
set chokers To work as a choker setter. (P.C.F.)

Setting chokers on a big **yellow fir** log at the Big Creek Logging Company operations near Knappa, Oregon, on the Lower Columbia about 1920. *Weister photo, Labbe collection*

set'em up 1. To turn up the retaining valve handles on a train air brake system before going down a hill with a heavy load of logs. This causes partial braking on the wheels so that the engineer can build up more pressure in the brake cylinders. 2. To set up the hand brakes before heading down the hill with disconnected trucks. (P.C.F.)
set gauge See spider.
set knobs To work as a choker setter. (P.C.F.)
set of sticks The boomsticks needed to build a raft of logs. (P.C.F.)
set pole A long pole used in holding a bateau in place. (N.F.)
sets Disconnected trucks were usually made up of a matched pair. Footboards and brakes were on opposite sides so that when made into a car, brakes would be on the same side of the train and footboards were at each end of the car. Thus, two trucks made a set.
setting 1. An area that is logged to one spar tree. Also side. 2. An area that is yarded by cats to one landing. 3. The smallest geographical unit to be identified for record keeping purposes for logging or silvercultural activity.
setting anvil See set block.
setting block See set block.
setting line An older term for a moving or transfer line on a skidder. (P.C.F.)
setting on his hands Said of a lazy man on the job. (P.C.F.)
setup 1. A job, arrangement or layout: "What's the setup?" 2. The important seating arrangement in a logging camp cook shack where the old-timers had their choice seats.
shack A brakeman on a logging train. Head-end shack is the brakeman on the engine; tail-end shack is the brakeman on the last log car, or on the caboose. (P.C.F.)
shacker 1. A man living in a small building of logs, or rough lumber. (L.S.) LL 2. A small farmer moving into the woods area. (P.C.F.) 3. A small subcontractor. (Canada)
shacking Living alone in the woods. (L.S.) LL
shackle The important clevis or heavy iron loop which is passed through an eye spliced in a wire rope to anchor a rope to a swivel, block or some other hold. The shackle is closed by a drive pin or screw pin. Also clevis.

A **shackle**, or **clevis**. *From Young Iron Works Catalog No. 52, 1952, Philip Schnell collection*

shackle yarding block A heavy yarding block with a yoke fastened by pins so that it could be opened to lay in a line.

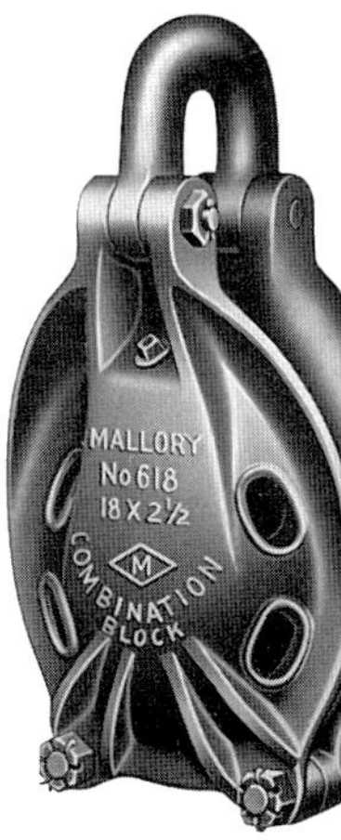

A **shackle yarding block**. *From Mallory Catalog No. 12, 1925, Glen Comstock collection*

shack up 1. To quit work and take it easy until times get better. 2. To leave the woods and go to town for the winter. 3. To sleep with a woman.
shade streak In turpentine orcharding, a streak so chipped that it shades the newly exposed surface from the direct rays of the sun. (S.F.)
shag A brakeman on a logging railroad. (P.C.F.)
shagpoke gut A very big eater; from the shagpoke (Shitepoke), a small heron which inhabits the mud flats and seems to eat anything and to eat all the time. (P.C.F.)
shake 1. A wood shingle made by splitting flat strips from a bolt. (P.C.F.) Also split. 2. A lengthwise separation of the wood, which occurs usually between and parallel to the rings of annual growth; usually occurs in a tree when buffeted by the wind. Shake usually follows the annual rings, while checks are radial. Also split. See windshook.
shake bolt A block of wood from which shakes are split. See bolt.
shake hands 1. To put the knob of one choker in the bell of another, making a big loop. (P.C.F.) 2. Sometimes used to denote the coupling of the air hose between cars, or between truck and trailer. The fittings thus joined are called glad hands.
shake roof A roof on camp buildings made of cedar or pine shingles or shakes split with a shingle froe. LL
shake the lines over the iron mule To run the donkey engine. (S.F.)
shake the sieve Said when men were to be laid off or a crew to be cut down in size. (L.S.) LL
shake tongs To work as a loader. (P.C.F.)
shake tongue To move a loaded log cart by ox or horse team.
shaky log A log showing evidence of shake, split by swaying in the wind.
shank The handle of a tool. Also holder.
shanty 1. Originally in the Canadian woods, all the buildings of a camp complex. 2. A bunkhouse. (L.S.) Also called a State of Maine camp. 3. A small building made with boards. 4. Occasionally a term for a lumberjack. LL

A big three-truck **Shay** locomotive of the White River Lumber Company, of Enumclaw, Washington, pictured here in March, 1953. *Courtesy of M. Koch*

shanty boat Another term for wanigan. (L.S.)
shanty boss A camp worker who got the wood and water and did chores. (L.S.) LL
shanty-boy The original name for the lumberjack. See lumberjack.
shanty-boy's alphabet See alphabet song.
shanty-man A 19th century term for lumberjack. (L.S.) (Canada)
shanty-man's life (the) One of the oldest and most widely known ballads which was sung in the lumber camps of the Great Lakes and which probably came from New England or Eastern Canada. The first stanza is as follows:

Oh, a shanty-man's life is a wearisome life,
Although some think it void of care,
Swinging an ax from morning till night
In thc midst of the forest so drear,
Lying in the shanty bleak and cold
While the cold stormy winter winds blow,
And as soon as the daylight doth appear
To the wild woods we must go.

shanty men The woodsmen who live in shanties. (S.F.)
shanty nasty The person who stays on living in one of the buildings of an abandoned logging camp. (L.S.) LL
shanty queen A logger's wife living in a rough shack in the woods. (P.C.F.)
shape up or ship out To work at a job or to get out.
shark An employment agent. (P.C.F.)
Sharkey carriage A slack-puller carriage for use with a skidder. The slack was pulled by a hand line trailing down to the ground from the carriage. (P.C.F.)
shavetail A mule. (S.F.) (N.F.) LL
shaving the whiskers on big dick Said of the men going from woods work to the wheat harvest in the Dakotas. (L.S.) LL
Shaw A type of rigging plate used to attach chokers to the butt rigging or skidding line on early skidder shows. (P.C.F.)
Shaw socket A sliding socket used on a line bringing in several chokers on a skidder show. (P.C.F.)
Shay The most popular of the geared locomotives used in the woods. The cylinders were upright on the engineer's side of the frame, and the boiler was offset to balance them. A propeller shaft driven by cranks ran along one side of the engine and was geared to the axles. In the Lakes States called a Lima or Limey or sidewinder; in the Northern woods, usually a stemwinder; and in the Sierra Nevada Mountains, usually a screwdriver.

The first Shay was built by E. Shay in Michigan in 1874; the Lima Locomotive Works built 2,771 Shays, the last in 1945. The Willamette Iron & Steel Works built 33 geared engines of the Shay type.
Shay swivel A rig used to attach the slack pulling line to the skidding line on a skidder show. (P.C.F.)

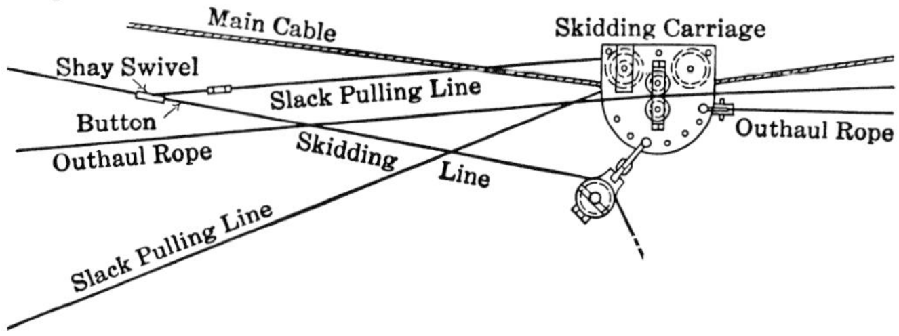

The rigging arrangement of the **Lidgerwood skidder carriage**, showing a **Shay swivel** used to connect the **slack pulling line** to the **skidding line**. *From "Logging," Bryant, Wylie & Sons, 1923, Philip Schnell collection*

shear skid See fender skid.
sheave The grooved wheel in a block to hold a line which runs through it. Also wheel.
sheave block A large, horizontal sheave mounted at ground level to a tree or stump; used to control line direction on a cypress pull boat operation. (S.F.) See pull boat.
sheaved wheels The grooved metal wheels used on various types of trucks on a pole road. See pole road. (L.S.) LL
sheave snubbing engine A donkey which had one big drum with

A **sheave block** attached to a **tail tree** on a cypress **pull boat** operation in the South. The X-shaped blaze on the tail tree is to indicate to **fallers** that this tree is not to be cut. *From "Logging," Bryant, Wylie & Sons, 1923, Philip Schnell collection*

very large flanges so it could be heavily braked. It was used as a lowering engine on an incline. See gypsy-type lowering engine. (P.C.F.)

she-bitch 1. A nasty woman. 2. An engine that refuses to respond. (P.C.F.)

she-boilers Women cooks. (App.) LL

shed The roof over a donkey or a cat. See house, lid.

sheepherder 1. A worthless man. 2. A logger who wears overalls. (P.N.F.)

sheer arm A boom set slantwise across a railroad track and above the bunks to shove logs off a car. See jill-poke.

sheer boom A small boom, one end fast to shore, and the other swinging out into a stream to force logs into a main holding boom. Also fender boom, glancing boom, side boom. LL

sheer log 1. A log placed on a turn in the skidroad so as to push the logs away from the bank to prevent them from digging into the ground. 2. Applied sometimes to the first log to be yarded, where it was placed against any obstacle on the road so that following logs could slide by.

sheer rail A guide rail for wheeled vehicles on a plank road or at a landing. See fore and aft road.

sheer skid See sheer log.

shell The outer framework of a block.

shellacker A cook. (Maine)

shelter house In turpentine orcharding, a small shed erected in the woods in which workmen may take shelter during storms. (S.F.)

shelterwood cut A harvesting method in which the mature timber is removed in a series of cuttings, encouraging the establishment of natural reproduction under the partial shelter of residual trees.

she pulls Said of the downriver floating action of the logs after the dam was broken. (L.S.) LL

she's going to haul Said of a log jam about to break. (L.S.) LL

she's made Said of a job that is finished. (P.C.F.)

she-stuff Women. (P.C.F.)

shifter A switch engine. (P.C.F.)

shikspoke A man who picks at his food; from the shagpoke, a small heron. (P.N.F.)

shim 1. A small piece of anything used to pry something apart or to level something up. 2. In felling, the wide steel plates of varying thickness often placed above and below the falling wedges to prevent them from squeezing into the wood. 3. Blocking placed on cross ties to level up the railroad track; generally used on soft wood ties, and also used to keep the track from sinking into the mud. (L.S.) LL 4. To raise or support.

shim hoe, shin hoe An adz.

shim up To place shims under a railroad track.

shin cutter An adz tool used in hewing logs to form the floors of stables and shanties. (N.F.) LL

shingle A thin wedge-shaped slat made as a covering for roofs and sides of houses.

shingle bench A bench used for shaping shingles or shakes after blanks were cut from a block with a shingle froe. LL

shingle bolt A bolt from which blocks are cut to be made into shingles. See bolt.

shingle camp A camp where shingles were shaved. (Maine)

shingle crib A system of a corral of poles used in the water in which shingle bolts are loaded for towing to the mill.

shingle froe A tool for splitting shingles from a block of wood. See froe. LL

shingle horse A long log, one end resting on two legs, the other on the ground, on which lumber was laid to be worked into shingles. (Maine)

shingle logs Short or defective logs. (App.)

shingle weaver 1. Originally the man who made shingles by hand with a froe. 2. The man who packs shingles. 3. The big worms that bore into cedar logs.

shingle weaver hair cut A crude bunkhouse hair cut. (P.C.F.)

shingling logs To sink one end of a floating log by putting another log on top of it. A boom was put around a unit of these logs to transport them. (L.S.) LL

shinny Liquor. (L.S.)

ship To float logs down a flume. (P.C.F.)

ship knee A big chunk of wood taken from the area where the trunk of a tree turns into a large flat root, forming almost a right angle; used at one time as a brace in framing sawmills and in shipbuilding. LL

ship knee camp A number of camps on the West Coast during World War I and II were engaged only in getting out ship knees for shipyard use in making wooden ships.

shipworm A destructive marine borer which eats the logs or other wood in salt water.

shittim See shitum bark, chittam.

shit out To force a lower log outward by a heavy top log on a logging truck or railroad car. A popular expression: "It shit out the small ones." (P.C.F.)

shitum bark The popular expression for cascara bark. See chittam.

shiv, shivy 1. A splinter which comes out of the chipper too long to be pulped. 2. Anything that is no good. 3. A common pronunciation of sheave.

shock absorbers Hotcakes. (P.C.F.)

shoe 1. A donkey sled. 2. To put a shoe on an animal. 3. To attach something to something.

shoe a chew To spike up a chew of tobacco with snoose and whiskey. (P.C.F.)

shoe a springboard To fit a dog iron on a springboard. (P.N.F.)

shoeing hammer A tool for driving the horseshoe nails into the horse's hoof. Also farrier's hammer. LL

shoeing tree Usually a maple stump made for the camp blacksmith to use in shoeing horses. (L.S.) Also shoeing buck. LL

shoemaker A very poor logger. (P.N.F.)

shoe pack pie A pie made of vinegar, cornstarch, sugar and water, and sometimes flavored with lemon or vanilla, like the modern butter scotch pie. (L.S.) LL Also larrigan.

shoe packs 1. Rubber bottoms with leather tops worn in cold weather. (L.S.) LL 2. Cowhide boots with moccasin soles worn by early woodsmen before the appearance of spiked boots and, later, lumbermen's rubbers. (Canada) See larrigans.

shoes 1. The tires on sled runners; usually made of cast iron on tote sleds, and thinner steel on logging sleighs. (N.F.) LL 2. Often used by truckers to refer to tires. (P.C.F.)

shoes with angry bottoms Cork boots. (P.C.F.)

Shingle bolts for the Grandy Lake Shingle Company at Concrete, in Northwestern Washington, were brought to the **dump** over the company's little railroad. The trains were handled by a Fordson locomotive from the Skagit Iron Works. *D. Kinsey photo, Labbe collection*

shoofly 1. A temporary track or road built around a washout. 2. A route which winds around the hillside instead of using a trestle or fill across a canyon.
shoofly rear To take only the logs that are easy to get in sacking the rear. (L.S.) LL
shoofly sacking Not getting all of the logs thrown up on the river bank back into the river during a river drive or following a log jam. (L.S.) LL
shoot See log chute. (P.C.F.)
shoot a jam To loosen a log jam with dynamite. LL
shore hold The attachment of the hawser of a raft of logs to an object on the shore. (N.F.) LL
shore up To reinforce anything: trestles, piers, bridges, excavations.
short-coupled 1. Said of anything hitched up with a short line. 2. Said of a small man. (P.C.F.)
short crew Said of not enough men for a job. (P.C.F.)
short ground A break in the surface of the ground which will cause a redwood to break in falling. (Red.)
short haul 1. An easy haul of a short distance. 2. To cut a man out of his fair share of something. 3. To take the easiest way when doing a job. (P.C.F.)
shortleaf pine (pinus echinata) One of the four important pines in the Southern states which has a wide botanical range. It grows more or less abundantly in 24 states from Long Island, New York, and parts of Pennsylvania southward to Northern Florida, and westward to Eastern Texas. Shortleaf pine, in common with longleaf, slash and loblolly, is called "yellow pine" or "Southern yellow pine" by the lumber trade. (Grant W. Sharpe, Clare W. Hendee and Shirley W. Allen, Introduction to Forestry, 4th edition)
short line skidding See jammer logging.
short log 1. The 16- or 20-foot log usually cut in the Pine Country. 2. The 32- or 40-foot or longer log cut in the fir belt; includes the 8-foot logs cut for stud mills. See standard log.
short log cat A pond man. (P.N.F.)
short log country The Pine Country. (P.C.F.)
short logger 1. A logging truck hauling short logs on the truck bed with no trailer. 2. A logging truck that hauls only logs of shorter lengths. 3. A logging truck that makes short runs. (P.C.F.)
short log loader A new loader who works out on short logs before trying to handle the big ones. (P.C.F.)
short log logger 1. A logger in the Pine Country. 2. A gyppo who handles small logs. 3. A pulpwood cutter. (P.C.F.)
short log show 1. A pine operation. 2. A tie operation. 3. A gyp of some kind. (P.C.F.)
short of cars Said of one who is out of work. (P.N.F.)
short road See go-back road.
short scale 1. The disparaging term of the fallers and buckers who think that they were cheated or should have gotten more money: "We must be working on a short scale." 2. The disparaging term of the stumpage buyers or log buyers who usually felt that they were getting short scale on general principles. (P.C.F.)
short splice A splice with tucks closer together than in a long splice; not to be used with heavy loads.
short staker The boomer who works only long enough to make a short stake and then moves on to another job. See boomer.
short sweet'nin' A lump of sugar to sweeten coffee. (L.S.) LL
short ton A U.S. weight measure equal to 2,000 pounds.
short tongue A tongue used between bobs in place of cross chains, to hold the rear bob at a distance. (N.F.) LL
shot 1. A dynamite or powder blast. 2. Said of a machine or tool which cannot be repaired. If fixed may be half shot. A machine is half shot if in need of repairs but still running.
shot gun bunk See muzzle loader.
shot gun system See duplex flyer.

shot hole 1. A hole dug to take charge for blasting. 2. A hole made in wood by boring insects. (App.)
shot in the brush (a) 1. Said of building a railroad into deep timber. (P.N.F.) 2. A method in the late 1800s of locating a logging railroad. (P.C.F.)
shoulder In turpentine orcharding, the uppermost corner of a face. The streak starts at the shoulder and terminates at the peak. (S.F.)
shoulder wear 1. The cutting action of a wire rope caused by a sheave being out of line with the direction of pull. 2. Wear on a block.
shove his nose down To thrash a person. (L.S.) LL
shovel A power shovel used for digging or for loading.
shovel loader A shovel converted to a loader.

A rail-mounted **shovel** and **Shay** belonging to Big Creek Logging Company, near Knappa, Oregon, on the Lower Columbia. *C. Kinsey photo, courtesy of Roy McCourry*

A tracked **shovel loader**. *From "The Timberman," October, 1949, Philip Schnell collection*

shovel runner The operator on a power shovel.
show 1. A logging operation, such as a high lead show, a cat show, a winter show, etc. 2. Also said of operating conditions, such as a good show or a bad show. See chance.
show boy (Quebec) See chore boy.
showing the white feather Said of a river pig or lumberjack who stopped work because of bad weather. (L.S.) LL
shrinkage 1. The decrease in size of wood due to decrease in moisture content. 2. Log loss during a river drive. (L.S.) LL
shroud A blanket. (L.S.) LL
shuffle the brogue An old bunkhouse game. (L.S.) LL
shutdown 1. Work stoppage in a lumber mill for maintenance work, usually in July and the latter part of December. 2. Work stoppage in the woods when there is a surplus of logs or because of weather conditions, such as wind, rain or low humidity. 3. Work stoppage because of a strike. 4. To close down the lumber mill or woods operations.
shut the hole Said of a river driver who disappeared after falling into the river. (L.S.)
siamese A fire hose with a Y coupling to give two outlets.
siamese cat Two cats linked by a heavy cable, working it ahead in a semicircle to pull down small timber inside the loop. (P.C.F.)
side The logging unit, including all men and equipment, operating around a single landing; called a cat side, high lead side, etc. See setting.
sidebinder A tree or log that binds from the side when being cut.
side block 1. Any haulback block along the side of the yarding, other than the corner or tail block, as used when yarding under a skyline. 2. See Sutherlin side block. 3. To move back the side blocks beyond the normal width of the skidroad in skyline logging to reach a few logs just outside the usual line of travel. Used also with a slackline.
side-blocking To chop a notch in a tree or log to prevent splitting when felling or bucking.
side boom See sheer boom.
side boss The foreman of a side. (P.C.F.)
side camp 1. A temporary small camp away from the main camp, such as for timber cruising. Also stub camp. 2. In turpentine orcharding, quarters for laborers at some distance from the main camp where the still is located. (S.F.)
sidecast To shove dirt over the side of the grade in road construction.
side delivery bunk A bunk built parallel to the deacon seat so that a lumberjack could crawl in from the side, in contrast to the muzzle loader which had to be entered in from the end. (L.S.) LL
side door pullmans Boxcars on the railroad. (L.S.) LL
side-heavy Said of a load of logs that is too heavy on one side.
side-hill gouger An imaginary creature found in the stories of the loggers. The legs are longer on one side than the other to keep it level on the slopes.
side-hill load A load with big logs on one side and small logs on the other. (P.C.F.)
side-hill mechanic In the late 1800s, a skinner on a steam or gas log-hauling tractor. (P.C.F.)
side-hill ranger A mythical animal in the lumber woods. (Maine)
side hill salmon 1. Deer killed out of season. 2. Bacon. (P.C.F.)
side jam A jam that has formed on one side of a stream, usually where the logs are forced to the shore at a bend by the current, or where the water is shallow or there are partially submerged rocks. (N.F.) See log jam. LL
side line 1. To roll a log around a stump or rock when yarding. 2. To bring logs up to the mainline using long chokers as short skidding lines. See side block.
side line man The man who carries the side lines from the main cable of a pull-boat and attaches them to the logs that are to be skidded. (S.F.)
side-line logs See siwash.
side-loader See bunks.
side mark See bark hack.
side-nip To pinch cable tightly against the side of the donkey drum because of a heavy pull or a foul lead.
side-notch 1. In crosscut saw felling or chain saw felling, to cut out part of the outside of a tree so that a short chain saw can cut through a tree. 2. To chop a notch in a tree or log to prevent splitting when felling or bucking. (P.C.F.)
side pier See wing dam.
side pole See sway bar.
side push The boss or foreman of a crew on one logging side.
side rod 1. The boss of a yarding crew or side, or one unit of a logging operation. 2. An executive's assistant.
side spool A gypsy spool on the outside frame of an old donkey engine. Also overhung gypsy. See outboard drum.
side sticks The boomsticks used at the sides of a log raft.
sideswiper A limb or tree which lashes out sideways when cut, or when knocked loose in yarding. See widow maker.

Weyerhaeuser Timber Company's Longview, Washington operation's No. 110. A **Baldwin** 2-6-6-2T **Mallet**, she was a **side tank** engine, as the water tanks were carried on either side of the boiler. Note the **rerailing frogs** on the pilot deck. *Courtesy of W. Clearman*

side tank A tank locomotive on which the water tanks were carried on either side of the boiler. See saddle tank.

side wheeler(s) 1. A low-geared engine. 2. Lice, crumbs or livestock. (S.F.)

sidewinder 1. A falling limb or snag that may fall from above in felling or yarding. 2. See sideswiper. 3. A Shay locomotive. 4. A man who cannot be trusted. 5. A tree which knocks down another in felling or in yarding.

sight 1. To look down the end of a gunstick to determine the line of fall of a tree. (Red.) 2. To look down the handle of a double-bitted ax to determine the line of fall of a tree. (P.N.F.)

sight gun A wooden triangle made of light sticks, used to determine the direction in which the tree will fall. The base of the triangle is set in the under cut and the top points along the line of fall. (P.C.F.) See gunstick.

sights A timber cruiser's compass bearing. (L.S.) LL

signal boy A young whistle punk. (P.C.F.)

signal man The man who signals the operator of a machine. See whistle punk.

signals Various signals are used in cable logging. Although there are variations in different camps, a more or less standard signal system is used in high lead and skidder operations. Hand signals are used around the landings, and whistle signals are used in yarding. Signals are also used in the sawmills.

silver tip 1. A young Shasta red fir. 2. A grey-haired man. (P.C.F.)

silviculture The art of producing and tending a forest; the theory and practice of controlling forest establishment, composition and growth.

simple locomotive 1. An articulated locomotive in which all of the cylinders receive steam directly from the boiler. 2. Any steam locomotive with multiple cylinders not compounded. See Mallet.

simplex donkey A donkey where all the gears are driven directly through a pinion on the crankshaft; has no train of gears.

singe To make a torch of a tree by burning off the dry moss. (P.N.F.)

single barrel An old fashioned wood splitter powered by a small steam cylinder. It was either fed off a steam line from the donkey or from a small stationary boiler.

single bit 1. An ax with one cutting edge. 2. A machine working at half its capacity.

single-bitted ax A pole ax. (L.S.) LL

single cord A pile of wood, 8 feet long, 4 feet high and 2 feet wide. (C.H.F.)

single coupler Single coupling grabs joined by a short chain or cable, used for fastening logs together. (App.) Also tail grab.

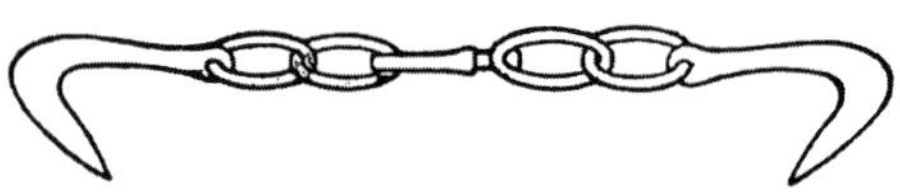

A **single coupler**. *From "Logging," Bryant, Wylie & Sons, 1923, Philip Schnell collection*

single-cut To float logs, usually cypress, one at a time from the woods to the front road. (S.F.)

single drum The early loading engine that did not have a haulback drum; the loading line had to be pulled out.

single drum wide face One of the earliest road engines; it had no haulback.

This view shows a **single drum wide face road donkey**. The **line horse** can be seen to the left, needed to take the line back out. Logs are being **parbuckled** onto a four-wheel **log wagon**. *D. Kinsey photo, Labbe collection*

single drum yarder The old donkey which was an improvement over the first donkey, the vertical spool donkey. A line horse carried the mainline back to the woods.
single-haul To yard logs direct from the stump to the landing with one donkey. Opposed to double and triple haul, using more donkeys. Also single-yard.
single-jack 1. A chopper who works alone in the woods. 2. To work as a single-jack: "I was single-jackin' yesterday." "I had to single-jack when my partner got hurt."
single line A long log chain used to send up logs when loading a sleigh without a top loader. (N.F.) LL
single line system An incline which simply lowers and raises a car on the end of a line; no use of endless cable or a block car. See one-way incline.
single slackline The usual slackline system of yarding, with the skyline passing from the drum out to a tail tree; opposed to the double slackline where the skyline was received through a block on the tail tree and brought back to a second skyline drum on the donkey.
single-speed A direct or compound geared yarder with only one speed; no low or high gear.
single stick boom A single pole used as a loading boom, from which the tongs were hung as in a hayrack boom, one from either side of the pole in order to line the log up with the car.

A **single stick boom**, devised by the Benson Logging & Lumbering Company at Clatskanie, Oregon, for loading long logs. The **ears** at the top of the boom were used to attach the various lines used in the loading process. *From the Benson collection*

single tong loader Another name for a heel boom loader.
singletree The bar to which the traces were fastened when skidding with one horse.
single-yard See single-haul.

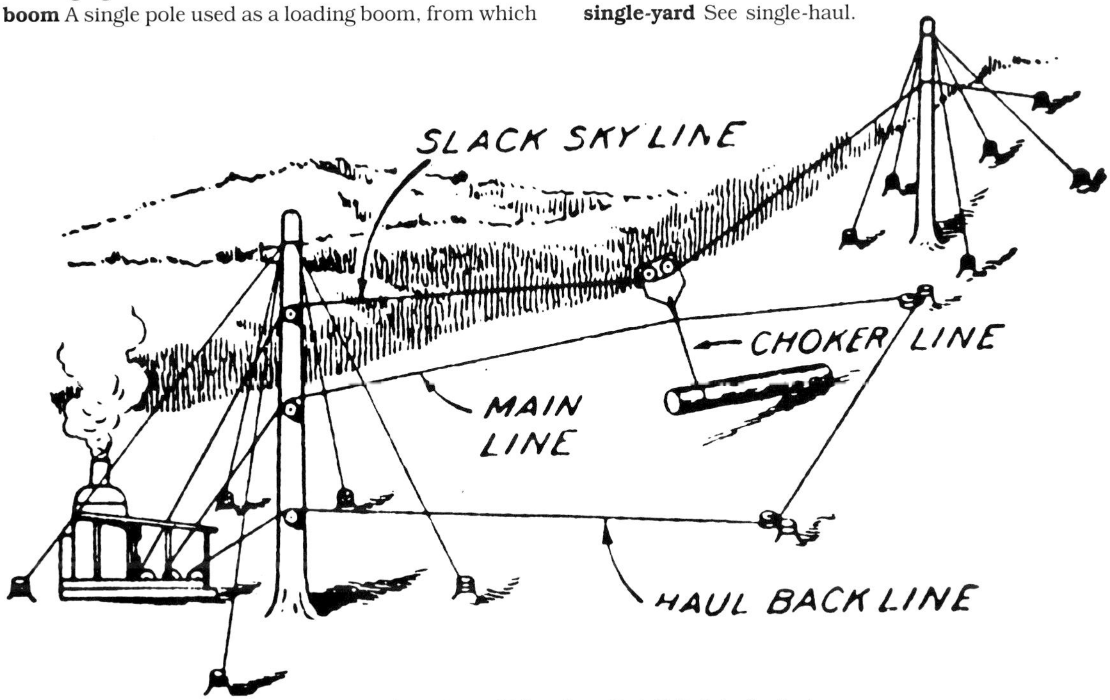

The **single slackline** system. *From "Logging, Principles and Practices," Brown, Wylie & Sons, 1934, Philip Schnell collection*

sinkage 1. The loss of logs in the water. 2. The loss of logs being driven downstream to the mills, causing shrinkage. (L.S.) LL
sinker 1. Usually a heavy butt log that sinks in a log pond. 2. Any log that sinks when dumped in the water. 3. A doughnut or a biscuit.
sinker loading A system of loading railroad cars in the ground lead days before power loading was developed. A track was sunk across the bottom of a log storage pond, and logs to be loaded were pike-poled into place as the car was hauled out of the pond.
Sioux pants Trousers of heavy wool having a very long nap. (L.S.)
sister hook A quick connecting link, two pieces pinned together so that they lap over when closed; will not open under a pull. (P.C.F.)

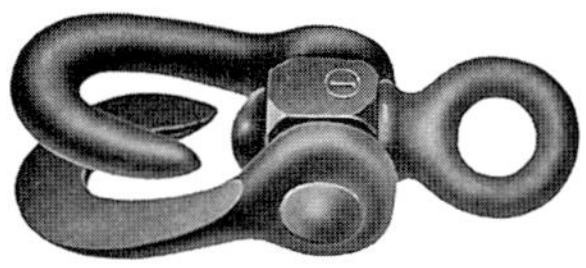

A **Swivel** with **sister hooks**. *From Young Iron Works catalog, Glen Comstock collection*

Sitka spruce (picea sitchensis) Also called tideland spruce and Western spruce, Sitka spruce is the largest of the whole family of spruce trees, attaining a height of two hundred feet with a fifteen-foot diameter. Its habitat ranges from Mendocino County, California, northward along the Pacific Coast to Kodiak Island, Alaska. It has thus a north and south range of 2,000 miles, but east and west the strip is a mere fifty-mile ribbon along the coast. It grows abundantly on the western slope of the Coast Range in the states of Oregon and Washington, and thrives in the fogs and rains from the Pacific Ocean. Posessing qualities of lightness and toughness, no other tree, probably, has been put to such varied uses, including aircraft construction, piano sounding boards, stringed musical instruments, boats, boxes and trunks, caskets, and paper pulp, to name a few.
siwash 1. An Indian (from the Chinook jargon, from French "sauvage"). 2. A line running around a stump or tree instead of a block. Also Oregon block, rub tree. 3. Slack or unproductive: "A siwash outfit." (P.C.F.) 4. To drag logs up to the main hauling cable. (L.S.) LL
siwashing The tangling of a line on a stump or other obstruction.
siwash logger A man who got his logs by beach-combing along the coast, not by logging. (P.N.F.)
siwash tree A tree used in place of a block to deflect a cable to change the direction of a line in yarding.
six feet and seven ax handles Said of a very tall person. (Maine)
six by six A truck with six wheels, two in front and four in back, with power applied to all of them. Originated in the Army and was borrowed by the loggers who found the multiple drive units ideal for logging.
six pack The six passenger pickup trucks often used by loggers and foresters. (P.C.F.)
sizzler A fry cook. (L.S.)
sizzle-sozzle A hard rain that floods. (S.F.)
Skagit A trade name for a gas donkey made by the Skagit Steel & Iron Works. (P.C.F.)
Skagit gadget A light mobile yarder and loader put out by the Skagit Steel and Iron Works. (P.C.F.)
Skandihoovian See Scandihoovian.
skeleton car, skel A connected log-hauling car on a railroad that has a center beam and bunks, but no floor. See connected truck.
skewers Round wooden pins of varying length and diameter, and pointed at one end.
skid 1. One runner of a donkey sled. 2. A log or pole, commonly used in pairs, upon which logs are handled or piled. 3. A log or pole laid transversely in a skidroad. 4. To haul logs over the surface of the ground to a landing. See snake, twitch and yard. 5. To reinforce a road by placing logs or poles across it. (N.F.) (App.) LL
skid camp A logging camp which was built on skids and could be moved easily to a new location. See skid shack.

A **Skagit** gasoline **yarder**. *From "Logging," Brown, Wylie & Sons, 1949, Philip Schnell collection*

A typical **skeleton car** used on Western logging railroads. *From Seattle Car & Foundry catalog, Labbe collection*

skid dauber See grease monkey.
skidded Said of a man who passed out from too much drinking. (West Virginia)
skidder 1. A big sled- or rail-mounted donkey that powered a skidder system, which could have three sets of engines if the loading and line-changing engines were mounted on the same unit as the yarding engine. Some of the skidders weighed over 300 tons. 2. In Mendocino County, California, a roading donkey or donkey used for ground logging. 3. A logger who kept care of the skidroad. 4. A teamster skidding logs out of the woods. (L.S.) LL
skidder car A car designed specifically for carrying a skidder and which was part of the machine. It consisted of the heavy frame and the trucks on which the skidder was mounted.

A steel **skidder car** built by the Pacific Car & Foundry Company at Renton, Washington, near Seattle. Equipped with **swivel trucks** that could turn a full 360 degrees, the car could be reversed by using two diverging tracks as shown here. The car, shown at the Willamette Iron & Steel Company shops in Portland, Oregon in 1923, is ready for a **skidder** to be mounted to it. *Willamette Iron & Steel Company photo, Labbe collection*

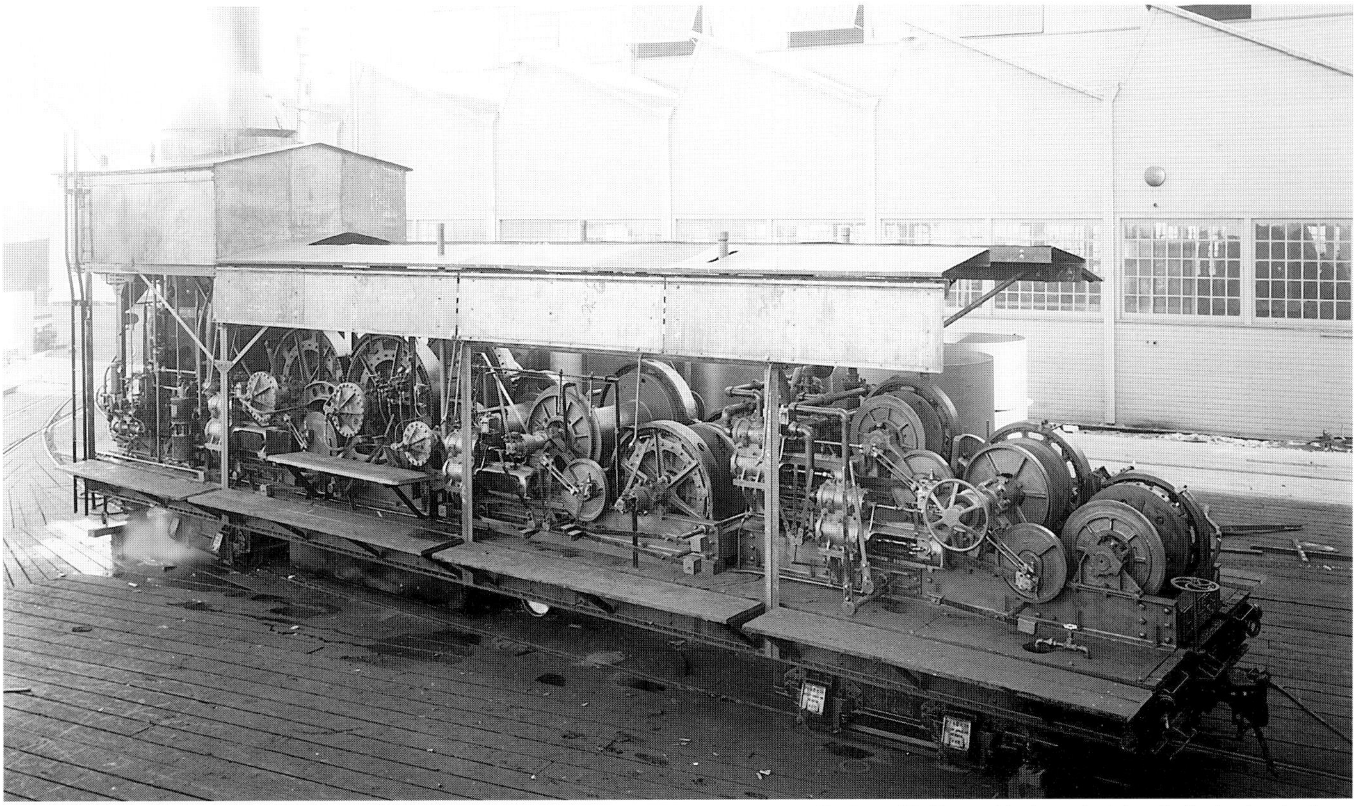

A Willamette tree-rig **skidder** being tested before shipment to the Nooksack Timber Company operations in Northwest Washington. The forward unit is a **duplex loader**, the drums being powered directly by steam engines. The rear drums are equipped with steam **frictions**. The machine is mounted on a **skidder car**, supplied by Pacific Car & Foundry, of Renton, Washington. *Willamette Iron & Steel Company photo, Labbe collection*

skidder carriage A skyline carriage designed specifically for a skidder system. See skidder system.

A **skidder carriage**. *From Young Iron Works catalog, Glen Comstock collection*

A **skidder head tree block**. *From Young Iron Works catalog, Glen Comstock collection*

skidder head tree block A skidding line (main line) block on a skidder system. The sheave grooves were designed to allow passage of the slack puller line swivel. See Shay swivel, slack puller, skidder system.

skidder punk A whistle punk on a skidder show. (P.C.F.) See whistle punk.

skidder receding line block A receding line (haulback) block on a skidder system.

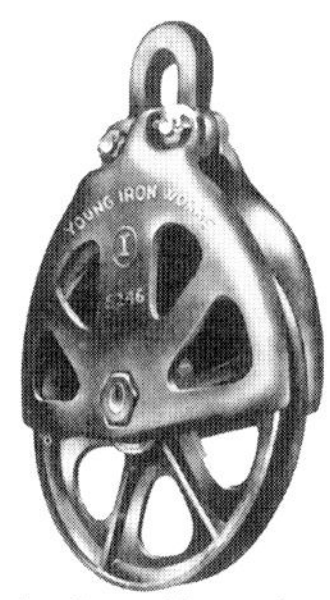

A **skidder receding line block**. *From Young Iron Works catalog, Glen Comstock collection*

skidder rigger The boss of the crew on a skidder. Could be compared to a hooktender on a high lead show. (P.C.F.)

skidder system A skyline system of logging which is the most complicated cable system to rig and uses the most men. Useful in rough country, this system depends on lots of power and mass production and uses a standing line, a big trolley and a slackpulling system which lowers the butt rigging to the crew. (P.C.F.)

skidder transfer block See transfer block.

skidding Hauling logs for short distances with oxen, horses or mechanical means. See yarding.

skidding chain A chain usually 16 feet long with a round hook on each end attached to a log when skidding or dragging a log. (L.S.) LL

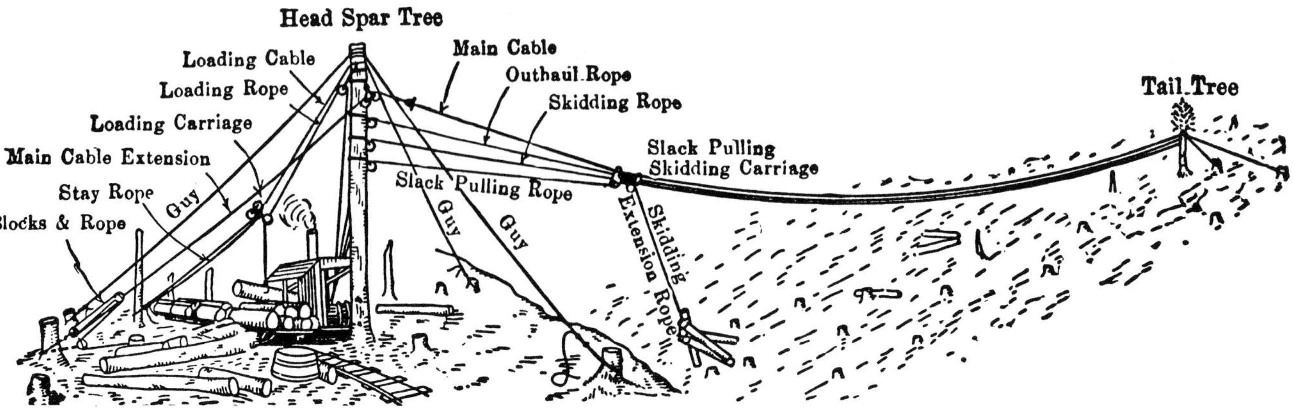

The **skidder system**. *From "Logging," R.C. Bryant, Wylie & Sons, 1923, Philip Schnell collection*

skidding extension rope A wire rope on the end of the main skidding line to which chokers were attached in the early days of skidder logging.

skidding grab A grab used in snaking logs with horses; consists of two hooks joined by short lengths of chain, with a ring in the middle.

skidding hooks See skidding tongs.

A **skidding grab**. *From "Logging," Bryant, Wylie & Sons, 1923, Philip Schnell collection*

skidding leverman The man who operates the skidder. (P.C.F.)

skidding line drum The drum that holds the line which hauls the trolley in a skidder system.

skidding pan A plate of heavy steel, rounded up in front, and placed under the front ends of logs being skidded to prevent the logs from digging into the ground. Used mostly in skidding with a tractor. Also pan.

skidding sled See dray.

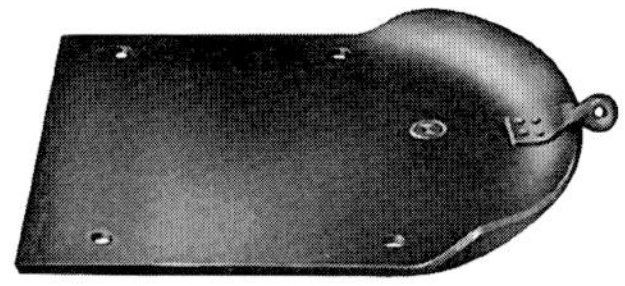

A **skidding pan**. *From Young Iron Works catalog, Glen Comstock collection*

skidding team The team used in dragging logs out of the woods. (N.F.) LL

skidding tongs 1. A pair of hooks attached by links to a ring used for grabbing and holding logs while skidding them. Used in the Pine Country instead of chokers. 2. The early Lidgerwood skidders used tongs instead of chokers on medium-sized logs. Also called chain grapples, coupling-grabs, dogs, grabs, grips and skidding hooks in the Lake States. LL

Skidding tongs. *From Mallory Catalog No. 12, 1925, Glen Comstock collection*

skidding trail See gutter road.

skid flat See flat spot.

skid grease 1. In the bull team days the grease or tallow that was used on the skids. 2. Butter.

skid greaser, skid monkey The man or boy who swabbed tallow on the cross skids. As he walked, he dusted the grooved skids with a broom in one hand and swabbed the tallow with the other.

skid oil Oil used instead of skid grease. (P.C.F.)

skidroad 1. A road on which logs are skidded. In the 1800s, bulls and horses dragged logs over skids laid on the road. At the present time, logs are skidded by a cat, and the skids have disappeared. Also run, snake trail. 2. The lower or cheaper part of town frequented by vagrants and alcoholics. The term originated from Yesler Way in Seattle, Washington, a former skidroad which became an area frequented by out-of-work loggers and bums. In recent years the term has been corrupted to skidrow.

A **skid shack** being loaded onto a **flat car** on the operations of the West Side Lumber Company. West Side operated an extensive 3-foot gauge logging line out of Tuolumne, California. *Philip Schnell collection*

A **skidroad** on the operations of Henry Colvin, near Marshland, Oregon. *Courtesy of Marshall Morgan*

skidroad queen A whore.
skidrow, skid row A "corruption" or shortened form for skidroad, popularized by ignorant reporters. This new form has spread throughout the country, much to the dismay of the Pacific Northwesterners who emphatically state that there is no such word and that the term was and always will be skidroad.
skids 1. The logs that were placed across a road to keep the logs from nosing into the ground. 2. The logs at a landing on which logs are cold-decked before loading them onto a truck or railroad car. 3. Short poles laid against a car to aid in topping off a load.
skid shack A bunkhouse mounted on skids for easy moving to a new location. See skid camp.
skid tender An early logger who maintained the skidroad. (P.C.F.)
skid timbers 1. Heavy timbers used at landings to hold logs before they are loaded. (L.S.) LL 2. Timbers used as skids when loading logs with a crosshaul. (P.C.F.)
skid up 1. To level or reinforce a logging road by the use of skids. 2. To collect logs and pile them on a skidway. (L.S.) LL
skidway 1. Another term for skidroad. 2. Another term for skids in which two skids are laid parallel at right angles to a road on which logs are piled for loading. The Canadian lumberjacks contributed the skidway to logging in the Lake States. 3. A loading dock built of logs. 4. Sometimes used for a pile of logs waiting to be loaded. (L.S.) LL (Canada) See rollway.
skidway man 1. A chaser at a landing. 2. A handyman.
skidway tongs Log grabs. See skidding tongs.
skimming off the cream Originally said of cutting only the choicest white pine near the drive river banks with logs at least 18 inches in diameter. (L.S.) LL
skin around a tree To move a yarder from one side of a tree to the other in order to log the other half of the setting. (P.C.F.)
skin cat To work as a tractor driver. (P.C.F.)
skin'er back To go ahead on the haulback to take the butt rigging back to the woods. (P.C.F.)
skinner 1. In the 1800s, a mule driver, a freighter or a horse teamster. 2. A machine operator, usually a tractor driver. Cat skinner is the popular term. (P.C.F.) 3. A teamster who has abused his horses and made them thin. (Maine)

A **skidway**, or loading dock, along a logging railroad in West Virginia. The logs in the structure were later loaded and hauled away. *From "Logging," Bryant, Wylie & Sons, 1923, Philip Schnell collection*

skin out To get out the timber. (P.C.F.)
skin timber, squaw wood 1. Pitchy wood used for starting fires, especially in the early days of wood stoves. 2. An attractive woman. Also squaw timber.
skipper 1. A hammer used to drive a dog into logs. It was a sledge on one end and the other end could be used to drive a hook out of a log. (L.S.) See grab skipper, grab maul. 2. One of the poles placed on a skidroad to facilitate the haulage of logs. (App.) 3. A boss. LL
skipper road 1. A skid road on which poles are placed zigzag across the road, the angle between skids being about 60 degrees. 2. A road on which poles are placed transversely at intervals of from 4 to 6 feet. (App.)

A **turn** of logs at the **dump** along a **skipper road** in West Virginia. The logs are fastened together with **single couplers**, or **grabs**. *From "Logging," Bryant, Wylie & Sons, 1923, Philip Schnell collection*

skip the grabs To release the skidding grabs from the log by means of a grab skipper. (App.)
skookum A Chinook jargon word meaning strong, good or powerful. (P.C.F.)
skookum block 1. A trade name for a block. 2. A big, heavy mainline block. (P.C.F.)
skookum camp A good camp to work for. (British Columbia)
skookum chuck Dangerous, rapid waters, such as falls or tide rips. In the Chinook jargon, "skookum" meant strong, and "chuck" meant water, a river or a stream. (P.N.F.)
skoovie A snubbing hitch to stop and tie moving rafts or boats on the bay or in log ponds. (L.S.)
skulch Refuse. (Maine)
skunk 1. A gasoline car in the woods. 2. A gasoline schooner.
skunk-bit Said of one who is very drunk. (P.C.F.)
sky bird A top loader. (S.F.) See sky hooker.
skybound 1. Said of a tree that will not fall after it has been cut. 2. A falling tree hung up in another. 3. Butt rigging fouled in the high lead block at the top of a spar tree. 4. Rigging that is motionless in the air.
sky hook 1. A term used whenever difficulty arises: "We need a sky hook." meaning to drop a hook down from the sky to solve the problem. 2. A practical cable system for carrying logs or other materials on a self-powered carriage, where operator, engine and load all ride along on double cables. The machine was invented by Phil Grabinski of Portland, Oregon. (P.C.F.) 3. The term has been used to describe balloon logging. (P.C.F.)

A **skyhook** in operation for the Glen Corning Logging Company out of Deming, Washington. *From Pacific Logging Congress "Loggers Handbook," 1948, Philip Schnell collection*

sky hooker 1. A head loader. (P.N.F.) 2. A member of a loading crew who stood on the logs as they were sent up. (L.S.) LL
sky-hooking The worker arranging logs on top of a car or sleigh. (L.S.) Also top-loading.
skyline 1. The main heavy steel cable between the head tree at the landing and tail tree or stump back in the logging area. Also track cable, trackway cable. 2. A logging system in which a traveling block or carriage rides on a skyline. The line which does not move is attached to a donkey drum or an anchorage at the landing, and to a tail tree or tail hold on the far end. In some systems it is fixed, but in the slackline system it can be lowered to the ground for loading, then could be raised up to free the logs from the ground when pulling in. The first skyline on the Pacific Coast was supposed to have been rigged by Bob Barr at the Bridal Veil Lumber Company in Oregon in 1899.
skyline block 1. A large block on a head tree through which the skyline passes. 2. A heavy block used as a carriage in a tight line system of logging.

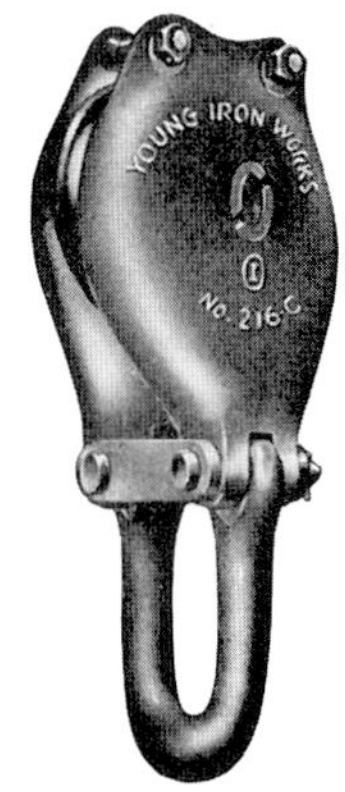

A **skyline block** for use with the **North Bend system**. *From Young Iron Works catalog, Glen Comstock collection*

skyline carriage A carriage specifically designed to be used on a skyline system.

Close up look at a **skyline carriage**, shown here on the Clark & Wilson Lumber Company operations in the hills north of Portland, Oregon. *C. Kinsey photo, Labbe collection*

skyline logging Logging with a cableway skidder. (P.C.F.)
skyline sheave A hardwood block grooved to fit the line; it was fixed and did not turn. See tree jack.
skyline stub A short piece of skyline run through the tail tree and made fast to a tail hold when the tail tree is rigged before the main skyline is set up. See admiralty shackle.
skyline tail stump The stump or stumps used to anchor the end of a skyline when a tail tree is not used.
skyline yarder A four-drum donkey used in skyline systems. One drum was used for the skyline, one for the mainline to the carriage, one for the haulback and one for the strawline. (P.C.F.)

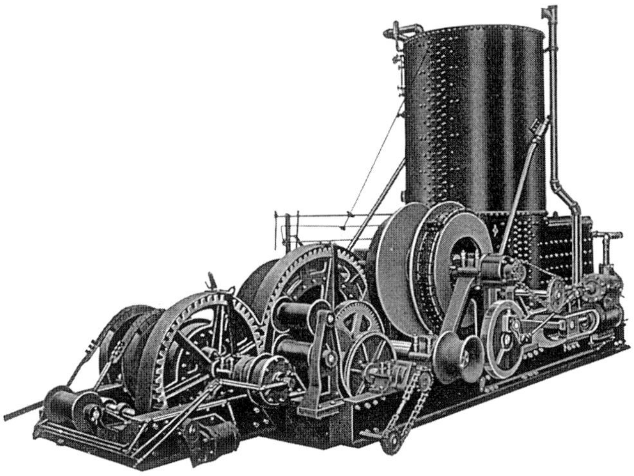

An S&W (Smith & Watson) 12x13 convertible, 4-drum **skyline yarder**. *From Smith & Watson Iron Works Logging Equipment Catalog No. 9, Merv Johnson collection*

sky-loader The man who loaded logs at the skidway and on the load. His job was dangerous and required much skill. (L.S.)
sky piece 1. The top log of a load in a horse camp. 2. A hat. LL
sky pilot 1. A preacher who walked from camp to camp in the late 1800s. 2. Any logger who takes religion seriously. (P.C.F.)
sky route The way to heaven. (L.S.) LL
slab To rip a big chunk or slab off the side of a log by careless use of the tongs in loading. Also caused by the use of old-style end hooks. Cup hooks helped to prevent slabbing. See end hooks, cup hooks.
slabbed 1. Said of a log with a chunk broken off, caused by careless felling, bucking or loading. 2. Said of a log which has been slabbed on the carriage. See slab. LL
slab streak In turpentine orcharding, a streak so chipped that the entire surface of the newly exposed wood is turned outward toward the direct rays of the sun. (S.F.)
slab tie The third tie made from a stick of timber too small to make four ties and too large to make two ties. (S.F.)
slack 1. A line without any tension. 2. On railroad cars the play between couplers on cars which takes up the shock in starting and stopping. 3. To slack a line to remove tension.
slacker A slackline yarder. See skyline yarder. (P.C.F.)
slacker show The slackline system of yarding with the skyline passing from the drum out to a tail tree. (P.C.F.)
slackline artist A disparaging term for a donkey puncher who runs his machine without keeping the lines taut. (P.C.F.)
slackline carriage A carriage designed specifically for a slackline yarding system.
slackline system A system of overhead cable logging in which the mainline and skyline are lowered to the ground to bring in the logs. Also slack skyline.
slackline yarder A special donkey used with a slackline system. (P.C.F.) Also called a slacker. See skyline yarder.
slack off To lower a line.
slack-off time Quitting time. (P.C.F.)
slack on the pin To back up to release tension on couplers so that a car or trailer can be uncoupled. Also slack the pin.
slack puller 1. A logger who pulls slack on a skidder. 2. The slack-pulling drum on a skidder and the series of blocks through which the slack-pulling line is attached to the haulback. When pulled, slack is hauled on the tong line, lowering it to the choker setters. See skidder system. 3. A log or chunk to counterweight the tongs on a loading rig. See squirrel. 4. A woman of easy virtue.
slack rope system 1. An older term for the high lead. 2. A slackline system. (P.C.F.)
slack skyline See slackline system.
slack water 1. In river driving, the temporary slackening of the current. 2. Low water or dead water. (N.W.) LL
slant dam See rafter dam.
slap-happy loading rig A type of crotch line loading outfit. (P.N.F.)
slapjack Bread cooked in a frying pan in the early 1800s. Cook stoves and box stoves were introduced in the logging camps in the 1850s. (N.W.)

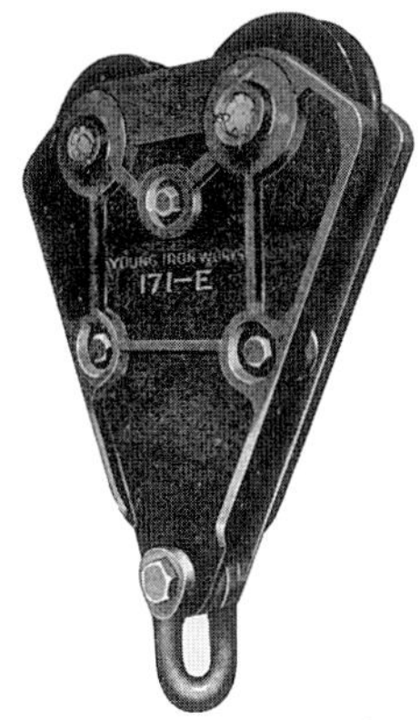

A **slackline carriage**. *From Young Iron Works catalog, Glen Comstock collection*

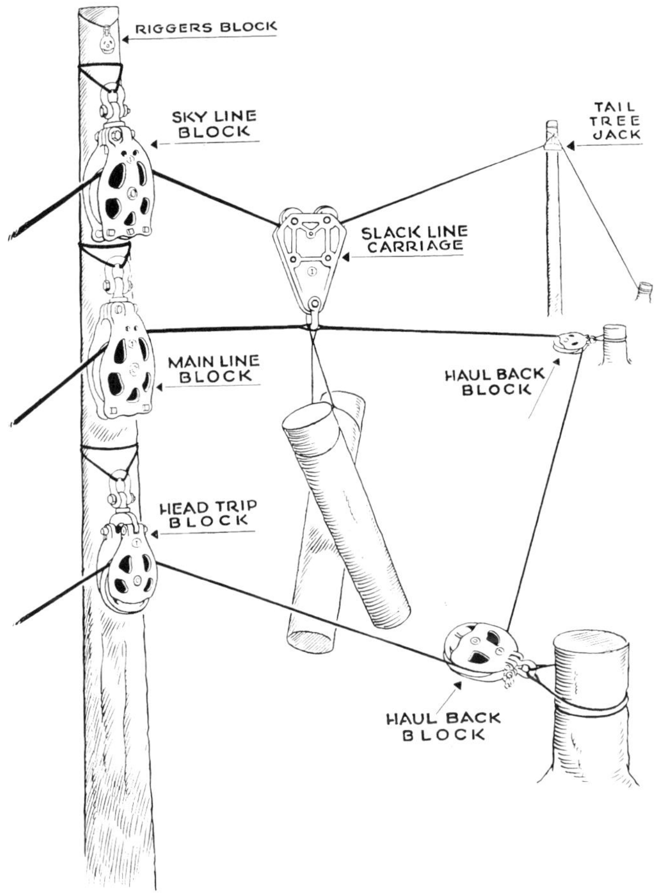

The **slackline system**. *From Young Iron Works catalog, Glen Comstock collection*

slash 1. Branches, bark, chunks, cull logs, tops, uprooted stumps and broken or uprooted trees left on the ground after logging. 2. A logged off area. 3. To cut down trees in order to make a right-of-way, or for other clearing.
slash burn A fire set in logging slash, usually in the fall, to get rid of dangerous fuels before the next fire season.
slasher A man who cleans up the slash after a logging operation.
slashings 1. The area of recently cut timber. 2. A land area full of slash. See slash.
slats A man's ribs. (L.S.) LL
slave Any working man.
slave driver A high-ballin' foreman.
slave market The employment office. LL
sled 1. The frame used as a foundation for a donkey or pile driver, consisting of cross timbers bolted to two heavy log runners. See donkey sled. 2. A logging sled to haul logs on snow or ice. (N.F.) See sleigh.
sled builder The skilled workman who used to make the donkey sleds. (P.C.F.)
sledge hammer two-speed A convertible compound-geared two-speed yarder built by Smith & Watson Iron Works of Portland, Oregon. In order to change from simple gearing to compound gearing, the machine had to be stopped. Keys on the pinion gears were removed, and the pinion gears were slid on their shafts (sometimes with a sledge hammer) to align with the desired gears. These machines were designed with the intention of changing the gear ratio according to the type of logging encountered.

The gear arrangement of a Smith & Watson two-speed convertible yarder, often called a **sledge hammer two-speed**. *From Smith & Watson Iron Works Catalog No. 9, Merv Johnson collection*

sled irons 1. The iron work used to bolt the cross timbers to the two huge timbers of a donkey sled. 2. All the iron work used in sled making. (P.C.F.)
sled plate A heavy plate that was fastened on a donkey sled for moving the donkey. (P.C.F.) See bridle iron, log plate, rope plate.

A **sled plate** for a **donkey sled**. *From Mallory Catalog No. 12, 1925, Glen Comstock collection*

sled runners The two large timbers on a donkey sled. (P.C.F.)
sled shoe A curved metal shoe that was fitted over the nose of a sled runner to protect it when moving a donkey.
sled tender 1. Similar to the P. F. Man, he accompanied the turn of logs to the landing, collected the dogs and grabs and returned them to the yarder. (P.C.F.) Also frogger. 2. One who assists in loading and unloading logs or skidding with a dray. (N.F.) LL
sleeper 1. A fire, apparently out, but comes to life again. 2. A misfire when setting off a blast. 3. An idler car used under the overhang when hauling extra logs on railroad cars shorter than the logs.
sleeping shanty A bunkhouse. (L.S.)
sleeping tent A tent set up on a river bank for the river drivers during a log drive. It held from eight to twenty men, sleeping on balsam boughs laid on the ground. (L.S.) LL
sleeping wanigan A houseboat used solely for sleeping during a river drive. (L.S.) LL
sleeve 1. A wide-throated shackle with a push pin, used to hold guy lines on a spar tree in some types of rigging. 2. An early rig to replace the butt hook, a sleeve with a link on one side to which the butt chain was attached. The sleeve slide along the mainline and could be spotted alongside any log to be yarded.
sleigh A logging sled for hauling logs on snow and ice. (N.F.) Although, as a rule, called a "sleigh" west of Michigan, "sleigh" and "sled" were used interchangeably. As a rule "sled" was used in the North Woods. Also logging sled, turn sled, two-sleds and wagon sleds.

A **sleigh** of Moore Keppel & Company at Port Wing, Wisconsin, about 1903. The single team of horses is handling more than 16,000 feet of logs over the frozen road. *Courtesy of W. Casler*

sleigh iron All the various pins, bolts, rods, plates, washers and other iron parts used in construction of logging sleighs. (N.F.) LL
sleigh shoes Heavy, iron, curved facings on the bottom of sleigh runners. (N.F.) LL
slept-in Said of Sunday, the day of rest in camp. (L.S.) LL
slew skid A skid set at an angle on one side of the road to warp logs around a bend in the skidroad, and to prevent the turn of logs from sliding off the skids. (P.C.F.)
slick A razor-sharp chisel, two to four inches wide. The blade had a handle twenty-five inches to thirty-five inches long. (L.S.) Also paring chisel. LL
slick as a chicken's tit Said of something very slick. (L.S.) LL
slicker than a schoolmarm's leg 1. Said of a surface so smooth that cork shoes will not hold. (P.C.F.) 2. Said of something very pleasing or pretty. (Maine)
slick-shod Said of a logger who comes to work with smooth shoes instead of corks. (P.C.F.) See barefoot.
slide 1. A rough chute for transporting logs. 2. A rough road cleared down a steep hill or a crude trough made of parallel poles. 3. A derrick for loading logs onto cars. (L.S.) LL See slide loader. 4. A V-shaped trough built of logs used to transport logs down a slope. (West Virginia). Also dry slide.

A **slide ass**, or **slide loader**, of the Cascade Lumber Company at Casland, near Yakima, Washington. These loaders were designed to slide from car to car as it completed a load. Logs are being handled with **end hooks** with **tag lines**. The loaded logs are secured with **car stakes**. *Brown photo, from Labbe collection*

slide ass A common term for the slide loader, used chiefly in the Northwest pine areas of Washington, Oregon, Idaho and Montana. The slide loader was far more popular in the East than in the West. It was the main loading machine in Pennsylvania, West Virginia and in the South. The reason was that the small timber and short logs were logged to trackside, thus eliminating the usual landing. The train loaded as it passed along the track, and the horses had a much shorter haul from the woods. See jammer, kahallek special, rapid loader, slide loader.

slide ass jammer A jammer that pulled itself along. (L.S.) LL

slide gate A dam gate which slides up and down, hoisted open against the pressure of backed-up water by a donkey engine. (P.C.F.)

slide hook A choker hook. (P.C.F.)

slide loader Any steam loader designed to be carried on the cars. As each car was loaded, the loader would slide ahead along the empty cars. Used chiefly with short logs, such as in pine and hardwood logging. See jammer, kahallek special, rapid loader, slide ass.

slide tender One who maintains a slide or log chute. LL

slide valve flunky A waitress. (P.N.F.)

sling 1. A strap hung from the tong and used to loop around a log too large to be picked up by the tong. 2. Oxen were often shod in a sling. See strap.

sling chains In early skidder operations, short chains used to pick up a number of logs instead of hooking one big one with tongs.

slingin' head To work as the head rigger or head slinger: "I used to sling head." "I was slingin' head at McKay's." (P.C.F.)

sling-rigging To work as a rigging man. (P.C.F.)

slip 1. Any dock that projects out into the bay or river. 2. An early dirt scoop pulled by a horse. 3. An early name for the pig which carried the tools back to the woods. (P.C.F.) 4. A part of a log sorting works. 5. The edge of a rollway. 6. To make a spar tree out of a standing tree, then to move or to slip it to a better location nearby, instead of raising a tree at this spot. After the guy lines are rigged, the tree is sawed off and slid along in an upright position to the new location. Also jump a tree. (P.C.F.)

Loading a big **fir** log on a **motor truck** in the Northwest in 1939. The log was too large for the **loading tongs**, so **straps** were hooked onto the tongs to make a **sling**. *From "The Timberman" collection*

slip grab A pear-shaped link attached by a swivel to a skidding evener or whiffletree, through which the skidding chain is passed. (L.S.) Also grab link. LL

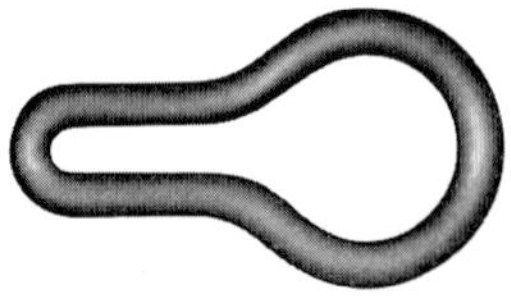

A **slip grab**. *From Young Iron Works catalog, Glen Comstock collection*

slip grab hook A hook used to choke a wire rope sling, as around a bundle of ties.
slip hook A grip hook used on a cable which releases when the line is slacked off.
slipper A bar used to couple two logging sleighs. (N.F.) See gooseneck and rooster. LL
slip skids See glisse skids.
slip-tongue wheels The big wheels used for yarding logs in the Pine Country. A 36-foot long tongue was slipped through hounds on the axle. As the animals pulled the tongue forward in starting the load, it also pulled ahead an iron rod running back to a short lever called a Jacob staff. As the staff swung forward in an arc, it turned two eccentrics from which hung the chains holding the logs. The chains were thus pulled up, lifting the logs. The iron rod was then latched to secure the load in place. Going downhill the skinner could unlatch the rod, or let the wheels run ahead on the tongue. This reversed the action and the logs were let down to act as a brake. Four horse teams were generally used with the large slip-tongue wheels, up to 12 feet in diameter.

Skidding logs in Central Oregon with horses and **slip-tongue wheels**, a method generally used on the flat ground of the Oregon and California pine regions until the advent of the **tractor**. The **teamster** rode the left wheel horse and operated the **Jacob's staff**. *From "The Timberman," October, 1949, Philip Schnell collection*

sliver cat 1. Splinters on stumps standing up two to four feet which whistle in the winter wind. Also screech cat. 2. A splinter, usually about a yard long, attached to a pole by a piece of leather. The lumberjacks twirled this splinter by swinging the pole rapidly to make a howling, screaming sound that frightened greenhorns. It was often used while icing roads at night. (L.S.) LL
slivers The soft inner bark of trees, especially birch, which children often eat as a delicacy. (Maine)
Sloan's linament A favorite linament. LL
slobber stack Said of a saturated steam engine by trainmen. Such engines, without the later superheaters, would blow condensed steam or water from an over-filled boiler out the stack. Also soak.
sloop Two runners with a bunk to haul logs out of the woods. (N.F.) See dray. LL
sloop logs To haul logs down steep slopes on a dray or sloop equipped with a tongue. (N.F.) LL
slop 1. Water loss out of a flume caused by current, curve or bunching and bumping of logs. 2. Excessive movement in machinery, due to wear or poor fit. (P.C.F.)
slopover A wreck caused by the dispatcher failing to hold one train for a meet. (P.N.F.)
slop trough A filthy cookshack. (P.N.F.)
slough pig A poor river man who is sent back to work logs out of the sloughs. (N.F.)
slow as an ox Said of a very slow person.
slow bell show A poor bunch of loggers; the opposite of a highball show. (P.C.F.)
slow elk A cow. (P.N.F.)
slower'n a jill-poke Said of a man who is extremely slow. (Maine)
slue line A line used to swing a loading boom hung from a spar tree. (P.N.F.)
sluice 1. A log flume. 2. A short stretch of heavy flume built to carry logs around a bad stretch of river. 3. To transport logs by water. 4. To sort logs into pockets from the bull pen where they are dumped. 5. To smash something. 6. To flood something.
sluice crew The crew who work on the booming grounds, moving logs into pockets.
sluiced Said of a team that has accidentally gone off the road and piled up. (L.S.)
sluice dam See splash dam.
sluiced horse A horse that has been injured by having a loaded sled run onto his heels as a result of poor braking on down grades. (N.S.)
sluice gate, sluiceway The main gate of a splash dam which was fixed to be kicked open in a short time. Also sluiceway. (N.F.)

The **sluice gate** has been opened on this **splash dam** in McKean County, Pennsylvania, and the logs are started on their way down the stream. *Courtesy of W. Casler*

sluicing Guiding logs through the sluice gate of a dam. (L.S.) LL
slumgullion 1. Very loose hash. 2. Any unappetizing dish.
slush Coffee. LL
slush cook An assistant cook. (L.S.) LL
slusher A horse-pulled earth scoop with twin handles manipulated by the teamster on road and railroad right-of-way construction work. (L.S.) See Fresno. LL
slush-handler A cookee. (L.S.) LL
small bore A little man who acts like a big man. (P.C.F.)
small game Bedbugs. (L.S.) LL
small saw size Pole timber trees from 11 inches in diameter, 4.5 feet above the ground. (Idaho)
small square See government square.
smear Butter. (P.N.F.)
smells like a shithouse turned over Said of something very smelly. (P.C.F.)
smells like the bottom of a log pond Said of something that smells horrible. The worst odor is a drained pond that has been used for hemlock. (P.C.F.)
smilo Hard liquor. (L.S.) LL
Smith & Watson See S&W.
smoke box The part of a boiler where the smoke collects before going up the stack; the front end of a locomotive boiler, or the conical cap of a steam donkey boiler; designed to induce draft.
smoke boy A fireman on a locomotive, especially one that is not steaming properly.

smoke eater A fire fighter. (P.C.F.)
smoke hole A large, square hole in the roofs of the early camp cambooses, directly over the fireplace below, that permitted the smoke of the open fire to escape. (L.S.) LL
smoke meat An Indian. (P.C.F.)
smoke pole A rifle. (P.N.F.)
smoke signals 1. Smoke signs that another train might be coming. This is what a man looked for when he took a logging train out without authority from the dispatcher. If he saw smoke, he ducked into a side track or backed up quickly to keep out of trouble. 2. Some smaller outfits did not use a dispatcher, and the trains ran by smoke signals. (P.C.F.)
smokestack reamer Similar to a left-handed monkey wrench, and what the greenhorn is sent after. (P.C.F.)
smooth as a schoolmarm's thigh Said of an exceptionally smooth board or other smooth object. (L.S.) LL
smooth-faced Said of a donkey drum with a barrel not grooved to lead the rope as it is spooled on.
snag 1. A dead standing tree. 2. A standing section of the stem of a tree broken off at a height of 20 feet or more. See ram pike. 3. A partly sunken tree or log in a river. 4. To fell snags. 5. To clean up a river.
snag boat A boat designed to raise snags and deadheads in navigable streams; primarily to clear streams for river boats. (P.N.F.)
snag faller A man who cuts snags; not part of the usual felling operation and generally paid on a bushel basis. (P.C.F.)
snag pusher Framework out in front of a cat to push over pine snags. Sometimes the framework would be attached to an arch. (P.C.F.)
snake 1. A drag at the end of a cable or chain. 2. The movement of a saw weaving up and down to make an uneven cut when felling timber. 3. A wavy cut in sawing. 4. To skid logs over a crooked road. 5. To wind a line around stumps and logs when dragging them out of the woods. 6. To horse-drag a log to its destination without using a sleigh. (L.S.) LL
snake block A clamp, shaped like a logging block, used on a guy line in order to pull slack and tighten the line. (P.C.F.)
snake hunter A person who came to a lumber camp wearing city clothes and pointed shoes. (L.S.) LL
snake in logs To yard logs.
snake oil A drink of unknown origin in prohibition days. (P.C.F.)
snake room A room off a saloon, usually two or three steps down, into which a barkeeper or bouncer could slide drunk lumberjacks headfirst through swinging doors from the barroom. (L.S.) LL
snake ranch A bootlegging joint. (P.N.F.)
snake trail A skidroad. (N.F.) LL
snakey Said of a log being dragged, i.e., a "snakey movement."
snaking line 1. A mainline. 2. A skidding line on a tree-rig ground skidder. (P.C.F.)
snaking machine A donkey used in ground lead yarding.
snaking system A system of power logging in which the main cable is returned to the woods by an animal. (P.C.F.)
snap To give a quick pull on a line in order to jerk a log over a stump or to avoid a hang-up.

Yarding a large log, using a **block-and-a-half** purchase. The log has been **sniped** to prevent **hang-ups**. Scene is of Oak Point Piling & Lumber Company operations, on the Lower Columbia River. *Courtesy of R. Reid*

snap-back team A team used to pull the big wheels back over the load when yarding with a slip tongue. One snap-back team usually worked with two four-horse wheeling teams, and bunched logs for them when not helping in the loading.
snap line 1. A tight line holding the A-frame of a loading rig to the lean, preventing it from tipping over backwards. 2. A line attached beneath a boom which prevents the boom from flying up and over the operator when a load breaks loose or the line breaks. 3. A line to keep the machinery erect.
snap truck A second truck, or helper, in the early days of truck logging that was used to help get a truck over a hump.
snatch 1. The vagina. 2. A woman. 3. A whore.
snatch block A block which can be opened on one side so that a cable can be laid in the block. Also Bouse block, open side block. See team snatch block.
snatcher A man on a lumber raft who manipulated the snatch pole. (L.S.) LL
snatch pole A long pole on a lumber raft to lift the front raft out of the water in going over falls. Also used to hold onto in rough water. (N.F.) LL
snatch team See tow team.
snatch yoke A yoke on a block which may be opened up to permit a line to be laid in on the sheave.
snib 1. In river driving, to be carried away accidentally on purpose on the first portion of a jam that moves. 2. To ride away from work under the guise of being accidentally carried off. (L.S.) LL
snipe 1. A bevel hewed on the ride side of the end of a log, making it easier to pull over the skids. Also barked. 2. An angle chopped or sawed on the undercut side of a stump to help the tree slide off and clear an obstacle. 3. Trim allowance on a log. 4. A section man (S.F.). 5. To round off the end of a log in order to facilitate skidding in the early days of logging. 6. To chop or saw an angle on the undercut side of a stump to help the tree slide and clear an obstacle in felling.
snipe for the ride To put the snipe on the side of the log which would ride on the bottom, saving the work of sniping a bevel around the entire end of the log. (P.C.F.)
sniper 1. In the early days, the logger who sniped the ends of logs for skidding. 2. The logger who flattened one side of a log (sap side) so that it would ride easier on the skidroad.
snooper The man who works hard to find logs, getting the edge on other buyers. (P.N.F.)
snoose, snuss 1. The once very popular Copenhagen snuff. Some loggers and mill workers moisten the snoose with whiskey to improve the flavor. Also condition powder, idiot powder, spit-or-puke. In the Lake States also called Swedish brain food, Swedish condition powder and Scandinavian dynamite. 2. Unusual power, as in "She had plenty of snoose."
snoosern A Scandinavian. (P.C.F.)
snort A drink of whiskey.
snorting pole 1. A pole that was nailed across the foot of the loggers' bunks to push against in order to get the cramps out of the loggers' legs. (P.C.F.) 2. A sapling laid under the hay in muzzle-loading bunks to keep the sleepers apart. (L.S.)
snotty end 1. The business end of a machine. 2. The worst part of a job.
snow a road To cover bare spots in a logging road with snow to facilitate the passage of sleighs. (N.F.) LL
snow break A swelling in a tree indicating a break when the tree was small, often caused by snow at higher elevations; almost always rotten inside.
snow brush Any type of brush. Also buckbrush.
snow-down Trees that have been knocked down by a heavy snow. (P.N.F.)
snow knocker A small hammer which has a pick on one end of the head. (N.F.) See ball hammer.
snow on the roof Said of grey hair, meaning that the man is too old to hire. (P.C.F.)
snow plow A buxom woman. (P.C.F.)
snow roller A huge, heavy roller used to pack down snow on roads. (L.S.)
snowshoe tracker A man with big feet. (P.N.F.)
snow show A very cold job. (P.N.F.)
snow slide A temporary slide on a steep slope, made by dragging a large log through deep snow when it was soft or thawing; when frozen solid it could be used to slide logs to a point where they could be reached by sleighs. (N.F.) LL
snow snake See travois.
snub 1. A plank slanted across a flume to kick out logs floating down the flume. 2. A patented friction device used as a brake on small inclines. See Barienger snubbing device. 3. To hold back logs, sleighs, cars on an incline, with a cable. Also breast line. 4. To check the movement of anything with a line wound around a stump, tree or braking device.
snubber 1. An engine used to brake log cars going down an incline. See lowering engine; incline. 2. An engine used to hold back logs going down a pole road or skidroad. 3. Sometimes just a drum and brake used for snubbing on an incline. 4. A device for checking the speed of a sleigh down a steep hill by winding a rope around a stump and letting it down slowly. (N.F.) 5. An old biscuit. (N.W.)

A snubber. *From Willamette Iron & Steel Works catalog, 1913, Glen Comstock collection*

snubbing guy A line that is used to hold the base of a spar tree while it is being raised.
snubbing line A car line that is used to spot empty cars for loading at a landing. See car line, spotting line.
snubbing machine See snubber.
snub check A wrap or two around a stump or tree to reduce the speed of a running line.
snub down the slack To slack off or lower the butt rigging down to the crew on the ground. (P.C.F.)
snub line An anchor that is used to hold down a donkey so that it will not lift or pull over under a heavy load.
snub rope A rope attached to timber and played around a stump or tree to check the speed on heavy downgrades, or to stop a raft. (App.)
snub yoke The wheelers in an ox team. (S.F.) (App.) See butt team.
soak 1. In river driving, the time off that a checker had when the gap in the river was closed. The checkers worked in pairs. (N.W.) 2. See slobber stack.
so big it takes three days to walk around it Said of a large donkey or cat. (P.N.F.)
socket To attach a piece of logging equipment to a wire rope by pouring melted zinc or babbitt around the end of the rope which has been fitted into the socket on the piece of logging equipment.
sod-buster A part-time logger who farms the rest of the year. (P.C.F.)
so dead the buzzards wouldn't touch it Said of a very dead logging show. (P.C.F.)

soft fir Yellow fir. (P.C.F.)
soft gum See crude turpentine.
soft hammer Usually a brass hammer, used with an old ax blade, to cut wire rope.
softness The property of wood indicated by a relative lack of resistance to cutting, denting, scratching, pressure or wear.
soft pine Old growth ponderosa similar to the wood of the famous "cork pine" of the Lake States. (P.C.F.)
softwoods The botanical group of trees that have needles or scale-like leaves and are mainly evergreen. The term has no reference to the actual hardness of the wood.
soldier To loaf on the job. (L.S.) LL
solid jam 1. A jam formed solidly and extending from bank to bank of a stream. 2. A drive is said to be in a solid jam when the stream is full of logs from the point to which the rear is cleared to the mill, sorting jack or storage boom. (N.F.) LL
solo truck A truck used without a trailer.
so muddy it would mire down a saddle blanket Said of very soft ground. (P.N.F.)
son of a bitch The log that squeezes out those below and tumbles a pile. (L.S.)
soogan, sougan A heavy blanket used by loggers and lumberjacks.
sort 1. To sort logs for grades, length, etc. 2. To sort lumber on the green chain or dry sorter according to grades.
sorter A man working on a rafting ground, particularly a sorting pocket. (N.W.)
sorting bay A boomed off area in a lake or stream where logs were sorted for size, species or grade. (N.F.)
sorting boom 1. A strong boom used to guide logs into the sorting jack. (N.F.) 2. Any sorting ground where logs are boomed and sorted.
sorting gap A port through which logs are passed on down a river after being separated by ownership.
sorting grounds A water area used for sorting logs.
sorting jack 1. The throat of a log pond through which logs are steered into the sorting pockets. 2. A raft, secured in a stream, near an opening in which logs pass to be sorted by their marks and diverted into pocket booms or the downstream channel. LL

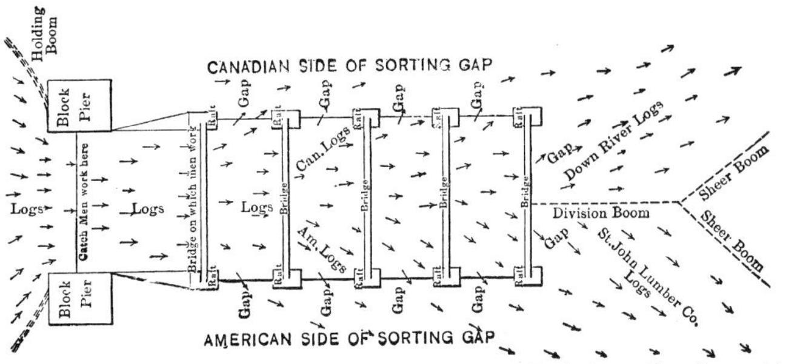

Diagram of a **sorting gap** on the St. John's River near Fredericton, New Brunswick. *From "Logging," Bryant, Wylie & Sons, 1923, Philip Schnell collection*

sorting pocket A division of a sorting grounds for holding logs which have been separated out of a boom or drive.
sorting pond See reload pond.
sorting works An enclosure on a river where logs were sorted as to ownership. (N.F.) Also boom works. LL
so steep the chipmunks have to double-head to get up the hill Said of very steep ground. (P.N.F.)
so steep you have to look up the chimney to see if the cows are coming home Said of a logging show on rough ground. (P.N.F.)
so strong he pulled himself right off his shoes Said of a good, strong ox. (P.C.F.)
sound 1. To hit a tree or log to listen to the sound to tell whether it is good or rotten. 2. Free of decay, as a sound tree, sound log, etc.
Sound (the) Puget Sound in the Pacific Northwest which was and is important in log rafting.
sound as a nut Said of something good. (P.C.F.)
sounds like a buzz saw in a knot Said of a heavy snorer. (P.C.F.)
soupbone Said of a very thin logger. (P.C.F.)
soup jockey One who performs menial duties in a cookhouse or lumber camp. (P.C.F.)
soup shanty 1. A cookshack. 2. A building where employees of boom companies were boarded near the booming grounds. (L.S.) LL

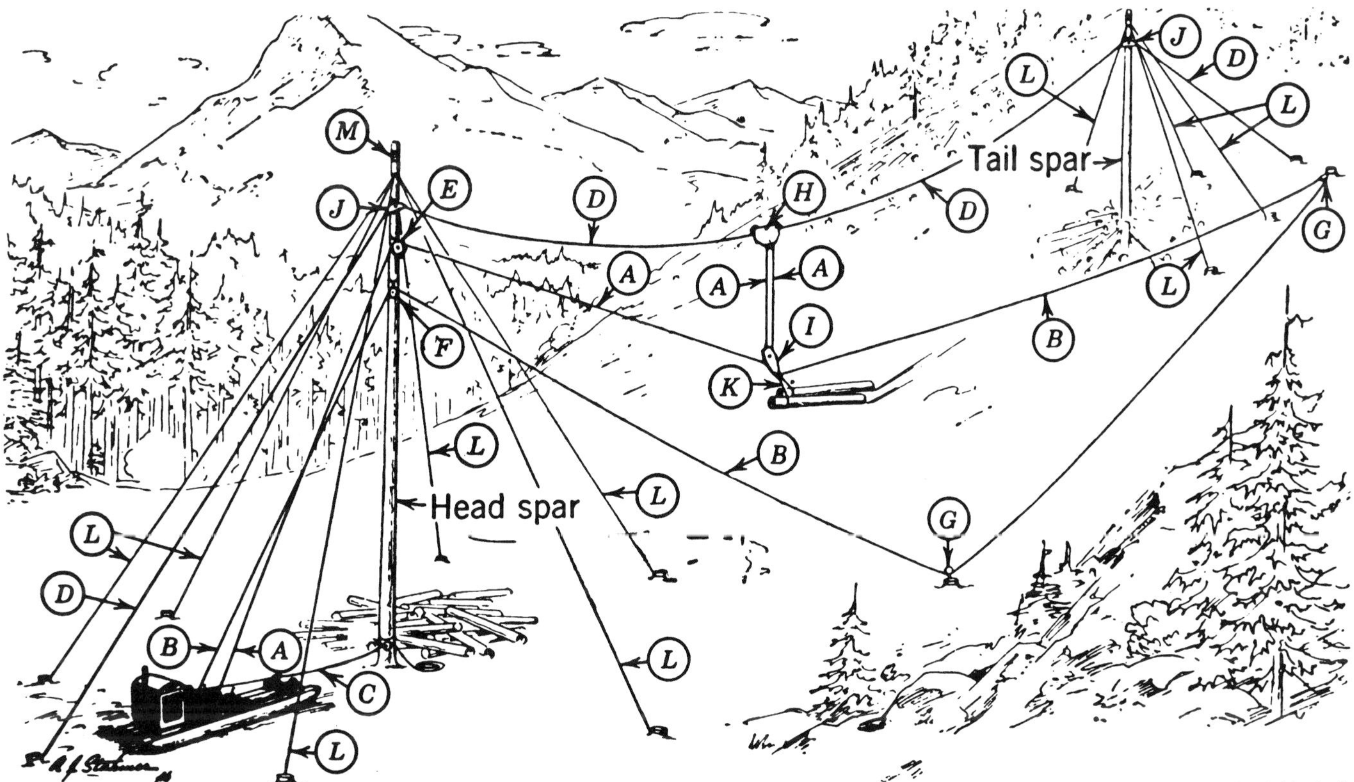

The **South Bend system**, also called modified or downhill **North Bend system**. A.-**Mainline**; B-**Haulback line**; C-**Strawline**; D-**Skyline**; E-**Mainline block**; F-**Haulback block**; G-**Haulback corner** or **tail block**; H-**Carriage**; I-**Fall block**; J-**Tree shoe**; K-**Chokers**; L-**Guy line**; M-**pass line block**. *From "Logging," Brown, Wylie & Sons, 1949, Philip Schnell collection*

sourdough 1. A partly fermented dough used as a starter for bread, flapjacks, biscuits in woods cooking. In the early days before yeast cakes, every cabin in the woods had a carefully kept can of sourdough which was taken to bed on cold nights to prevent loss by freezing. 2. A mountain man or prospector.
sourdough cook A poor bread maker. (L.S.) LL
South Bend A type of spark arrestor. (P.C.F.)

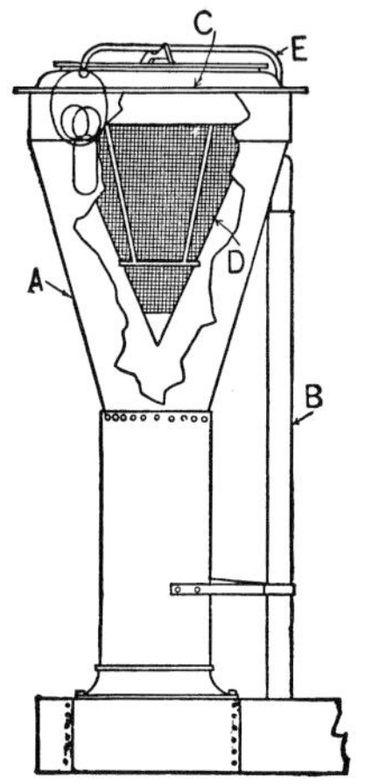

The **South Bend spark arrestor**. *From "Logging," Bryant, Wylie & Sons, 1923, Philip Schnell collection*

South Bend system A modification of the North Bend system, used for swinging down steep hills. It was rigged to give a greater lift to the turn of logs. The end of the mainline was fastened to the top of the fall block, then passed through a sheave in the bottom of the carriage, down through the sheave of the fall block, and forward to the mainline lead block. This gave more lift on the turn to overcome obstructions. Also downhill North Bend system, modified North Bend system. (P.C.F.) See North Bend system.
sow belly Salt pork.
sow bosom Bacon. (L.S.) (S.F.)
space blocks Small blocks spaced on a loading boom to guide the tongs.
spacing bar A bar used for spacing logs when starting a turn of logs. LL
span 1. The skyline distance from the head tree to the tail tree. 2. Two horses harnessed together. (Maine) 3. To stretch a line across a canyon.
Spanish windlass A device for moving heavy objects in logging. It consists of a rope or chain, within a turn of which a lever is inserted and power gained by twisting. (L.S.) Also twister. LL
spar, spar tree A tree or pole used in high lead logging.
spar cap One of different types of metal caps which were tried at times on the tops of spar trees to prevent chafing and cutting by the guy lines.
spare bunk A log bunk set on a landing for preloading, swung out of the way when not in use.
spar jumping See jump a tree.
spark arrestor 1. A screen fitted over the smokestack of a locomotive or a donkey to prevent sparks from escaping. Some of the well known arrestors were Adams, balloon, boomerang, cyclone, Ellsworth, Gerlinger, Glafke spark killer, gooseneck, Merry Widow, Monarch, never-clog, Mother Hubbard, Oregon, Pacific spark consumer, Radley Hunter, Sequoia, South Bend and Western. 2. A spark arrestor was used on mill stacks, especially when they were located in the woods. 3. Spark arrestors were also required during fire season on some gas and diesel rigs.
spark arrestor day The first day of fire season when arrestors have to be in place on woods machinery. (P.C.F.)
spark cap A spark arrestor.
spark catcher A fire watchman. (P.C.F.)
spark chaser 1. Usually an old logger at the landing who kept care of the spot fires caused by the donkey. 2. An old mill worker at the sawmill who watched the burner for sparks which could cause spot fires. (P.C.F.)
spar pole See spar tree.
spar tree The important key to cable logging, the tree on which rigging is hung for one of the various types of cable hauling systems. The spar tree is the center of operations for each side; here is the landing, the loading, the end of the run for the crummy (except for fallers), and the beginning and ending of the day's work. The location of each spar tree is very important to the success of logging. Also high lead tree, spar pole.

An unidentified **spar tree**. This may be a **tail tree**. Note the **high rigger** near the top. *From Philip Schnell collection*

Spaulding rule A log rule. See Humboldt log rule.
spear-edged bunk A sharp-topped bunk on a railroad car.
spearhead A smoking or chewing tobacco. (L.S.) LL
speeder A rail motor car.
speeder monkey The operator of a speeder. (P.C.F.)
spell one To relieve a person at work.
spider 1. The spokes on the bull wheel of a donkey. 2. A small metal tool for testing the set in the cutting teeth of a new saw in the sawmill and on felling and bucking saws. Also set gauge. 3. A big star-shaped wheel used on the carriage of the Opsal skyline system. (P.N.F.) 4. A heavy cast-iron frying pan. (L.S.) LL
spider web 1. A boom on a river or bay across the mouth of a feeder stream to catch logs which have escaped from booms farther up the stream. (L.S.) LL 2. Paul Bunyan's helper whose job was to see that Babe, the Blue Ox, stayed healthy.
spike To add snoose or whiskey to chewing tobacco. (L.S.) LL
spike bar A bar with a claw used to pull out railroad spikes that are used to fasten guy lines to a stump, and also used on the railroad for the same purpose.

Crew **speeders** of the Crown-Zellerbach Corporation near Cathlamet, Washington, on the Lower Columbia, in June, 1957. Built by the Skagit Iron Works, they were equipped with war surplus Cadillac motors. *Labbe photo*

spike camp A temporary camp away from a main camp used by the Forest Service. (P.C.F.)
spiked bell hook An end hook with a large spike in the center of the cup; used for loading. (P.C.F.) See cup hook.
spiked skid A skid on which spikes are inserted to keep logs from sliding back when being loaded or piled. (L.S.) Also skid. LL
spike'er 1. To spike guy lines to a stump. 2. To stop. 3. To fasten anything securely.
spike knot In the woods, a knot that is deep in the wood. Also horn knot.
spike money A bonus to keep a good man on the job.
spike peddler The man who delivers spikes to spikers in a railroad track-laying crew. (S.F.)
spiker A lumberjack who has quit his job. (L.S.) LL
spike top A tree with a top which resembles a spike.
spile 1. A heavy stake or timber driven into the ground as a foundation or support. 2. A small log used as a pile in dam construction. (N.F.) LL 3. A spout inserted in a tree to draw off the sap.
spile driver Another term for pile driver.
spiling A series of spiles driven in front of a dam. (L.S.) LL
spill 1. A log dump where one track is raised to spill logs off a car or truck. 2. A wreck. 3. An overturned load of logs. 4. The leaf of a softwood evergreen tree. 5. To have a wreck. 6. To let water through a dam.
spill the ink To write out a check because the man is quitting. (P.N.F.)
spillway The overflow of water over a dam in a stream or lake. (N.F.)
spiral 1. Twisted grain in a log or tree. 2. A gradual turning in and out of a curve on a logging road or a railroad.
spiral auger An auger with a down-cutting lip and an open, spiral shank. (L.S.) LL
spiral grain A type of growth in which fibers take a spiral course about the bole of a tree instead of the normal vertical course. The spiral may extend right-handed or left-handed around the tree trunk.
spirits See turpentine.
spit-or-puke Snoose. (P.N.F.)
spitter A short piece of notched fuse causing the fire to spit out of the holes in the casing; used to set off dynamite blasts by hand.
splash 1. To move logs down river by use of a splash dam. Water is held back while logs are yarded into the river. When the gate is kicked loose, the flood of water will wash logs down the stream for miles when ordinarily the current would not move them. This system of driving logs was used extensively in the North Woods and the Lake States. It was brought west by the loggers, but never proved as efficient on this side of the Rockies. Big River in Mendocino County, California was the best driving stream in the redwoods, and the only one used for an extended period of time. 2. To work on a splash operation. Also flood logs.
splash boards 1. Planks added to the top of a splash dam to hold back an extra head of water. 2. The section of the water tank on a locomotive which extends above the top of the tank.
splash dam In the 1800s and early 1900s, a dam was placed across a stream to build up a head of water for moving logs. Also flood dam, flush dam, sluice dam.

An unidentified **splash dam** under construction, probably somewhere in Idaho. *Courtesy of William Hayes*

splatter sheet A stretcher. (P.N.F.)
splayfoot An awkward logger. (P.C.F.)
spleeny A person lacking in stamina and courage and who is always complaining about his health. (Maine)
splice 1. A joint or joining strands of wire rope. 2. To weave or join together the strands of wire rope.
splicer In the early camps, a logger who spent all his time splicing wire rope.
splice shed The rigging headquarters in the logging operation.
splinter draw Splinters pulled out of the butt log and left sticking up on the stump. (P.N.F.)
splinters A disparaging term, used by the loggers of the Redwood Country, to refer to other trees cut in other areas. (Red.)
split 1. The lengthwise separation of the wood due to the tearing apart of the wood cells when drying lumber in the lumber yard. 2. A large shingle made from very straight-grained pine, fir or cedar trees. (N.F.) (Eastern Canada) See shake.
split a quarter To begin the first road in a setting half-way between the fair lead and the square lead on a diagonal line from the donkey.
split ear A mean man. (P.N.F.)
split guy A guy line split in two pieces and joined by a shackle; a time-saver in rigging.
split roof A roof of a logging camp or barn made by laying strips split from straight-grained timber. (N.F.)
split shingles Shingles four or more feet long and made thin on one edge; used for roofing camps and sometimes for barns. (Maine) LL Also long shingles.
splitter 1. A cordwood maker. 2. The man who made the top cut in a hemlock barking crew. (L.S.) LL
split the blankets To get a divorce. (P.C.F.)
splitting gun A piece of tubing, or a piece of shafting bored out, with one end closed. The tubing was filled with powder and driven into a log to split the log when it was too big to handle in one piece.

split yoke A loading system in which two loading loops are suspended from the rigging between two equal-sized loading trees.
sponger 1. A moocher. 2. To beg or sponge.
spool 1. A drum for holding cable. 2. An unright spool along the skidroad to keep the wire rope from hanging up. See road spool. 3. A small drum used as a capstan; several turns of line around the drum gave it pulling power for use in loading or small jobs around the landing. See gypsy head, nigger-head. 4. To wind cable onto a drum.
spool cart 1. In the late 1800s, a small cart holding a spool of cable which was paid out as a horse hauled it from the landing out to the bucked logs. 2. A low two-wheeled cart with wide heavy wheels and a bunk; used as logging dray during the dry summer months.
spool donkey One of the first donkeys with an upright spool or drum. See Dolbeer.
spool off To coil rope on the ground as it came off the side of the spool as the rope was reeled in from the woods or to run line off a drum by hand.
spool-off man The logger who coiled the line on the ground as it came off the vertical spool of the early Dolbeer donkey. See spool tender.
spool on To guide a cable or rope as it was wound on the spool or drum so that it would wind evenly.
spool-on man The logger who guided the incoming line onto the spool on the first donkeys when the line came in at an awkward angle; later fairleads were designed to guide the line onto the drum.
spool stuff Spaghetti. (P.C.F.)
spool tender 1. The logger who coiled the line on the ground as it came off the vertical spool of the early donkeys. Also drum tender, wire tender. 2. A man who tends the spool on a pile driver in construction work. 3. A logger who used the spool of an old-time roader or yarder to load logs.
spool wheel 1. A log cart with solid wooden wheels. 2. The wheels themselves.
spool-wright The man who took care of the skidroad and greased and repaired the spools and blocks.
spoon To turn over together in bed. In the earliest and crudest camps, the lumberjacks slept close to one another to keep warm. (L.S.) LL
sporting house A whorehouse.
sporting lady A whore.
spot 1. Used with a locomotive number to designate the number, as One-Spot, Two-Spot, etc. 2. To help or signal an operator of a rig or engine for placing of loads. 3. To spot a car for loading, signalling the operator to place the car at the exact spot for loading. 4. To put a blaze on a tree. (N.F.) LL
spot-burning A controlled burning confined to a relatively small area for removing potential fire hazards.
spot fire A small fire started from the main blaze by wind-blown sparks.
spot plate The number plate on the smoke box of a locomotive.
spot rot The spots of rot in a log.
spotted dog A steamed pudding with raisins. (P.N.F.)
spotter 1. A man who signals the operator of a crane or lift truck as to the proper movement and placing of loads about a yard and buildings. He moves bunks, bolsters, kiln trucks and places them properly under a load. 2. One who locates big trees and spots them for cutting.
spotting drum A small drum on loaders that is used to pull in a line that is attached to the empty railroad cars to spot them for loading. See car line.
spotting line A cable by which a log loader or power skidder moves itself for short distances. Also car line, snubbing line.
sprag See gooseneck.
spread The rate of travel of a forest fire. (P.C.F.)
spreader 1. A short piece of the mainline between hook-ups to spread the butt hooks apart when using two chokers. 2. A man who talks too much. 3. A wooden hitching device that held apart a crotched chain used in skidding in place of a doubletree. (L.S.) LL 4. A piece of steel rail used to separate the loading hooks in loading with a gin pole or crotch line loading. (P.C.F.)
spreader bar A bar that is used to keep the tongs apart on a crotch line loading system.

A **spreader bar** for use on a **crotch line loading** system. *From Young Iron Works Catalog No. 52, 1952, Philip Schnell collection*

spreader chain A loading safety chain in the position of a gut chain to prevent the peaker from crowding down to break the bunk chains and force the bunk logs over the cheese blocks.
spring 1. To widen the bottom of a hole by an explosion to make a bigger pocket for the shot. 2. To pull a line to one side to increase its tightening effect.
springboard 1. An older term for the rigging in felling a redwood. 2. In the Pacific Northwest fir country, a strong plank on which fallers stand when felling in rough country or when the tree is too swell-butted for easy sawing. Springboards are disappearing with the wide use of power saws. 3. A stout timber attached to the axle of a big wheel logging cart. When pulled down it raised the tongue in the air. After the log was chained on, the tongue, being pulled down by the team, acted as a lever to lift the log at least partly in the air, which made for easier hauling. 4. The support on which shingles rest while being jointed, so called because it springs back into place after having been pushed down so that the shingle is brought into contact with the saw.

Two **fallers**, working for the Cathlamet Logging Company, near Cathlamet, Washington, pose on their **springboards**. They are making their **undercut** fairly high up the tree to avoid the **swell** of the **butt**. The four springboards, set into **springboard notches**, form a makeshift ladder. A **falling saw** hangs from the undercut, to be used to make the **back cut**. *J.F. Ford photo, courtesy of Frank M. Potter*

springboard notch In the Pacific Northwest fir country, a narrow deep notch cut in a tree to hold the iron on the inner end of a springboard.

spring break up 1. The time when logging camps closed for the season because transportation became too difficult because of the thawing of snow and ice in the woods 2. The time when the ice in the river broke up and floated downstream so that the drive could start. (N.F.) LL

spring burn Slash burning in the spring. (P.C.F.)

spring guy 1. A guy line to tighten another guy line. 2. A guy line to pull another out of the way temporarily.

spring heel A corked shoe in which a single piece of leather is used from the toe to the heel and is curved in the arch. (P.C.F.)

spring hole A hole in which a small amount of dynamite is placed to widen the hole so that a bigger charge can be placed in it. (P.C.F.)

spring link yarder A yarding block with one cheek plate hinged so that the block can be opened up. See Bouse block, snatch block.

spring pole 1. A prop used in one-man bucking with a crosscut saw. See Dutchman; rubber man. 2. A stick used to guide the whistle punk's jerk wire to give a straight pull on the donkey whistle; used to keep the whistle wire taut. 3. A springy pole attached to the tongue of a logging sled for holding the weight of the tongue off the horse's necks.

spring pole falling A vine maple or other spring used in one-man felling with a crosscut saw. The pole on the other end of the saw pulled the saw away from the man. See rubberman.

spring set Saw teeth bent alternately to one side and then the other.

springwood The portion of an annual growth ring which forms during the early part of any season's growth; usually less dense, weaker mechanically, and lighter colored than summerwood.

sprinkle 1. To water down a logging road to keep down the dust. 2. To water down around a landing to decrease fire danger.

sprinkler 1. Some locomotives were equipped with sprinklers which were used primarily to dampen bridges and trestles during dry weather, but were sometimes used to sprinkle the rails to keep the trucks from riding up. (P.C.F.) 2. A large water-tight box built on a sleigh. It sprinkled the logging road with water which froze, making an ice road for a sleigh haul. (N.F.) LL Also ice box, icer, ice tank, tank, water box, water tank, water wagon. See ice a road. 3. The man who "sprinkled." Also tank teamster.

A **sprinkler**, mounted onto **sprinkler sleds**, being filled with water, some place in the Adirondacks. *From "Logging," Bryant, Wylie & Sons, 1923, Philip Schnell collection*

sprinkler sleds The sleds on which the sprinkler was mounted. It consisted of two sleds whose runners turned up at both ends, fastened together by cross chains, and each end having a pole so that the sprinkler or water tank could be hauled in either direction without turning around. (N.F.) LL

spruce Common name for a genus of about 40 species of coniferous trees, indigenous to the Northern Hemisphere, nearly half being natives of North America. The genus, of the pine family, was formerly combined with a fir genus, from which it differs in having pendulous cones, persistent woody leaf-bases, and four-angled leaves, scattered and pointing in every direction. Spruces occur farther north than most trees, forming forests within the Arctic Circle, and extending south, especially in the mountains, as far as the Pyrenees in Europe and the Himalayas in Asia; in the United States, spruces occur as far south as North Carolina and Arizona. The white spruce and black spruce cover extensive areas in Canada, almost to the total exclusion of other trees.

In North America the prevailing species over the greatest extent of territory are the white spruce and the black spruce, although in lumbering the two species are seldom distinguished. The former is found from New York to British Columbia and northward to Newfoundland, Hudson Bay, and Alaska. The white wood is light in weight, soft, compact, nearly odorless, and with scarcely distinguishable sapwood. It is used for general building purposes and also yields some of the spruce gum of commerce.

The black spruce has a distribution somewhat similar to that of the white, except that it descends along the mountains to Virginia. It is a somewhat smaller tree, but otherwise similar to the white spruce. It forms immense forests in Maine, New York, Canada, and elsewhere and is used largely for wood pulp and paper. A third species, the red spruce, a valuable timber tree, is limited essentially to New England but is also found along the Appalachians as far south as Georgia. See Sitka spruce.

Spruce Production Division During World War 1, a division of the U.S. Army Signal Corps. The Spruce Production Division was established in order to dramatically increase the production of spruce lumber required to satisfy an urgent military need for airplanes. Begun in the Spring of 1917, the program was soon placed under the command of Colonel Disque (later General Disque), who was charged with the duty of substantially increasing the quantity being produced.

It soon became apparent that in order to produce the enormous quantities (up to twenty million feet per month) of airplane-grade spruce required, it would not only be necessary to secure sufficient manpower to get the job done, but to provide a means of getting the spruce logs to the giant government-owned cut-up mills.

Early on, primitive logging methods (such as riving the giant logs into crude cants) were employed, and motor trucks were used to transport the logs to the mills. It was soon realized that railroads would be needed, and some thirteen of them in the coastal Sitka spruce regions of Oregon and Washington were laid out, four of which were of permanent construction.

To secure adequate manpower, loggers were enlisted as "soldier/loggers," and were subject to standard military discipline. Logging camps under the Spruce Production Division often resembled military encampments more that the typical logging camps of the day. The army offered working and living conditions greatly improved over what had been the custom, and established a new standard, which continued after the war.

The Spruce Production Division was a huge success—As an example, in October, 1918, less than a year after the program had been instituted, the cut-up plant in Vancouver, Washington put out 28,681,239 feet of spruce, of which 20,282,933 feet was of airplane quality.

spruce up 1. To dress up. 2. To clean up, to get things in order.

spud 1. A tool for peeling bark. 2. A peavey with no hook. 3. A block placed under a skidder or jammer. 4. A potato 5. To peel bark. LL

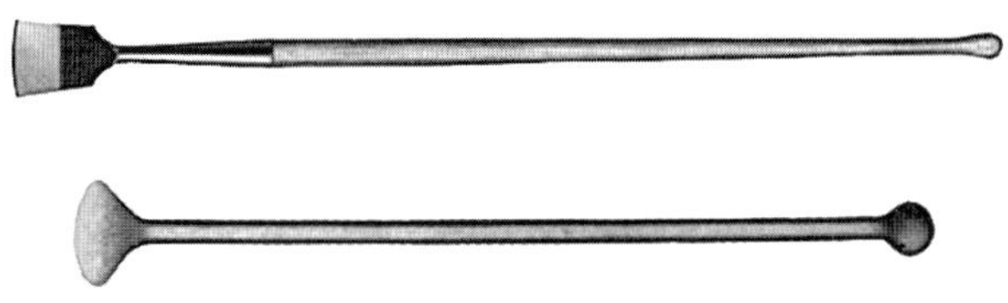

Two types of **spuds** used for peeling bark. *From Mill & Mine Supply Company catalog, 1916, Glen Comstock collection*

spur 1. A branch logging road. 2. A climbing iron. 3. A sticker or jagger on a wire rope. 4. A small ridge extending off a main divide. 5. Another name for a board iron used on a springboard in the Pacific Northwest.

spur wheel The large toothed gear wheel on donkeys, generally used to drive the main and haulback drums.

A tent camp of the U.S. Army's **Spruce Production Division** during World War 1, location unidentified. *A.M. Prentiss photo, from "Pictorial Review of World's War Activities, Spruce Production Division," Philip Schnell collection*

square 1. A basis for paying buckers or snag fallers, so much per square, meaning per square foot of stump or log cross section. 2. A measure of shingles or shakes to cover 100 square feet. 3. To square a timber by using a broad ax to hack off the round outer part.
square around (away) 1. To get equipment lined up ready to go to work. 2. To line up logs at the landing parallel to the track or road for loading.
square dress See swage set.
squared stuff Hewed timber. (P.C.F.)
squareface An engine. (P.C.F.)
square foot A basis for paying timber fallers, not by scale but by the area of the cross-section of the stump.
square head 1. A Scandinavian. 2. A green man. (P.C.F.)
square lead A line at right angles from the fair lead. Yarding begins on a line in front of the donkey or fair lead and moves to a line about 45 degrees called the square lead. Yarding beyond the square lead is called the diamond lead.
square rigger An old man with a square beard. (P.N.F.)
square the breaks To cut the broken ends of logs that have been broken in falling. (P.C.F.)
squaring cord Chalk line used in hewing logs. (L.S.) Also marking cord. LL
squatter 1. A person who lives on cutover land. 2. A person who builds a cabin on another man's timberland. (P.C.F.)
squawk box The radios carried by all rolling stock of a company so that the men can communicate with one another and with headquarters.
squaw dance 1. There were few white women living near the lumber camps, and the Indian women came to the lumberjack dances in the late 1800s. 2. Stag dance (L.S.) LL
squaw hitch Said of the hitch when a choker is tightened over just a corner of the end of a log, when it is impossible to get it under the log. (P.C.F.) Also donald.
squaw link A double-ended grab link used to connect the loose end of a broken chain as temporary or emergency repair. (L.S.) LL
squaw log A log which is rotten on the big end. (P.C.F.)
squaw man 1. A logger who goes to the reservation at night. 2. A man married to an Indian woman. (P.C.F.)
squaw timber, squaw wood See skin timber and trading stock.
squeak heel Extreme pain in the heel because of very cold weather or working in snow or water.
squeegee A logger's or lumberjack's homemade washing machine; a tin can nailed to the end of a stick, open side of the can downward, to make a vacuum.
squeeze The boss or bull buck. (P.C.F.)
squeeze box A concertina.
squint eye A saw filer.
squirrel 1. A counterweight to pull a loading boom into position. The squirrel might be a chunk of wood which rides up and down the spar tree, a separate tree or a guy line. Also dummy log, plumb-bob, squirrel chunk, slack puller. (P.C.F.) 2. A rough house game in the bunkhouse. (L.S.) LL
squirrel block 1. A small block through which a line passes to the squirrel. 2. A small block used to handle the haulback when using a jammer for yarding. (P.C.F.)
squirrel cage 1. A safety fence built around the spot where the squir-

rel comes down close enough to hit a man walking under it. 2. When a strand of rope on the squirrel line breaks, the rope at the spot frays out, resembling a squirrel cage. Thus one hears "A squirrel cage on the squirrel line." (P.C.F.)

squirrel line Short sections of small line that are used to transfer the position of the heavier line. Also ratline.

squirrel man A high climber. (P.C.F.)

squirrel porch A conk on a tree. (P.C.F.)

squirrel tree A tree with many branches, good only for squirrels. (P.C.F.)

squirrel whiskey A cheap brand of whiskey sold by the early saloon keepers. (L.S.) LL

squirt dam A dam built on a small creek tributary to the driving stream and opened to give a "squirt" at strategic moments. (N.W.)

stab a log To punch a hole in a log with loading hooks or tongs. (P.C.F.)

stack 1. An abbreviation for a smokestack. 2. A pile of stacked logs. 3. A pile of pancakes.

stack blower A rig for exhausting steam in the stack of a locomotive or donkey engine to increase the draft.

stacker See gobbler.

stack ground The location where logs are stored. (S.F.)

stag 1. An ox. 2. To shorten work pants by cutting them. 3. To cut a large slice off a pay check for deductions.

stag camp A logging camp only for single men. Also called bachelor camp. (P.C.F.)

stag dance The lumberjacks danced among themselves to the tunes of the camp fiddler. The "ladies" wore a red handkerchief around one arm or tied a sack around their waist. (L.S.) LL

stage The depth of water in a river, specifically for river driving. LL

stage cut To fell timber in several rounds or successive cuts to reduce breakage. (P.C.F.)

stage of water The amount of water in the river for driving logs. In a good stage, the banks were full of water. If too high the logs would catch on obstructions; if too low, the logs would not float. (N.F.) LL

stage up To put the staging boards in position before beginning the undercut above the swell at the base of a redwood: "Stage up to it." (Red.)

stagged Said of pants that have been cut off. See stag pants.

staggered settings A system of logging alternate settings so as to leave timber for seed and for a fire break for the cutover areas. Also block-cutting, patch logging.

stagging trousers Said of trousers cut off at the bottom for easier working in the snow, on river drives and from hanging up in the brush. LL

staghead A tree with a dead top. (P.C.F.)

staging boards 1. The boards that are laid across the drivers to make a platform on which the fallers stand. (Red.) Also driver boards. See board up a tree. 2. Boards laid across springboards to provide a chopping platform when working on big trees. (P.N.F.)

stag pants The work pants that are stagged to remove the dangerous cuffs that may trip a man. (P.C.F.) Called stags in the Lake States.

stags See stag pants.

stag shoes Shoes with tops cut off to make slippers. (L.S.) LL

stag tree A tree whose top has been broken off. (L.S.) LL

stake 1. A corner post, such as a survey stake. 2. A car stake used for loading on flat cars, usually 4x4's trimmed to fit the stake pockets. See car stake. 3. A grub stake. 4. Enough money saved to quit. 5. To stake out or claim ownership. 6. To stake a man.

stakebound 1. Said of logs which are jammed so tightly against built-in stakes that it is difficult to loosen them quickly for unloading. 2. A logger or lumberjack with too much money who will not return to work until broke.

staked bunk A bunk with built-in stakes.

stake-maker The man who cuts car stakes and blocks to hold logs on railroad cars.

A **stag camp**, or **bachelor camp** of the Wisconsin Logging & Timber Company, near Oak Point, Washington, on the Lower Columbia River. Note the small **Shay** in the background.
Courtesy of Alvin Larson

stakey 1. Said of a man who has saved up money and who is hard to get along with because he can quit any time. 2. Said of a man who has saved up money and who is restless because he wants to go to town. (P.C.F.) (British Columbia)
stamp 1. A mark on a log showing ownership. LL 2. To stamp a brand on a log.
stamp ax A branding ax. Also branding hammer, log stamper, marking hammer, marking iron or stamping ax.
stamper A timber or log marker.
stamping hammer, stamping iron See branding ax.
stand Standing timber.
standard gauge 1. A regular gauge railroad that is 4 feet 8 1/2 inches wide. 2. Anything up to standard.
standard log The standard pine log was usually 32 feet, and the standard fir log was 40 feet. Anything shorter was usually a short log, and anything longer was a long log.
standing boom Permanent boomsticks used as a frame when making up rectangular rafts. The sticks are kept in place by piling, and are used as walkways by the boom men. LL
standing line A fixed cable during logging operations.
standing line drum The skyline drum on a tight line logging system.
standing shot An unshot charge of dynamite. (P.C.F.)
standing skyline See tight line.
standing timber The merchantable timber that is uncut.
standing tree A live tree standing at a landing used as a spar tree; opposed to a raised tree, one brought in and raised at the landing.
Stanfield's half inch Thick woolen underwear favored by British Columbia lumberjacks.
star chief A camp cook. (L.S.) LL
star light The skylight in the roof of the sleeping shanty. (L.S.) LL
star ring A pitch ring in a log.
start A pin or pins fastened to the runners of a dray and holding in place the upper removable bar or bunk. (N.F.)
starting bar 1. The bar which was used to start the old "Sixty Cats," to turn one compression stroke at a time. The bar fitted into holes in the flywheel, which were tapered on the back side for quick release. 2. A heavy bar used to jar loose a sleigh which has frozen to the icy road and used as a lever to raise the runner. (L.S.) 3. A heavy bar used to move railroad cars short distances. (L.S.) LL
starvation outfit (show) A logging camp that offers poor food or little pay. (P.C.F.)
starved fit to eat the Lord's supper Said of a logger who is very hungry. (Maine)
starveout 1. A poor show for logging. 2. A poor place to work. (P.C.F.)
station 1. In surveying, a hundred-foot long section of a grade. 2. A point along a flume where a flume tender is stationed. 3. A place below the head of a flume where more logs are dumped in.
stationary engine An engine used for skidding or uphill hauls. Same as a donkey. (L.S.) LL
station man 1. A small contractor building grade by a station. 2. The man who built logging roads or railroads for so much a lineal unit. (L.S.) LL
stay boom A boom fastened to a main boom and attached upstream to the shore to give added strength to the main boom. (L.S.) LL
stay rope A line used to hold a loading carriage at a desired spot on a guy line.
steadying guy See buckle guy.
steam bucker A stave bucker operated by steam power. (S.F.)
steam bucking saw A portable steam-driven saw used for bucking logs at the landing. Also deck saw, log pond saw. See drag saw.
steam cat A steam hauler. (L.S.) LL
steam choker See donkey puncher.
steam dago A power-drive log-bucking device. (P.C.F.)
steam donkey See donkey.
steam drag A portable steam-powered drag saw that was used before the gas engines; such saws were often mounted on the log deck in case a log needed bucking at this point, or to remove a bad section.
steam dummy See dummy.

An unidentified **steam bucking saw** or **steam drag**. *From Philip Schnell collection*

steam freighter A three- or four-wheel steam traction engine once used to haul log carts in the Pine Country. The Russell was a popular one. The Best and Holt were also popular in the West and were the forerunners of the Caterpillar, and ultimately became the Caterpillar tractor. Also road locomotive.
steam hauler A wood-burning steam engine, the forerunner of the modern tractor; used to haul trains of sleighs of loaded logs over iced roads. The term was applied primarily to the Lombard and Phoenix steam haulers, which were steam engines mounted on caterpillar-type tracks. (L.S.) LL Also steam cat.
steam jam (jamb) A steam-powered brake on a donkey or locomotive. (P.C.F.)
steam jammer A machine operated by a steam engine and used for loading logs on cars. (L.S.) Also loader, steam loader. LL
steam loader Any one of several large railroad log loading machines.
steam pot A steam donkey.
steam power haul An older term for a steam donkey.
steam skidder See skidder.
steam splitter A one-cylinder steam engine with a piston attached to an enormous "ax" blade; used to split wood for donkeys and locomotives. Also used in stave mills, shingle mills, etc. for spitting bolts; usually in mills that bought logs, rather than bolts.
steam up To fire up a donkey or a locomotive.
steam wagon A heavy log wagon hauled behind a steam tractor.
steel The rails on a railroad.
steel crew, steel gang The men who put down the steel rails in building a railroad.
Steel King An early donkey built by Columbia Engineering Works of Portland, Oregon. (P.N.F.)
steel man A worker on the steel gang.
steel moving car See moving car, bull car.
steel spar skidder See Lidgerwood, tower skidder.
steep as a cow's face Said of very steep country. (P.C.F.)
steep grade engine A lowering engine used on an incline. (P.C.F.)
stem 1. A lumberjack's leg. 2. The main street in town. (L.S.) LL
stemwinder 1. A Shay locomotive. 2. A geared locomotive in the East. 3. A very good machine. (P.C.F.)
step-and-a-half Said of man who is crippled on one leg. (P.C.F.)
step-and-fetch-it 1. Said of a good handy man. 2. Said of a good relief man. (P.N.F.).
sterno stiff A drinker of denatured alcohol. LL
stew A toot; to get drunk. LL
stew builder A cook in a logging camp. (L.S.) See gut robber. LL
stew bum A camp cook. (L.S.) LL
stewed Said of one who is drunk.
stick 1. A log or tree. 2. A measurement used in the Lake States and Northeast forestry, where a "stick" of wood is a 100-inch length log; usually used in pulpwood economy areas.
sticker A hung up log raft. LL

sticks 1. Long, slender logs used for piling. 2. A set of boom sticks. 3. In the 1700s and 1800s, logs used for masts. (N.W.)
stiff Any working man who does not work in an office.
stiff boom On the early skidders, a fixed yarding boom reared up at about a 45-degree angle to the base of the machine. Also used for loading and unloading cars. (P.C.F.)
stiff leg 1. A rig for unloading cars by pushing logs off as the train moved by the dump. Also stiff leg crane and stiff legs. See jill-poke. 2. A boomstick used to fend a raft off shore or to hold outer boomsticks in place off an anchorage. 3. The Donovan unloader. (P.C.F.)
stiff-tongue wheels Big wheels with a fixed tongue, used for yarding in the Pine Country.

A set of **stiff-tongue wheels** or **stinger-tongue wheels** in use in the Pine Country of Oregon in the 1880s. *Labbe collection*

stiff tug road An iced road not level but just slightly uphill; so-called because there was never any slack in the harness tugs. (L.S.)
stillwater That part of a river having such slight fall that no current is apparent. LL Also deadwater.
stinger 1. The hitch at the rear of a logging truck to which the reach is fastened. 2. The reach connecting the logging truck and trailer, which sticks high above the cab when the trailer is loaded for the return trip to the woods. 3. The chain end of a power saw frame. 4. Any pole that sticks out. 5. The handle on the end of a chain saw when used as a two-man saw. Also tail piece. (P.C.F.)
stinger-tongue wheels See stiff-tongue wheels.
stinkpot Any internal combustion contraption used in the woods.

This small, Fordson-powered **donkey**, working in an unidentified location in the snow, might well have been called a **stinkpot** by loggers. *Gregory J. Maxwell collection*

stock The handle of a cant hook or peavey. (App.)
stocking 1. The quantity of young trees coming up in a logged-off area. 2. An indication of the number of trees in a stand relative to the desirable number for best growth and management, such as well-stocked, over-stocked or partially-stocked.

stock logs To deliver logs from stump to mill or railroads. (S.F.)
stock widths Lumber cut in even widths from 4 to 112 inches.
stog See chink.
stogged up Said of one who is stuffed. (P.N.F.)
stomach hollow as a coon log Said of one who is very hungry. (P.C.F.)
stomach organ A concertina. (P.N.F.)
stomach robber A cook.
stomach robber beatin' the gut hammer Said of the cook beating the triangle in the lumber camps to announce that the chow was on the table. (P.C.F.)
stoneboat 1. A small flat wooden sled used in skidding by some horse loggers; probably originated from the "clod masher," usually loaded with rocks to give it weight, which was used in farming. 2. A sled to haul stones away when clearing fields.
stone heavy Said of anything that is very heavy.
stool pigeon 1. An informer. 2. The tattle-tale or meter on a truck that determines the running time. (P.C.F.)
storage boom A strong boom used to hold floating logs in a storage at a sawmill. Also receiving boom. (L.S.) See holding boom. LL
store-bought Not homemade. LL
stout 1. Strong. 2. Rugged.
stove Said of a raft striking a rock and breaking up. (App.)
stove in Said of something dented or smashed. (P.C.F.)
stovelid A pancake or griddle cake. (L.S.) Also washer. LL
stove up 1. Said of an injured man. 2. Said of something out of order. (P.C.F.)
stow To place logs in position in a boom, within a pocket.
stowing gang The crew that puts logs in position in a boom when building rafts.
stow logs In rafting, to place logs together and parallel within boom sticks which mark the outside of the raft section. (P.C.F.)
stows his grub with a number nine scoop Said of a hearty eater. (P.N.F.)
straight dozer A straight bulldozer blade on a cat which could not be angled, and remained square with the cat.

An International Harvester TD-24 diesel **tractor**, equipped with a **straight dozer** blade, brings a **turn** of logs to the **landing**. The **arch** was manufactured by Isaacson. The location is unidentified. *From Pacific Logging Congress "Loggers Handbook," 1948, Philip Schnell collection*

straight-falling The system where fallers fell the trees, but do not buck them. (P.C.F.)
straight-grained 1. Said of good timber. 2. Said of a good man. (P.C.F.)
straight gut 1. A long narrow channel favored by tug captains with a long tow of logs. 2. Any straight haul. (P.C.F.)

straight haul A straight pull on a cable. Also straight line haul.
strain a gut 1. To over-tax a machine. 2. To lift or pull too hard; to hurt one's self.
strain a tree 1. To fell a tree on a low layout which has a curve, causing checks on one side of the tree; the top half (running lengthwise) will usually be merchantable. 2. To force a tree to fall where it does not want to by holding wood on one side or the other. (P.C.F.)
strain your milk 1. Usually used as a warning in lifting; "Don't strain your milk." 2. To hurt one's self physically by over-exerting.
Straits (the) 1. The Straits of Georgia between the British Columbia mainland and Vancouver Island where much rafting has been done. 2. The Straits of Juan de Fuca. (P.C.F.)
strand a cable 1. To unwind the strands of a cable to make a splice. 2. To break strands of a cable in use.
stranded Logs which are hung up on the bank or on gravel bars by a sudden fall of the water level in a river.
stranded line 1. The wires separating or unstranding in a wire rope. 2. A line with broken strands.
strap 1. A short steel cable with an eye on each end which has many uses in loading, rigging and yarding. See sling. 2. Flat steel bands used with special crimping tools to bind or strap together small pulp poles for rafting. 3. Bands used to bind whole loads of logs to reduce handling and to help contain sinkers in the water.
strawberries Beans. (L.S.)
straw boss The foreman under the general foreman or super. Also straw push.
straw drum A small drum on a donkey that works the strawline.
strawline A light wire rope that hauls the heavy logging cables into position. Also grass line, guinea line, haywire, lumberman's thread.
strawline fairlead A light fairlead bolted on the front head block of a donkey sled, the line feeding through a hole in the timber.
strawline hook A small hook on a strawline used instead of a shackle.
straw push See straw boss.

A detailed view of the front end of a **donkey**. This **Willamette yarder** sports a **sled** with three **fairleads**: at left is a large four-roller fairlead for the **mainline**; at right is a **bullseye fairlead** for the **haulback line**; bolted to the front face of the **head block** is a **strawline fairlead** with the **strawline** passing through the timber. Moving **straps** can be seen passing through the **bridle irons**, or **sled plates**, bolted to the **sled runners**. *From Willamette Iron & Steel Company collection, Labbe collection*

stray 1. An unbranded log floating loose. 2. A marked log passing through the sorting gap of a boom company and no instructions given about it. (L.S.) 3. A log which has passed the mill where it should have been taken from the water. (N.W.) (Eastern Canada) 4. See prize log.
stray logs Logs that had been misdirected at the sorting works. (L.S.) LL
streak In turpentine orcharding, the incision made when a tree is chipped. (S.F.)
stream jam See center jam.
strength The properties of wood which enable it to resist different forces or loads.
stretcher A bar to keep the singletree spread apart when skidding with horses.
strike (the) The redwood strike of 27 months which began January 14, 1946 and ended April 12, 1948. There were other strikes in the Redwood Country in 1903, 1919 and 1935, but the long strike is called "The Strike."
strike on the job An I.W.W. scheme to get paid while striking by pretending to work, stalling on the job. Used in a number of camps in the fall of 1917 in the Pacific Northwest, and one of the reasons behind the formation of the Four L.
striker A helper.
string 1. A trail of logs. 2. To hook up to a set of boomsticks. (P.C.F.)
stringer 1. In bridge construction, a timber laid lengthwise to connect the caps of the bents. 2. A 2x6 or 2x8 piece used in making foundations for loads in a lumber yard. 3. A long thin stand of timber. (P.C.F.)
stringer road 1. A fore-and-at road. 2. A tram road with sawed wooden rails, used for hauling logs. (App.)

A Class A **Climax** locomotive and train of logs on a wooden **tram**, or **stringer road**, of the Yellow Poplar Lumber Company, in Buchanan County, Virginia. *Courtesy of D. Richter*

stringhalt A man with a limp (from the disability which affects horses). (P.N.F.)
string of flats 1. A train of empty flat cars. 2. Flapjacks.
string of logs A section of a log raft usually 16 feet wide, 400 feet

long. Rafts were made up of ten or more such strings. (L.S.) LL
string of sticks A set of sticks to hold a boom together.
strip 1. An area of timber assigned to fallers for working on contract. 2. A narrow one-inch board. 3. An area, usually 22 yards wide, marked out by foreman or strip boss, where a piece-work cutter felled and stacked his pile, or cord cradle, of pulpwood. (Canada)
strip adjustment An agreement in which a faller gets a better rate because of a rough show.
strip a tree To remove the rigging from a spar tree. Also take the tree down, unrig. (P.C.F.)
strip burning 1. Burning slash across the face of an operation, gradually working the fires downhill. 2. Burning strips through a big slash area instead of broadcast burning the entire slash..
strip cruise The old system of timber cruising in which a compass line was run across the country, and the trees were counted on each side of the line.
strip logging A system under which certain strips were logged, while alternate strips were left as seed trees.
strip road In a cross-tie operation, a road cut out by the tie hacker or a given strip so that the haulers can reach the ties. (R.M.F.)
strong enough to float a wedge Said of good coffee. (P.N.F.)
structural lumber That part of a log which comes from the middle portion of the log between the sap and the pith.
struggle A dance. (L.S.) LL
stub 1. A standing section of a tree broken off at a height of less than 20 feet, from which leaves and most of the branches have fallen. 2. Any snag. 3. A broken-off branch. 4. A dead-end spur road or railroad.
stubbing In horse logging, the action of a horse stepping on a stick, which pops up and jabs him. (Maine)
stubble jumper A part-time farmer who works in the woods. (P.N.F.)
stub camp See side camp.
stub line A line spliced to the end of a skyline. In some skidder rig-ups it runs through the skyline block to the heel tackle. See extension.
stub track A dead-end track.
stud monkey jungle A place where drifters assemble. (P.N.F.)
studjo In the lumber camps in the early 1800s, rabbit, venison or any meat plus potatoes, all stewed together. (Maine)
stump 1. The lower end of a tree remaining after the trunk has been cut off. 2. To break a tree across a stump when it is felled. 3. To blow stumps from a right-of-way. (P.C.F.)
stumpage 1. The price paid for standing timber. 2. The price paid for timber, alive or dead.
stumpage inspector A man who sees that the purchaser of standing timber cuts the right trees. (L.S.) LL
stump-busted A tree which has been broken by careless falling across a stump. (P.C.F.)
stump deduct The deduction from a faller's scale of the wood left above four feet in the stumps, which were scaled to discourage high stumps. (P.C.F.)
stump detective A forester who measured the waste in stumps and tops of trees. (N.F.) LL
stump height The height at which trees are to be cut in timber sale contracts.
stumping powder Dynamite or black powder used to blow up stumps. (P.C.F.)
stump jumper 1. A person who farms logged-off stump land. Also stump farmer. 2. A choker setter. (P.C.F.) 3. A high-wheeled log wagon.
stump logging See hot-log.
stump ranch A farm built in cutover land. Also stumpstead.
stump roller A vertical spool fastened to a stump and used to guide the mainline in ground lead yarding. (P.C.F.)
stump rot The decay in the base of a tree which the bucker cuts off the lower log. Also butt rot.
stump scale In disputes over cut timber or trespass, a fair count can be obtained by tallying the stumps from which the logs were taken.
stump scaler The man who figures the fallers' and buckers' scale at the stump in the woods.
stump shot The splinters that pull out on the stump and the butt end of a tree when the tree is felled. (Red.)
stump spud A tool with a crowbar point on one end and a small spoon-like shovel in the other end, used in digging holes under stumps to place a blasting charge. (P.C.F.)
stumpstead A farm built in cutover land. Also stump ranch.
stumpwood Wood cut into short lengths and ricked at the stump in pulpwood operations. (N.W.)
succor An uprooted tree that has fallen into the stream. (L.S.) Also sweeper. LL
sucker 1. A tree growing out of a main trunk. 2. A general term for the willowy young trees growing around the timber to be felled. 3. A small log floating alongside a larger log into which a driver put his peavey to help balance while riding on the larger log. (L.S.) LL
sucker line The rope that tied lumber cribs together on a river drive. (L.S.) LL
sucker sweep See succor.
sugar berth A maple sugar grove. (Maine)
sugarloaf A small round hill on top of a range of hills or mountains. (P.N.F.)
sugar pine (pinus lambertiana) The tree of the Western mountains and the largest of all the pines, it grows at an altitude of from 1,000 to 7,000 feet above sea level from the Coast and Cascade Ranges of Southern Oregon, along the Coast Range and Sierra Nevada of California, and through the southern part of the state in scattered stands into Lower California. It attains its largest size and heaviest stands in California from Tulare to El Dorado Counties, where it is found on cool moist sites of the west slope of the Sierra at elevations ranging from 4,500 to 6,000 feet. Botanically, sugar pine belongs to the white pine group, along with Eastern white pine and Western white pine. Almost the entire cut of sugar pine goes into lumber and its products.
sugler The man who accompanied a load or train of logs down to the landing, throwing or slinging water ahead of the train to make the logs skid more easily. He supervised the logs as they were brought in by the bull teams. (Red.) See water slinger.
suicide show A dangerous piece of ground to log. (British Columbia)
sulky A two-wheeled carrier used in yarding behind a cat in place of an arch.

Logging **tree length** jack pine in the Superior National Forest, in Minnesota, with a **Cat D8 tractor** and a **sulky**. *From "Logging," Brown, Wylie & Sons, 1949, Philip Schnell collection*

summer cigarettes Snoose or chewing tobacco, which is not outlawed by fire regulations. (P.C.F.)
summer show 1. A logging operation conducted on dirt roads; rock roads are needed during wet weather. 2. An operation at low altitude to avoid a shutdown by snow. (P.C.F.)
summerwood That portion of the annual ring which is dark and hard compared with the light-textured springwood portion.

A small **donkey** built by **Sumner** Iron Works. *Courtesy of Will Davis*

Sumner A steam donkey used in the Pacific Northwest woods in the early 1900s; built by Sumner Iron Works of Everett Washington. (P.N.F.)
Sunday night train In the Redwood Country, some of the lumber companies used to run trains from the woods camps to First Street or the nearest point to Eureka on Sunday night to pick up loggers to make sure that there would be a crew on Monday morning. Old-timers tell many stories of the times that the trains had to stop along the way to let the drunken loggers fight before arriving back to the camps from the skidroad. The best known on the West Coast was the one run by the SP&S down the Columbia River from Portland to Astoria, which served many camps. See drunkard and whiskey special.
Sunday school Poker. (S.F.)
sun kink A bend in a railroad track that warps in the hot sun, resulting from the light ballast.
sunshine logger A logger who does not come to work when it rains in the summer. (P.N.F.)
sun toast Bread. (L.S.) LL
supe, super A logging superintendent. (P.C.F.)
super heat Time and a half or overtime. (P.N.F.)
superstitions See folklore.
support tree A tree which holds up a hung tree or leaner. (P.C.F.)
supported line A skyline with a middle gin pole or middle tree to hold up the line where high ground prevents a free wing between the head tree and the tail tree.
surface yarding See ground-lead.
Surrey-Parker loader A timber-framed, steam-powered tunnel loader used in the South; sometimes used for skidding short distances as well as loading. After being positioned at a landing, the machine would be jacked up, and then supported by adjustable posts at each corner. Unloaded cars could be rolled beneath the frame as each car was loaded. Loading was usually done with end hooks. (S.F.)

A **Surrey-Parker loader** in use somewhere in the South. A Butler **log wagon** brings a load in for loading onto railroad cars. This type of wagon had a trip device to permit the wagon to drop its load without stopping. *From "Logging," Brown, Wylie & Sons, 1934, Philip Schnell collection*

survey See grade.
surveying Before the Civil War, this term was used to describe the measurement of lumber. The term "scaling" was substituted so that the word might not be confused with the measurement of land. (Richard G. Wood, A History of Lumbering in Maine, 1820-1861)
surveyor 1. A state official who inspects and tallies cargo lots of lumber. 2. See marker. 3. A scaler in the late 1700s and early 1800s. (Maine)
surveyor's ax A small light ax used by survey crews to cut small brush and trees to clear a line of survey. Also cruiser's ax.
suspended log skidder An early name for a Lidgerwood skidder. (P.C.F.)
suspended skyline One supported by trees along the line, or by a secondary skyline at right angles to the first line; required an openside carriage to get past the hangers. (P.C.F.)
suspension carriage system See Tyler system.
sustained yield The practice of replanting, or reforestation, on a large enough scale to provide a constant fixed output of timber in perpetuity.
Sutherlin side block A special type of block used in yarding behind a cat. The mainline from the arch is first run out to several logs. The block has an opening yoke formed by sister hooks. The opened block is slipped over the line, and also takes the eye of a choker, so that several logs may be strung on one line. Also side block. (P.C.F.)
swabber A skid greaser. (P.C.F.)
swage 1. A tool used to form teeth from the material of a saw as the teeth are filed away. Usually pronounced "swedge." 2. To form new teeth. Also upset.
swage a saw To spread the ends of the teeth of a crosscut saw.
swage bar A blacksmith's tool for shaping hot metal by holding the swage and striking with a hammer. LL
swage-ground Said of raker teeth that are ground to pattern at the factory.
swage-set 1. Said of a saw fitted so that the teeth stick out to each side making a cut wider than the saw is thick, giving good clearance in the cut. 2. To spread the teeth of a saw to make a wider cut.
swagin, swogun Bean soup in the lumber camps. (Maine)
swallow tail stump A stump, badly cut, with wood split off the butt log left sticking up in the air. (L.S.) See barber chair.
swamp To clear the ground of underbrush, fallen trees and other obstructions preparatory to constructing a logging road, landing, skid trail or even to felling and bucking. LL
swamp angel See rut cutter.
swamp auger A mythical tool. A common joke among the lumberjacks was to send a new hand back to the office to get a swamp auger. Usually when the victim found out he was the object of a joke, he packed up his belongings and left camp. When the foreman wanted

to fire a man, he told him to go into town and get a swamp auger. (L.S.) LL

swamper 1. A man who clears out the brush and other obstructions preparatory to moving timber from its natural habitat to lumber manufacturing facilities. Also beaver, busher. 2. In the Lake States they may cut the limbs off trees after felling. 3. In the Redwood Country, after the trees were felled and brush burned, the swampers built roads, bridges and chutes. 4. In the Southern Forests, the swampers cut pine limbs from logs and cut roadways. 5. An unskilled lumberjack who swamped. (Canada)

swamper's guide The imaginary book where greenhorns are told to look up answers to logging questions. (P.C.F.)

swamp fire bucket A galvanized bucket with a stout wire handle used for fire fighting. Cone-shaped or round-bottomed, so it could not be used for any other purpose, it was filled with water and hung up in all camp buildings for fire protection. (L.S.) LL

A **swamp fire bucket**. *From Mallory Catalog No. 12, 1925, Glen Comstock collection*

swamp gaboon An imaginary animal that left snowshoe tracks. (N.F.)

swamp hook A large all-purpose open hook on the end of a line or chain; used in animal skidding or on a chunk-out operation. It has a link in the head and a very sharp point to dig into a log or stump, and can be pulled free very easily when the job was done. Also jam hook.

A **swamp hook**. *From Mallory Catalog No. 12, 1925, Glen Comstock collection*

swamping ax A wide double-bitted ax used in rough work.

swamp out To clean up or sweep out a building. LL

swamp sauger A mythical, swamp-dwelling animal of great strength. When a log stuck in the mud or in a swamp, the men said it would take a swamp sauger to get it out. (L.S.) LL

swamp the tongs To hook one point of the tongs over the tong line in order to roll a log on the load.

swamp wagon A wagon with wheels out of a section of a log held together with iron bands or hoops like tires on wheels. Used for hauling timber out of swamps. (L.S.)

swamp water Tea. Tea was the national drink of the woodsmen from Maine, and it was standard in the camps until the Swedes introduced coffee in the Lake States in the 1870s. (L.S.)

sway-back A load of extra long logs on disconnected trucks with a sag in the middle. (P.C.F.)

sway bar 1. Either of the two timbers used in coupling the front and rear sleds of a logging sleigh, and the skids rested on the bar in loading logs. Also side pole. 2. The bar used to couple two logging cars. (N.F.) Also rooster. LL

sweat pads Pancakes. (L.S.) LL

The Benson Timber Company was famous for its long logs. Here a **Shay** and a train of **sway-back** loads are at the **landing**. *Courtesy of Don Roberts*

swede See persuader.
swede car A hand-pushed car used for dumping fill dirt along the grade in early logging railroad construction. (P.C.F.)
swede fiddle See crosscut saw.
swede flag See swede level.
swede hooks See lug hooks.
swede level 1. Two boards nailed to make a cross; used as a line of levels for the shovel runner or cat skinner making a grade. 2. A keel mark on a tree or stake at eye level of the cat skinner; used as a line of levels for making a grade. (P.C.F.)
swede saw A short saw with a bow-type handle, generally used to cut pulpwood. Also fin saw. (N.F.) See bucksaw. LL
Swedish brain food, Swedish condition powder See snoose.
Swedish snowstorm Said of snow taken out of the woods and placed on bare spots in the road where the log sleds passed. (L.S.)
Swedish steam Sweat raised on a hand-felling or bucking job. (P.C.F.)
sweenied Said of a horse whose shoulder muscles are atrophied. (N.W.) LL
sweep 1. A curving section in a log, tree or line. 2. A string of short logs used in rafting. 3. The spokes of a wheel on old-time unloading rigs which swept logs off cars as the train slowly pushed past the unloader. (P.C.F.) See Donovan unloader. 4. To move a skidder from one road to the next by taking the far end to a new tail tree, and then tightening up to take out the bight, sweeping down any small timber left after felling. (P.C.F.) 5. To follow a log drive downstream in a pointer boat, salvaging logs snagged on the banks after the main drive had passed. (Eastern Canada)
sweeper 1. A sapling bent over in yarding and swept around by the incoming turn. 2. A tree which has fallen into a river, its roots still fast on the bank; the current sweeps it back and forth. 3. An uprooted tree which has fallen into the stream and can sweep men off a raft or log or cause a jam. Also succor. (L.S.) LL
sweeps 1. The long oars used on bateaus. (L.S.) LL 2. Trees overhanging a stream which impede log driving. (Eastern Canada)
sweep the bull pen To use short logs in a loop to pull other logs out of the bull pen when making up a raft. (P.N.F.)
sweep the rear See sack the rear.
swell 1. The flared-out end of a butt log. 2. A big bump on any log.
swell-butted Said of a tree greatly enlarged at the base. (P.C.F.)
swelling The increase in size of wood due to increase in moisture content.
swell-knobbed Said of the shape of the butt of an ax handle. (L.S.)
swifter 1. A cross timber joining the boom sticks on either side of a raft. See raft. 2. A cable or rope used for the same thing. 3. A stick used to tighten a rope by twisting.
swifter gang The crew which ties together the logs in boom sections ready for towing.
swiftering machine A winch used in tightening up swifters and for other jobs on a booming grounds.
swifter knot A half-hitch with the rope doubled through to act as a lock. (P.N.F.)
swifter line A line used in tightening up the swifters in making up a raft.
swindle stick See cheat stick.
swing 1. To take logs to a landing from a distant tree to which they have been yarded. 2. To take logs from a road donkey to the landing. 3. To move logs by cat from a yarder or cold deck to the landing. 4. An older term for swinging logs free of the ground in skyline yarding. (P.C.F.)
swing boom See gate boom.
swing brakeman A brakeman who rides in the middle of a long logging train. The regular brakemen are on the engine and the caboose, or on the last car in the train. Also swing shack. (P.C.F.)
swing chain A short chain used to attach logs to big wheels in pine logging. (L.S.) LL
swing dingle A single sled with wood-shod runners and a tongue with lateral play; used in hauling logs down steep slopes on bare ground. (N.F.) Also loose-tongued sloop. LL
swing dog A cant hook.
swing donkey A donkey used to swing logs from one road or tree to another, when more than one yarder is required to reach the woods from the landing.
swinging bitch See log wrench.
swinging block A fall block.
swinging bunk Any bunk that pivoted on a king pin; used on disconnected trucks, motor trucks, etc.
swing line A cable used to move a hayrack or heel boom from side to side. (P.C.F.)
swing road A road over which logs are swung by a donkey or cat between a yarder or swing tree, and the head tree at the landing.
swings In river driving, a series of rafts built up along the bank and cut away from the shore and tied end-to-end together, making a long raft. The long raft or "swing" was then attached by a hawser to a buoy in midriver. (N.W.) See joints, dropping off.
swing shack See swing brakeman.
swing team In a logging team of six, the pair between the leaders and the butt team. (P.C.F.) See swing yoke.
swing the block To change the bull block and haulback block from the side of the tree which has been logged to permit logging on the unyarded side. The donkey is then swung to the new lead.
swing tree A spar tree to which logs are yarded, then swung to the landing. If the country is rough, a skyline is used for the swing, since it is cheaper to swing than to build a road to a second yarder tree.
swing yoke An ox team of three or more yokes, the pairs between the leaders and the wheelers. (S.F.) (App.) Also swing team.
switchback 1. Originally a series of levels on a railroad whereby trains could saw back and forth to get over a steep hill. 2. A horseshoe turn on a truck road.
switching them in Keeping the logs moving with the current on a log drive. (L.S.)
swivel A universal joint used everywhere in rigging to prevent lines from twisting.

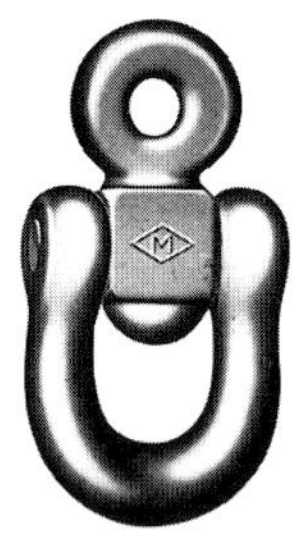

A **swivel**. *From Mallory Catalog No. 12, 1925, Glen Comstock collection*

swivel block See yarding block.
swivel chair logger The man in the office who handles the business end of the logging. (P.C.F.)
swiveled timber carrier See lug hooks.
swivel head 1. A type of fairlead used on the boom of some portable loading rigs. See universal lead. 2. A greenhorn who spends most of his time looking. (P.C.F.)

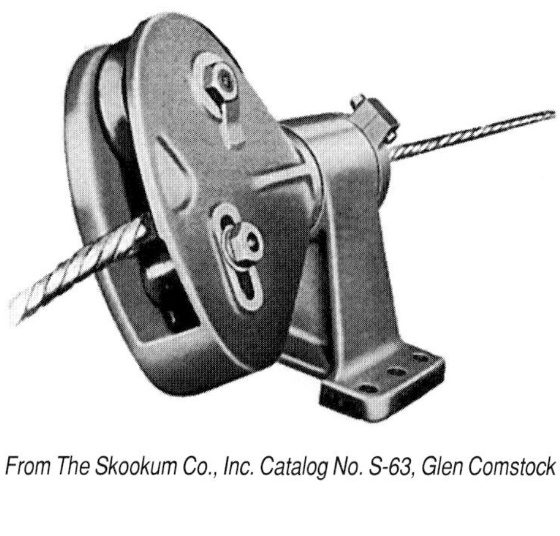

A **swivel head**. *From The Skookum Co., Inc. Catalog No. S-63, Glen Comstock collection*

swivelhips An active rigging man. This term, which was popular in football slang, can be traced back to 1914 in the woods. (P.C.F.)
swivel hook An early rigging hook with a swivel in the head.

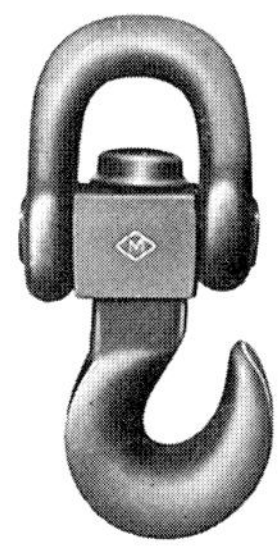

A **swivel hook**. *From Mallory Catalog No. 12, 1925, Glen Comstock collection*

swivel trucks Trucks under a skidding and loading railroad car permitting the unit to be run sideways off the tracks on temporary short rails at landings. These trucks also allowed the outfit to be turned end for end on any switch, thus saving time. Could turn a full 360 degrees under the car frame. (P.C.F.) See skidder car.
swivel yoke On some blocks, the heavy swivel-shackle arrangement through which a wire rope is run to attach a block to a tree or stump.

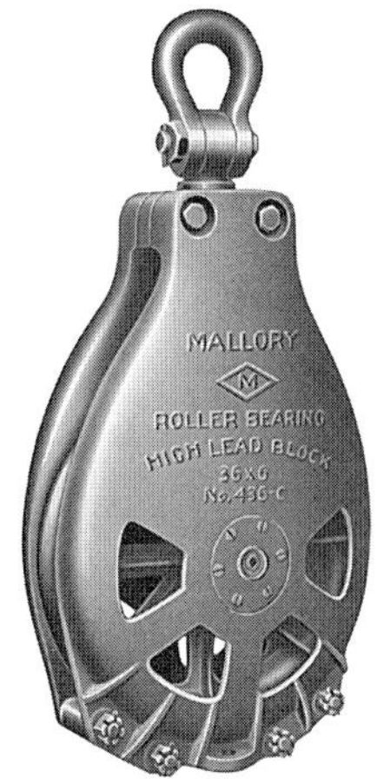

A **high lead block** equipped with a **swivel yoke**. *From Mallory Catalog No. 12, 1925, Glen Comstock collection*

T

tackle 1. All the rigging, blocks and cables, etc. hung on a spar tree. 2. Blocks and line rigged up to get a purchase, such as tightening a skyline. See block and fall tackle.
Tacoma A steam donkey built by the Puget Sound Iron and Steel Works of Tacoma, Washington. (P.N.F.)

A **Tacoma**; specifically a **Columbia River** 11x13 combination **swing** and **yarding donkey**. *From Mallory Catalog No. 6, Glen Comstock collection*

taffle A cook's helper. (L.S.) LL
tag 1. An older name for a butt chain. 2. A short line with Ds on either end, swiveling between butt rigging and choker. (British Columbia)
tag line 1. A short cable used between two sets of butt hooks to keep the chokers from tangling when more than one choker is used. 2. A short piece of line added to a mainline. 3. A rope attached to an end hook; used to swing logs into place when loading. 4. In the late 1800s and early 1900s, a short cable used from the road line to the grab hooks in loading. 5. A flexible cable used on the extreme end of the mainline on the South Bend system. (P.C.F.) See South Bend system.
tag sled A short sled which trailed behind the tote sled and was loaded with supplies. (L.S.)
tail 1. The rear end of anything. 2. To follow logs into a landing down a fore-and-aft road. 3. To ride the pig that's dogged to the last log in a turn. See P.F. man, pig.
tail block 1. See corner block. 2. Any block used on a tail tree in place of a tree shoe or tree jack.
tail chain A brake consisting of a heavy chain bound around the trailing ends of logs, used to check the speed of sleds on steep slopes. (N.F.)
tail crew The men who anchor the baloney on the opposite end of the donkey.
tail down 1. To snub the tail end of a turn of logs down a steep pitch when skidding with animals. 2. To shove the last rollway of logs into the river. 3. To take down a cold deck of logs and load it out. 4. To roll logs on a skidway to a point where they can be easily reached by the loading crew. (N.F.)
tail-down man A logger who works on a deck of logs being loaded out or rolled into a river. (P.C.F.)
tail end triangle A heavy steel link used with two shackles to connect a block to the tail tree strap on a duplex aerial yarding system. (P.C.F.) See duplex flyer.
tailer-in The man who tails down for a loading crew. (S.F.) LL
tail fastening A tail hold or anchorage for a tail tree.
tail grab See single coupler.
tail grip A cable grip hooked to a pair of grabs on the last log in the turn, then hooked to the haulback, thus serving as a brake on a fore-and-aft road or on a chute. (P.C.F.) Also tail hook.
tail hold, tail holt 1. Almost any purchase used to anchor something, for example, the stump or tree where the baloney is anchored or fastened. 2. An added line at the tail tree end of the skyline that is attached to an anchor. 3. A hold that is used to tie the back end of a cat or donkey to a stump. 4. The back end of a snub line, etc. 5. To rig a block to something to be moved, by running a line through the block from the donkey and out to a tail hold on a stump.
tail hook Another name for the dog attached to the rear log when skidding down a fore-and-aft road. See tail grip.
tailing down 1. Rolling logs down a skidway from the rear of the log pile. 2. Rolling logs to the deck pile. (N.F.)
tail jack A tree jack on a tail tree. See tree jack. (P.C.F.)
tail line 1. The line that was paid off a spool on a donkey. 2. An older name for a haulback.
tailor-made 1. Said of logging shoes made to order. 2. Said of commercial cigarettes.

tail over the dashboard Said of a logger on his way to town. (P.N.F.)
tail piece The handle attachment on the outer end of a two-man power saw. See stinger.
tail pole See tail tree.
tail rigger A logger who spends most of his time rigging tail trees. (P.C.F.)
tail rope An older term for a haulback.
tail sweep Load overhang behind a truck trailer.
tail track Usually the extension of the loading spur which was on a slight grade if possible. Empties were stored on the tail track to be fed past the loader, and the loads were dropped down still further to clear the landing. The returning locomotive usually had to switch out the loads before it could shove the returning empties into the tail track, although in some cases it was possible to incorporate a siding where the loads could be dropped.
tail tree A spar tree at the opposite end of the landing which supports the far end of the skyline in an overhead cable system of logging. Also tail pole. (S.F.) (P.C.F.)
tail(ing) up To collect derelict logs left stranded at the close of a river drive. (L.S.)
tailwater The upstream slack water behind a splash dam.
'taint weather, it's a disease A popular expression for a long period of foul weather. (P.N.F.)
takeaway The train that takes the logs to the mill. (S.F.)
take'em by the ears To slide logs into the river for driving by men using peaveys. (P.N.F.)
take five To take a short rest.
take his plate off the table To fire a man. (P.N.F.)
take the bridle off 1. To make a fast run. 2. To throw caution to the winds. (P.C.F.)
take the slack 1. To bunch cars in order to put slack into the couplers. As the locomotive started, it took out the slack from one set after another progressively, which allowed the engineer to overcome inertia more easily by getting a roll on the locomotive before it had all the weight to handle. 2. In rigging, the puncher had to take the slack out of the line before making a pull. 3. To tighten up on almost anything.
take the tree down 1. To remove the rigging from a spar tree. Also strip a tree. 2. To fell a tree. (P.C.F.)
take to the bushes To jump off a moving vehicle or train which appears to be heading for a wreck. (P.C.F.) Also hit the bush, join the birds.
take to the tall timber 1. To leave camp in a hurry. 2. To go back to camp after a short time in town. (P.C.F.) (L.S.) LL
takes two men and a boy to see clear to the top Said of a tall tree. (P.C.F.)
taking a drink standing up Said of one who falls into the water. (L.S.) LL
talking to his blankets Said of a man quickly leaving camp to avoid being fired, his blanket roll hung over his shoulder. (P.C.F.)
tallest tree For years the world's tallest known tree was believed to be Founders Tree in Humboldt Redwoods State Park in Humboldt County, California, which measured 352.6 feet. In 1957 a rival was discovered in nearby Rockefeller Forest, measuring 359.3 feet. Because of storm damage to the crown, the tree now measures 356.5 feet. In 1964 Dr. Paul Zahl and others from the staff of the National Geographic Society discovered four other giants in Redwood Creek Grove, east of Big Lagoon on the property of the Arcata Redwood Company: the world's tallest tree was measured at 367.8 feet; the second tallest at 367.5 feet; the third tallest at 364.3 feet; and the sixth tallest at 352.3 feet. The giant trees and the grove are now part of the Redwood National Park, which was signed into law by President Johnson on October 2, 1968.

The tallest redwood (367.8 feet) was named the Howard A. Libbey Tree in honor of the late president of the Arcata Redwood Company; the second tallest tree (367.4 feet) was named the Harry W. Cole Tree in honor of the late Mr. Cole, who managed the former Little River Redwood Company; the third tallest tree (364.3 feet) was named the National Geographic Society Tree in recognition of the part played by the Society in the discovery of these tall trees. The fourth and fifth tallest trees are the Rockefeller Tree (356.5 feet) and the Founders Tree (352.6 feet) which grow on opposite sides of Eel River, near the mouth of Bull Creek. The sixth tallest tree (352.3 feet) was named the Hill-Davis Tree in recognition of the Hill-Davis Company, which acquired the property containing these tall trees in 1905.
tallow In the bull team days, tallow was melted in tallow pots along the skidroad to be swabbed on the cross skids by the skid greaser.
tallowpot A fireman on a locomotive. In the early days he used to put tallow in the pots. (P.C.F.)
tallow pot 1. The pot or can along the skidroad in which tallow as melted for grease to be used on the cross skids. 2. The oilcan carried in a locomotive for applying oil to the main bearings. The tallow pot usually rode in a bracket on the backhead where it was kept warm, dating from the days when tallow was melted, and it was used to keep the engineer's regular oilcan filled. In later years, almost any oilcan came to be called a tallow pot by the engine crews. 3. A locomotive fireman.
tall uncut Said of virgin timber. (P.C.F.)
tally 1. A total count of logs, trees or footage in a unit of lumber. 2. Five chains or 330 feet, a standard measure of distance. 3. To mark down the scale of lumber or logs.
tally board A thin, smooth board used by a scaler to record the number or volume of logs. (L.S.) LL
tally card See tally sheet.
tallyman In the early days, four-man crews were used in big cruising jobs: two cruisers, a compassman and a tallyman to keep the figures as the cruisers reported the measurements on the trees.
tally sheet A card or sheet of paper on which is recorded the number of pieces and the grades of lumber. Also tally card.
tallywhacker A small mechanical counter generally worn on the index finger and activated by the thumb.
tame ape A logger, usually a rigging man. (P.C.F.)
tamp 1. To push ballast under ties. 2. To push dynamite gently into a hole. (P.C.F.)
tanbark 1. The barks of the hemlock and of several species of oak were for years the main source of the tannin supply in the United States. The eastern hemlock furnished the only coniferous tanbark in the East, and the western hemlock of the Pacific Coast were used to a limited extent. The eastern chestnut oak was used, and in the West the bark of the tanbark oak was extensively harvested. 2. A compound derived from hemlock bark and used in tanning leather.
tandem drum The most common type of donkey with two drums, one drum behind the other; the front drum handled the mainline, and the one behind handled the haulback. See road donkey, swing donkey.
tandem yarding Skidding logs down a road with a lineup of two or more donkeys. See swing, swing road.
tang The working end of a handled tool, such as a pick or hammer. (P.C.F.)
tangent A straightaway on a truck road or railroad. (P.C.F.)
tangle with a buzz saw To take a severe licking. (P.C.F.)
tank 1. A locomotive carrying fuel and water on the locomotive and requiring no tender. See saddle tank; side tank. 2. The water or oil tank of a locomotive. 3. A water or oil tank along the railroad track. 4. See sprinkler.
tank conductor One who has charge of the crew that operates a water tank or sprinkler, and who regulates the flow of water in icing logging roads. (L.S.) LL
tanker 1. A road-watering tank truck. 2. A fire-fighting tank truck. (P.C.F.)
tank heater 1. A sheet-iron cylinder that extends through a water tank or sprinkler, in which a fire is kept to prevent the water in the tank from freezing while icing logging roads in the extremely cold weather. (L.S.) LL 2. Oil-burning locomotives also had a tank heater, a steam pipe to the oil tank.
tanking Hauling water in a tank to ice a logging road. (N.F.) LL
tank teamster The man who drove the water tank for making ice roads in camps where logs were hauled by sleighs. See sprinkler. (L.S.) LL
tan one's hide To beat a person soundly.

taper The amount in inches in which a tree narrows in diameter from the butt to the top.
tap line A logging railroad that connected with a mainline road; could be a common carrier.
tap timber To tap a new stand of timber when extending a railroad or opening a new operation.
tar A low grade of coffee. (P.C.F.)
tar heel A logger from the South. (P.C.F.)
tarp 1. A canvas cover. 2. A tough flapjack. (P.C.F.)
tattle tale 1. A dangling chain to show the weight of a load by the distance from the chain to the axle. As logs are put on the truck, the chain goes down, telling the loader if he is violating the weight restrictions. 2. A meter on a truck to determine the running time. 3. A stick or limb propped in the frame of the last truck in a train of empties, which allowed the engine crew to determine at a glance if the train was complete. See tell-tale.
team 1. Any number of animals, horses or oxen, used in skidding in the 1800s. 2. Either a single horse or the wagon itself. (Maine)
team boss The man in charge of the skidding teams in a logging operation. (S.F.)
team snatch block A small light snatch block block with an attached hook. See snatch block.

Team snatch blocks; single-sheave, left; double-sheave, right. *From Mallory Catalog No. 12, 1925, Glen Comstock collection*

teamster The bull whacker. Newer inventions speeded the movement of logs in the woods. The steam log-hauler, which carried immense loads, increased the speed. Shortly came the Caterpillar tractor and then the steam skidder. These innovations relegated the oxen, the horses and that man of men, the teamster, to the past. Speed had changed the peaceful atmosphere of the old logging camp. The powerful team of oxen, their bows squeaking and chains clanking, and the "clear, profane voice of the ox teamster ringing through the forest" were gone. A donkey engine and the locomotive had replaced them. The teamster, missing his old world, sang:
Then I was king of the whole woods-crew,
And I ruled with an iron grip.
And never a slob on the whole dam' job
Dared give me any lip.
But, now, alas, my days are past;
There's no job for me here.
My bulls all killed and my place is filled
By a donkey engineer.
Instead of my stately team of bulls
All stepping along so fine,
A greasy old engine toots and coughs
And hauls in the turn with a line.
(Stuart Holbrook, "With Loggers in the Northeast," Travel, February, 1933, 35-39)
tear a hole in the timber To log in highball style. (P.C.F.)
tee A strip of iron about 6 inches long with a hole in the center, to which a short chain is attached; it is passed through a hole in a gate plank, turned crosswise, and used to hold the plank when tripped in a splash dam. (N.F.) See toggle.
telescope bag, telescope suitcase A fiber suitcase which held more gear. (L.S.) LL
tell-tale An indicator of any kind, such as a gauge on a water tank. See tattle tale.
tend To work at a specific job, such as tend hook, tend donkey, etc.
tender A logger who works at a specific job in the woods, such as hooktender. (P.C.F.)
tender skid See glancer.
tending out Stationing men at places along a stream where jams may form to keep the logs moving. (N.W.) LL
tend out To get logs moving down the river in good shape. (P.N.F.)
teredo A marine borer which destroys logs or timbers in salt water.
terminal The end, or terminus, of a lumber or log flume, near the mill, or where loading onto other forms of transportation was done. See elephant.
that's how the play come up An expression meaning "That's how it happened." (P.C.F.)
the baby's born Said when a job is finished. (P.C.F.)
the big bear walks 1. Said when it is time to eat. 2. Said when it is time for something to happen. (P.N.F.)
the cutover The land where the forests once stood. (L.S.) LL
the more you stir it the more it stinks Similar to "Let sleeping dogs lie." (P.N.F.)
there are two notches on the throttle Said to the donkey puncher who is working at half speed, meaning that a hooktender wants the donkey puncher to work at full speed.
thief stick See scale rule.
thieving rod A scale rule. (West Virginia)
thill A shaft of a vehicle.
thimble 1. A metal loop that forms an inside support for an eye in a line and helps to reduce wear on the line. 2. The nubbin on the end of a choker.

A **thimble** to support an eye in a line. *From Mill and Mine Supply Company catalog, 1916, Glen Comstock collection*

thin 1. To take out a small number of trees from a stand. 2. Said of a poor stand of timber.
thinning Partially cutting an immature stand to increase its rate of growth, to foster quality growth, to improve composition, to promote sanitation, to aid in litter decomposition, to obtain greater total yield, and so recover and use material that would otherwise be lost.
thinks he throws two shadows Said of a man who has a good opinion of himself. (P.C.F.)
third faller The low man on a power saw felling crew. He carries the tools, chips out the biscuit to make the undercut, watches for widow makers and does odd jobs. (P.C.F.)
third growth 1. The young growth. 2. Applied to a stand of young timber which followed the second growth in Coos County, Oregon. (P.C.F.)
third loader A third man who helps the first and second loaders by marking logs, setting up blocks, etc. He does not do the actual loading, and he is only used when there is much loading to do. (P.C.F.)
third rigger A second rigger's helper working on a skidder or slackline show. (P.C.F.)
Thompson loading boom An early one-tong heel boom loader. (P.N.F.)
thousand A thousand board feet, the unit of measure for timber, logs or lumber.
thousand legs A geared logging locomotive. (App.)
thread To reeve a line through a logging block.
three-donkey hold See block hold.
three-donkey show A logging operation with a yarder on the far end, a roader in the middle and a swing donkey at the landing.
three-drum loader A loading donkey with two mainline loading drums and a car-moving drum. Usually used with hayrack boom, heel boom, and crotchline loading systems.

A three-drum loader. *From Willamette Iron & Steel Works catalog, 1920, Glen Comstock collection*

three-drum road engine A road donkey with two haulbacks and one main line. The second haulback was used so that the donkey could haul alternately from one road to another when working with two yarders. (P.C.F.)

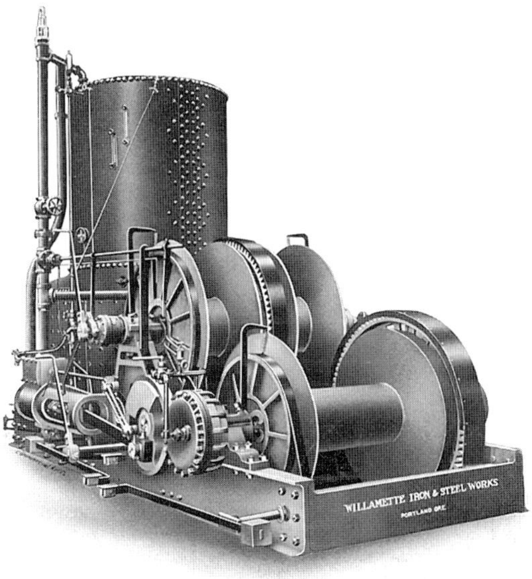

A three-drum road engine. *From Willamette Iron & Steel Company catalog, 1913, Glen Comstock collection*

three-gang system A trick worked by man-sharks who split the take with a grafting foreman. The loggers paid a fee to the man-shark for a job, then went out to the camp and were told to go back to the man-shark and pay another fee for another job. There were always three gangs of men, one going, one working and one coming.
Three L Ladies' Loyal Legion, the ladies' auxiliary of the Four L.
three-lick snipe A snipe hewed in three angles on the end of a log. (P.N.F.)
three-trip road Three round trips (three turns) each day. (N.F.) LL
throat 1. The narrow part of a block. 2. The far end of an undercut. 3. The gullet of a saw.
throttle jerker The engineer on a locomotive. (P.C.F.)
throw 1. The distance which a tree jumps off a stump in falling. 2. The thrust of water and floating logs against the outside of a flume on a curve. 3. The distance covered by the moving of a handle, lever or rig of any kind. (P.C.F.) 4. To fell a tree in a certain direction, particularly against the lean. See wedge a tree.
throw a sag into her To set a cant hook firmly in a log, bending the handle, in working hard at log rolling. (P.C.F.)
throw a whingding 1. Said of one who gets very angry. 2. To throw a party, usually a drinking party. (P.C.F.)
throwing the sieve To thin out the logging crew. (L.S.) LL
throw in with 1. To go to work for an outfit. 2. To agree with a man in an argument.
throw line See trip line.
throw-out 1. A timber placed at the mouth of a slide to direct the logs. 2. The junction of railroad tracks in railroad logging. (L.S.) See frog. LL
throw the bedroll in the bunkhouse To go to work for an outfit. (P.C.F.)
throw the book away 1. To do a job the way a man wants to instead of following orders. 2. To do a job in an unorthodox manner.
throw the rope over the top To work haphazardly as to cause injury or death. The expression originated from the high climber who accidentally threw his safety rope over the top of the tree and came down in a hurry. (P.N.F.)
throw up 1. To quit. 2. To give up work on a contract. (P.N.F.)
tick 1. A lazy man. (P.C.F.) 2. A fabric case which was stuffed with straw to make a mattress. 3. To strike lightly.
tide A freshet. (App.)
tie 1. To make a connection, such as to hook up, to hook together, to get hold of, etc. 2. To close up a survey line.
tie a can on To fire a man. (P.N.F.)
tie back To put a short line on a stump which may be too loose to be used alone as a tail hold or guy line stump. The strain is thus passed back to another stump which may be too unhandy to be used directly.
tie beam A timber across the sluiceway of a dam to hold the side walls riding at the top. (L.S.) LL
tie block The timbers that held two sleigh runners together in a bob and upon which the bunks were laid. (L.S.) LL
tie bucker A man hired to load railroad ties on a car. (L.S.) LL
tie cutter See tie hack.
tied Said of a logging railroad grade on which ties have been placed ready for rail laying. (P.C.F.)
tied on To work on a drive. (App.)
tied up 1. Said of a camp that is shut down. 2. Said of a train crew whose day's work is done.
tie'em down 1. To set the hand brakes on a railroad car. 2. To bind a load of logs to a truck.
tie'er down 1. To bind a load of logs to a car or truck. 2. To "close 'er down." 3. To fasten the end of a cat or donkey to prevent lifting. (P.C.F.)
tie hack(er) The man who spit or hewed ties by hand. (P.C.F.) Also hack, hacker, tie cutter, tie man (S.F.), tie whacker.
tie one on To get drunk.
tie piler A tool used for handling ties. (P.C.F.) See lug hooks.
tier A row of logs in a flat raft; usually two tiers make one section of the raft.
ties Lumber cut for the use of railroad ties.
tie up 1. To quit work. 2. To stop work because of a breakdown. (P.C.F.)
tie-up line A cable used to hold the base of a spar tree in place while the tree is being raised. (P.C.F.)
tie whacker See tie hack.
tie wire A small flexible wire used to tie up the ends of a wire rope before cutting it, or before fitting on an attachment.
tiger juice A powerful drink. (P.C.F.)
tiger shit Sticky mud. (P.N.F.)
tight as a growed bear in a cub's skin 1. Said of anything that is stuck. 2. Said of a timekeeper refusing an advance on wages. (P.N.F.)
tight bark pole A pole that is cut during the winter when the sap is down. (P.C.F.)
tightening block A block with up to four sheaves to give extra purchase in tightening the skyline in a skidder show; used in pairs. See block and fall tackle, heel block drum, heel tackle.

A pair of 2- and 3-sheave **tightening blocks**. *From Mallory Catalog No. 12, 1925, Glen Comstock collection*

A **standing skyline**, or **tight line**, somewhere in the Northwest. The **tightening line** is connected to the upper **tightening block**. It then passes through a block in the spar tree, and is anchored to what looks like two stumps. The tightening blocks are **reeved** with four parts of line to gain extra purchase; the lower block is fastened to a **tail hold** stump with a strap. It appears that this spar tree is in the process of **rigging up**. *From Mallory Catalog No. 12, 1925, Glen Comstock collection*

tightening line An extension on a skyline to which heel tackle was attached to pull it tight.

tight line 1. A standing skyline used in logging. 2. To hold a mainline taut by braking the haulback as the mainline is pulled in. 3. To keep a strain on a line holding an object so that men can work behind the object. 4. To reef on the haulback and mainline at the same time, thus tightening the line; often used to throw a log over a hangup. 5. To move rigging from one road to another by changing the tail blocks and then tightening the mainline to pull the cable to the new road. 6. To fire a man. 7. To yard or load with a tight line. (P.C.F.)

tight line show Difficult country to log. (P.C.F.)

tight rope overhead system An older name for a skyline system of logging; used in the early 1900s. (P.C.F.)

tight skyline system See North Bend system, skidder system, South Bend system, Tyler system.

Tillamook Burn 1. In Northwest Oregon, the total area of about 350,000 acres covered by the three great fires in Tillamook and neighboring counties in 1933, 1939 and 1945. 2. Any big fire. (P.N.F.)

tillicum A friend, a man, a person or people (from the Chinook jargon). (P.N.F.)

tilter 1. A log dump in which the platform is tilted toward the water. 2. A log dump in which the rail on the land side is raised to tilt cars. 3. A log dump with a balanced platform tilted by hand or machine. (P.C.F.)

timber 1. The uncut forest, merchantable logs and trees. 2. A large squared piece of wood used in construction, as dock timber, etc. "Timber" as "building material generally; material for the construction of houses, ships, etc. or of any manufactured article" is recorded in the Oxford English Dictionary for 900, and by 1100 was used as "wood for the building of houses."

tim-m-ber-r! 1. The logger or lumberjack warning to all in a nearby area that a tree is about to fall; often accompanied by "up the hill" or "down the hill." 2. The warning cry heard in a saloon that a man was buying drinks for the house. (L.S.) LL

timber baron 1. A derogatory name for a man who owns a logging outfit. 2. A large timber speculator or owner. Also pine hog.

timber beast A logger or lumberjack working in the woods.

timber berth 1. A timber claim. 2. In British Columbia, a railroad land grant granted by the Dominion Government.

timber-bound 1. Said of a bucker or faller who has his saw stuck. Usually if a cross cut tightened up on a saw, it was called a "pinched" saw, although "timber-bound" would fit. 2. Any tree, or log, containing interior stresses is said to be "timber-bound." 3. Anything hard to move.

timber brand Initials or characters made on the top of raft lumber with a hammer to denote ownership. (App.)

timber break A log that is broken in two when a tree is felled across gullies, rocks or stumps.

timber chute See log slide.

timber claim A tract of timber owned and logged by one outfit.

timber compass A sight gun used in felling a tree. See gunstick.

timber contract The right to purchase standing timber without getting title to the land. (L.S.) Also timber right. LL

timber fitter A man whose job was to size up a tree, to determine the exact spot where it was to fall and to notch it accordingly. (L.S.)

timber grapple See lug hooks.

timber hitch A hitch made by placing a line around a tree and over the running line and twisting the cable around itself. (P.C.F.)

timber hooks See lug hooks.

timber-hungry 1. Said of a logging show about to shut down because it has run out of timber. 2. A highball logging operator who has cleaned up a big show and is looking for a bigger one. 3. A highball outfit that is looking for production.

timber inspector 1. A bull buck. 2. A cruiser. 3. A big beetle. (P.C.F.)

timberland The uncut forest.

timber license In British Columbia, a license to cut and remove crown timber. The license is granted by legislative act.

timber limit In British Columbia, the land specified in a timber license.

timber looker A timber cruiser interested in speculation. (P.C.F.)

timber marking The selection and indication of trees to be cut or left in a cutting operation.

timber pirates The men who stole logs as they drifted downstream, were piled on landings or were grounded during a river drive. (L.S.)

timber plugger One who surreptitiously plugs knotholes and bad knots, especially on spar timbers. (S.F.)

timber right See timber contract.

timber road A fore-and-aft road.

timber sale The contract area when public timber is sold to a logging operator.

timber scribe A stylus used by a surveyor to mark witness trees. LL

timber shave A draw-shave with curved blade for removing bark from round wood. (L.S.) LL

timber shaver See barker, spud.
timber slide See log slide.
timber-tetched Said of a man slightly irrational from being alone in the woods too long. (P.C.F.)
timber tongs See lug hooks.
timber tramp A logger who moves from camp to camp, looking for an easier job. (P.C.F.)
timber trespass Cutting timber on someone else's land. (L.S.) LL
timber wheels See big wheels.
timber wolf A cruiser. (L.S.) LL
time Pay.
time checks A kind of script that was issued to lumberjacks and could be discounted at banks or business establishments. (L.S.) LL
tin coat A heavy water-repellent canvas coat worn in the woods. (P.C.F.) See tin pants.
tin dinner Lunch in a dinner bucket.
tin hat An aluminum hat that loggers and some mill workers wear for safety. (P.C.F.) See hard hat.
tinker 1. A poor mechanic. (P.C.F.) 2. A wood butcher or carpenter. (L.S.)
tinker shop A small log building in some logging camps where axes and saws and pots and pans were repaired. (L.S.)
tin pants Heavy canvas trousers that loggers coated with paraffin during the rainy season. (P.C.F.) (British Columbia)
tin-pants show 1. A period of rainy weather. 2. A logging show in very wet country. (P.C.F.)
tissue flipper A railroad conductor. (L.S.) LL
tit 1. The throttle of a locomotive or donkey. 2. A sharp growth on a log. 3. Any object that sticks out.
Titan 1. An early donkey made by the Smith and Watson Iron Works of Portland, Oregon. 2. A brand name of an early gasoline-powered chain saw.
toe 1. The bottom part of a slope or hill. 2. The end of something.
toe piling Sharpened poles or timber which are driven next to the upstream face of the mudsills of a dam to prevent water from getting under the foundations. Also toe spiling. LL
toe ring The heavy ring or ferrule on the end of a cant hook. LL
toe splitter An adz. (P.C.F.)
togged out Said of one who is dressed up.
toggle 1. A stick pushed through two eyes in wire rope to hold them together, or through one eye to make it fast to something. 2. A metal bar with a hole in the center by which it is attached to something. Turned lengthwise the toggle can pass through a hole; turned sideways it acts as a lock, as on a boom chain.
toggle and ring See boom chain.
toggle chain 1. A short piece of chain used to tie up a piece of line. 2. See boom chain. 3. A short chain with a ring at one end and a toggle hook and a ring at the other, fastened to the sway bar or bunk of a logging sled and used to regulate the length of a binding chain. (N.F.) LL
toggle hook 1. A hook passed down through the hole in a boom stick to fish the toggle out of the water. 2. A hook used in closing the chain which held the receding line block on the tail tree in early skidder shows. 3. A grab hook with a long shank, used on a toggle chain. (N.F.) LL
toggle knocker The man who knocked logs loose from the binder chain, particularly when working with big wheels. (P.C.F.) LL
toggler The man who fastened chains around logs on railroad cars and toggled them tight. (P.C.F.)
toggle with ring See boom chain.
tolerant Said of a tree capable of developing and growing in the shade of and in competition with other trees.
toll dam A dam by charter, built by a person or group who collected toll on passing logs until the original costs plus expenses and interest were paid for, when the toll was reduced so as to pay only for necessary repairs. (Maine)
tombstone A slab torn from the bole which adheres to the stump when a tree is felled. (S.F.) See barber chair. LL
tomcat A log-hauling cat with a wide tread and a built-on arch which is excellent for mud or winter logging. The cat was built in Portland, Oregon in 1943 and designed by Ted Flynn, Newell Wright, Axel Brandstrom and Tom Coldwell of the U.S. Forest Service.
Tommy Moore block 1. A mainline block with a wide throat and sheave used originally in ground lead logging which allowed the butt rigging to pull through without fouling. 2. Any block with a wide throat. Also butt chain block, open face, wide-face block.

This unidentified **riggin' crew** poses with a **Tommy Moore block**. The wide **throat** was developed to allow the heavy **butt rigging** to pass through the block without hanging up. The block is anchored with a **tail hold** on a stump, using a **strap**. The block was first used near Cathlamet, Washington, on the Lower Columbia River, about 1905. *Courtesy of Mrs. William Pringle*

tong To handle logs with skidding tongs. (N.F.)
tong block A block on a loading boom or hanging from a guy line, from which tongs are hung. See hayrack boom, McLean boom loader, heel boom, duplex loading system.
tonged Said of a man caught in a bad situation. (P.C.F.)
tong hooker 1. A logger who sets tongs in the logs to be loaded. 2. A choker setter on early day skidder shows when tongs were occasionally used with the butt rigging instead of chokers. (P.C.F.)
tonging Handling logs with skidding tongs. (L.S.) LL
tong line 1. The loading line on a loading boom or duplex loading system. See hayrack boom, heel boom, McLean boom loader, duplex loading system. 2. On a skidder the line that holds the head hook and second hook, and is connected to the mainline. 3. A rope that was attached to the tongs to help haul them into place. (P.C.F.)
tong line block See tong block.
tong line bumper A metal loop babbitted on the end of the tong line to prevent the shackle from jamming the loading block when the tongs are pulled up to the block.

A **tong line bumper**. *From Young Iron Works Catalog No. 52, 1952, Philip Schnell collection*

tongman See tong setter.
tong puller See tong shaker.
tong rigger A loader's helper. (P.C.F.)
tongs The big scissor hooks used for loading logs, and sometimes for skidding in the Pine Country. See loading tongs, skidding tongs.
tong setter In the early days, the logger who set the skidding tongs (forerunner of modern chokers) to bring in the logs. (P.C.F.)
tong shaker A logger who handles the tongs in loading logs. The second loader loosened the tongs on the load, but the first loader was the one who set the tongs on the ground. (P.C.F.) See second loader.
tongsman A choker setter. (P.C.F.)
tong unhooker 1. The man who unhooked the tongs when loading cars on a skidder show. (P.C.F.) 2. A man stationed near the power skidder who releases the skidding tongs or removes the chokers from logs which have been drawn alongside the railroad. (S.F.)
tonsil polish Booze. (P.C.F.)
too bad he doesn't have two heads so we could knock them together Said of a man who makes stupid remarks. (P.N.F.)
took water Said of a bully who challenged a man and then quit fighting. (L.S.) LL
tool boat Another name for the pig. (P.C.F.)
too low to oil the hinges on a skunk's stinker Said of a no good man. (P.N.F.)
too much brush to work in A common excuse given by a man quitting a camp. (P.C.F.)
toothpicks 1. A derogatory term given to the timber of the Great Lakes and the North Woods. (P.C.F.) 2. The smaller second growth or even small old growth.
top 1. The top diameter size to which the small end of a log may be cut. 2. To cut the top off a down tree. 3. To cut the top out of a tree to be used as a spar tree. (P.C.F.)
top a tree, top a spar tree To climb, limb and top a tree to prepare it for a spar tree. (P.C.F.) Also topping.
top barker In early logging, the man who barked all the log he could reach, down to the ground on both sides. (P.C.F.)
top chains Chains used to secure the upper tiers of a load of logs after the capacity of the regular binding chains has been filled. (P.C.F.)
top guy Any one of the guy lines that are attached to the top of a spar tree.

The **high climber** has just **topped a spar tree** for the Oregon-American Lumber Company, near Vernonia, Oregon. This is the moment of truth for the climber. As the top falls, the **spar** will sway violently and the climber must hold on tightly, lest his rope be thrown over the top of the spar. *Harold M. Brown photo, Labbe collection*

top load The logs piled more than one tier high on a car or truck.
top loader 1. The head loader who is in charge of loading operations at the landing. 2. Usually the head man on top of a load of logs while piling logs on the skidway or on the sled. (L.S.)
top loader platform A platform on a truck's cab guard for the second loader.
top-loading Stacking and arranging the logs on top of the car or sleigh. (L.S.) See sky-hooking. LL
top man The foreman. LL
topog 1. A topographic map, showing the contours of the ground. 2. To make a topog map. (P.C.F.)
topper A high climber. (P.C.F.)
topping See top a tree.
top sawyer The man who stands above the log and aids in operating a whip saw. LL
tops the spars Said of the high climber who cuts off the tops of spar trees.
torpedo 1. A strongly-built tank, shaped like a torpedo; used to carry fuel oil from the landing out to a yarder donkey on a swing show. The torpedo was designed to be carried into the woods by the yarding lines. 2. A charge of dynamite. 3. A loud explosive that was put on the track in case of an emergency to stop another train. (P.C.F.)
tote To haul supplies to a logging camp. (N.F.) LL
tote road A supply road to camp. (N.F.) Also pike road and portage road. LL Also fly road, hayroad.
tote sled A sled shod with wood; used for hauling supplies over bare ground into a logging camp. (L.S.) LL
tote team A team used in freighting supplies to camp. (N.F.) LL
tote teamster The driver who brought supplies into camp. (N.F.) LL
touch'er off 1. To set a fire. 2. To explode a charge in blasting. (P.C.F.)
tourist A logger who moves often when the job gets too difficult. (P.N.F.)
tow A boom of logs made up to be handled in one tow by a tug boat.

tow chain The big chain that was put down in the middle of an ocean-going log raft. Smaller chains ran out on both sides to bind the raft together and to spread the strain of towing. See ocean-going raft.
tower The steel tower that was used with a steel spar skidder. Later portable towers were designed for use with cats and truck-mounted yarders. See Lidgerwood.
tower skidder A skidder with a built-on steel tower; no spar tree was used. Also steel spar skidder. See Lidgerwood.
towhead A sandbar on an island with grass and willows growing on it. (L.S.)
tow line The cable from a raft to a tug boat.
town clown A small town policeman. LL
tow team An extra team stationed at an incline in a logging road to assist the regular teams in pulling the loaded sleigh up the hill. (N.F.) Also snatch team. LL
trace chain In the early days of animal logging, the short chain from the traces to the singletree or doubletree.
track 1. In railroad logging, the railroad which served the camp was known as the track. (P.C.F.) 2. The endless tread of a tractor.
track cable See skyline.
track layer 1. Any one of several rigs for taking up and laying track, such as the Clyde or Norby. It usually consisted of some type of overhead boom for passing ties and rail ahead or back to flat cars. Also Japanese skidder. 2. A machine mounted on tracks, as a track-laying shovel, used in the early days when most machines were rail-mounted.

A **Clyde** ridgid boom **track layer** in operation, carrying a bundle of ties forward for laying track. These machines required two flat cars—one to support the machine, and another behind it to carry the rails and ties. *From Clyde Track Machine Catalog TM, Glen Comstock collection*

track line See skyline.
track pads The individual sections of a crawler track.
track scale The scale of logs loaded on railroad cars or trucks.
track side The logging side operating at the landing or track in a railroad show; opposed to the yarder side back in the woods. (P.C.F.)
track toad A derail device. (P.C.F.)
track walker The man who patrolled the track, looking for wash outs, broken rails or any other dangerous road conditions.
trackway cable See skyline.
track wheels The endless treads on a logging arch or trailer. See Athey wheels.
traction An oil-burning or gasoline traction engine used in hauling log trucks. (California)
traction car A power car used on a logging incline.
traction line The pulling line in a tramway system.
tractor In early day woods work, a tractor was a three- or four-wheeled steam traction engine. There are records of such machines pulling log carts in the Pine Country as early as 1893. Later three- and four-wheeled gas machines appeared. Both types were used to pull sleighs or log carts at first; not for yarding. The crawler tractor, running on endless treads, made it possible to go into the woods after logs just as bulls and horses had done. One of the earliest loggers to use tractors was W. K. Dyche in the Pacific Northwest.
tractor hook See cat hook.
tractor skidder See lighthouse cat.
tractor wheels In the Pine country, a set of big wheels hauled by a cat.

Tractor wheels being hauled by small **tractors** on the Klamath Indian Reservation in South Central Oregon. *From Logging, Principles and Practices, Brown, Wiley and Sons, 1934, Philip Schnell collection*

trade blankets for a lantern 1. Said of one who works on a night shift. 2. Said of a railroad crew working all night to haul the last loads of the day out to the dump, to bring back the empties and to spot them at various landings ready for the next day's work. (P.C.F.)
trader rat See pack rat.
trading stock 1. Pitch wood used for lighting fires. 2. An attractive woman, so-called because pitch wood was sometimes traded to Indian women for their favors. (P.C.F.) Also squaw timber, squaw wood.
trail 1. A path in the woods. 2. A turn of logs being hauled down a skidroad. 3. To skid logs down a skidroad with horses or tractors. 4. To pull or skid logs between the rails by using a locomotive. (P.C.F.).

Trailing logs on a 9% grade on the 3-foot **narrow gauge** railroad of the Bridal Veil Lumber Company, near Bridal Veil, east of Portland, Oregon. The company began hauling with **flat cars**, but found them difficult to handle on the steep grades. They laid claim to the development of this system of handling logs, which was followed by a number of other operators in the Northwest with similar conditions. *Courtesy of D.L. Stearns*

trail chute A log chute where the incoming logs were pulled or trailed by team or donkey. (P.C.F.) See trailing slide.
trail dogs 1. Any type of dog used to connect up a number of logs for skidding. 2. The dogs used to tie together all the logs bucked from one tree and made up into a turn.

trailer 1. The tail unit pulled by a truck. 2. A flatcar pushed ahead of an engine to couple on to cars standing on the track which are too light to support the locomotive. Also idler 3. The last log in the turn. (P.C.F.)
trailer monkey The man who handles the steering trailer sometimes used with extra long loads. (P.C.F.)
trailers Several logging sleighs hitched one behind another and pulled by four or five teams or horses driven by one man, thus saving a teamster's wages. (N.F.) (Eastern Canada) LL
trailing down To bring logs from the piles in the woods. (N.F.) LL
trailing logs See trail.
trailing slide A slide on which the grade is so low that animals are required to move the logs. (App.) See trail chute.

Lumberjacks constructing a **trailing slide** down a mountain slope near Corbett, Pennsylvania. *Courtesy of W. Casler*

trail log The end log in a turn which held a dog which was attached to a tag line to the mainline, if trailing by donkey, or to the butt chain if horse trailing. (P.C.F.)
trail of logs In the days of animal skidding, a turn of logs being brought to the landing, arranged end to end.
trail slide An earth skidding trail, reinforced on the lower side by a fender skid. (App.)

The lower end of a **trail slide** in Idaho. A **fender skid** can be seen to the right of the slide, used to keep the logs in the **gutter road**. Note the **corduroy road** over which the **tow team** travels. *From Logging, Bryant, Wiley and Sons, 1923, Philip Schnell collection*

train 1. In the early days, a string of logs all set to be pulled from the woods. 2. In Mendocino County, California, a locomotive.
train delayer A dispatcher. (P.C.F.)
traineau A Quebec sleigh hauled by one horse. (Quebec)
tram 1. A wooden railway on which log carts were pulled by oxen or horses. The trucks were fitted with large concave wheels to fit the poles that served as rails. See pole road; pole tram road. 2. The railroad woods spurs. (Pennsylvania)

A horse **tram** of the Eel River Valley Lumber Company at Newberg, in Humboldt County, California, with two **redwood** logs. The wheels of the tram car are concave to fit the poles used as rails, which are spaced far enough apart to allow the horses to walk between them. *Freese & Fetrow photo, courtesy of M. Koch*

tramped Said of a man being fired. (P.N.F.)
tramp logger A lumberjack who changed jobs frequently. (L.S.)
tramp the road To pack the snow by tramping with men or oxen so that the road would freeze. (L.S.) LL
tramway 1. See tram. 2. See pole road.
tranes Sleighs. (Canada) LL
transfer 1. A system of preloading extra logging trailers to be used on a long haul. After the loading crew is gone for the day, truck drivers can come into the woods, leave the empty trailers and pick up the loads. See preload. 2. The reload where whole truck loads are transferred to rail cars. See reload.
transfer block 1. A block used with the skyline when changing roads in skidder logging. See transfer line. 2. A block with a wide throat to pass various pieces of rigging.
transfer drum A drum on a skidder which handles the transfer line.
transfer engine A donkey used to haul railroad logging trucks from the top of an incline along a piece of more or less level track to the loading point. (P.C.F.)
transfer line A cable used to haul the skyline to a new position when changing roads in skidder logging.
transport 1. A crew truck. (Red.) 2. A heavily built low trailer used for carrying cats. 3. To move or carry.
trapeze line In river driving, a second line suspended from a cable or checkline from which the checker clung to sort logs in a river. (N.W.)
trapezed Said of man being thrown in the air when caught in the bight of a line. (P.C.F.)
trap tree A system of beetle control used at one time in the Pine Country. A tree was cut down to attract beetles; when the bark was full of bugs, it was peeled off and burned.
traveler 1. A tramp logger. 2. A carriage used on some slackline systems. (P.C.F.)
traveling block 1. A block which travels on the sail guy supporting a heel boom loader; it carries the loading line. 2. An early block riding on the mainline in a swing operation. The cheek plates of the block covered only the lower half of the sheave as it hung from the line. Also running block. See heel boom.

A **traveling block** for use with a **heel boom**. The small sheave rides on the **sail guy**; the **mainline** of the **loader** is reeved through the large sheave; the **tong line** is attached to the **shackle**. *From Young Iron Works Catalog No. 52, 1952, Philip Schnell collection*

traveling dandruff Lice. (L.S.) LL
traveling lead roller An early fairlead that moved in a slot back and forth across a donkey frame to spool the line evenly on the very wide main drum. See cable saver. (P.C.F.)
travis pole A pole laid crosswise to connect two floating logs. Waterlogged logs were wired to the travis pole to float them to the mill. (L.S.) LL
travois, travoy A drag used to haul logs from the woods to the skidway, usually made from the natural fork of a tree with a cross piece bolted midway in the V. One end of the log rested on the travois and the other end dragged on the snow or ground. (N.F.) Also crazy-drag, ginnie, snow snake. (L.S.) See go-devil. LL
travois road A long skidroad used to drag logs from the woods to the skidway. (N.F.)
tread 1. The groove of a sheave. 2. The running surface of a road.
tree See spar tree.
tree controller A patented rig with jaws opening by a screw; used like a big wedge to fell trees by prying them against the lean. Also tree faller, tree jack.

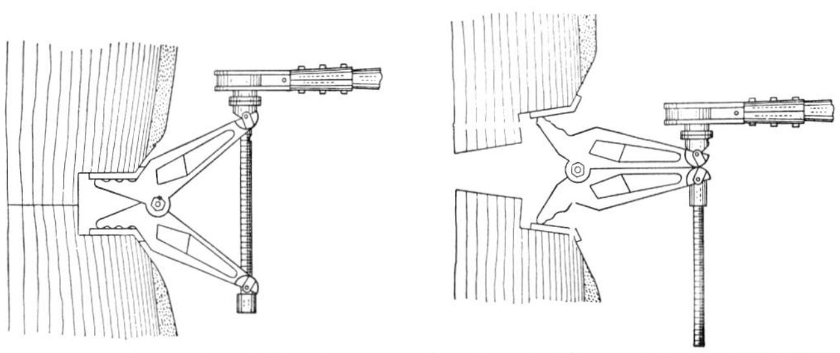

A **tree controller**, or **tree faller**. *From Logging, Bryant, Wiley and Sons, 1923, Philip Schnell collection*

treed Said of something stuck or of a person trying to solve a difficult problem. (P.C.F.)
tree faller See tree controller.
tree farm An area of forest land of any size, in private ownership, devoted primarily to the continuous production of commercial timber crops under good forest practices.
tree hanger A steel bracket supporting a skyline on a long haul skidder show. (P.C.F.)
tree improvement A general term describing the practice of producing genetically superior forest trees.
tree iron A heavy iron with one or two hooks which is bolted to a spar tree to make a support for blocks, straps, or other rigging, and to prevent cutting of the wood by cables where the guy lines and straps are hung. Also hook plate, tree plate.
tree jack 1. A support for a tight skyline at the head tree end, to allow frequent lowering of the skyline. Similar to a tree shoe, except that several small sheaves, placed in a large arc pattern, replace the hardwood block in the center, over which the line is carried. Also skyline sheave. See tree shoe. 2. See tree controller.
tree length log A long log usually containing the entire merchantable part of a felled tree. (P.C.F.)

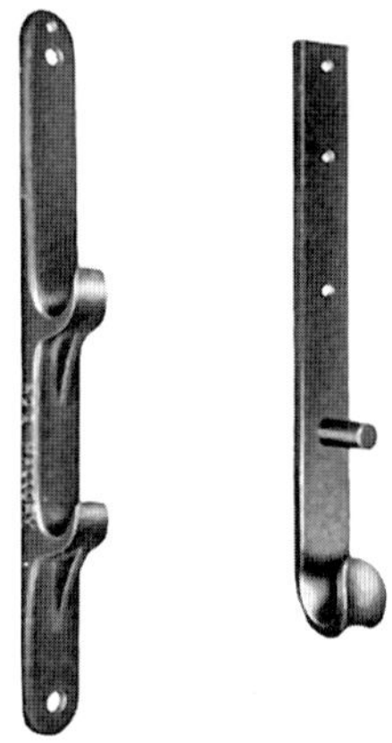

Two types of **tree irons**. *From Mallory Catalog No. 48, 1948, Philip Schnell collection*

A six-sheave **tree jack**. *From Mallory Catalog No. 12, 1925, Glen Comstock collection*

tree-length logging The operation of felling and transporting the trimmed bole in one piece for bucking at a landing or mill.
tree plate See tree iron.
tree rig 1. In the early high lead system, the ground lead raised into a tree at the landing. 2. See tree-rigged skidder.
tree-rigged skidder A skidder rigup using a tree for the spar; these skidders were contemporary with tower skidders. (P.C.F.)
tree roller See tree spool.
tree shoe A support for the skyline at either end. It consists of two heavy plates with a hardwood block in the center, over which the line is carried. Since the skyline does not move, it merely supports the line, which is then brought down to the ground and fastened to a tail hold. Also used to support the sail guy and anchored end of the loading line on boom loading systems. See tree jack.

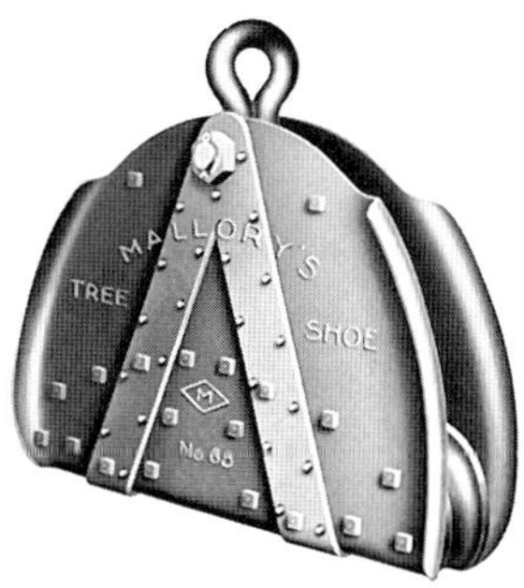

A **tree shoe**. *From Mallory Catalog No. 12, 1925, Glen Comstock collection*

tree-spar overhead cableway system The catalog's label for a tree-rigged skidder. See skidder.
tree spool A long deep roller, often grooved, mounted on a bracket which could be spiked to a tree to guide a haulback. A face was chopped on the tree or stump to fir the base of the roller. Also rollers. See road roller, road spool.
tree spotter 1. A timber cruiser. (P.C.F.) 2. A man who picks out spar trees suitable for landings. (P.N.F.)

A **Willamette** electric-powered **tree-rigged skidder** at work for the Sugar Pine Lumber Company, out of Central Camp, in the Southern Sierras of California. Power was transmitted through cables, or **baloney**, from the power house at the mill at Pinedale. **Duplex loading** was the standard method of loading cars on this operation, as is shown here. *Labbe collection*

tree squeak The noise made by two trees rubbing together was said to be an imaginary bird called a "tree squeak." (N.F.) LL

tree strap A short thick cable used to hang a block on a spar tree.

trespass To cut another man's timber.

triangle A triangular-shaped piece of steel that was used to call the men to dinner. The triangle was not closed on one end to allow it to ring more loudly. Also gut hammer.

trig A chock or block to stop or slow the motion of a wheel.

trig a load To block or stop the motion of a load.

trigger Another name for dog, one that could be knocked out quickly to release a log. (P.C.F.)

trigger block A block with a latch that opens and closes, allowing lines to be released from it.

trigger work Said of poorly laid plank roads. (P.C.F.)

trim See overlength.

trip 1. A single trip and return made by one team in hauling logs. (L.S.) LL See turn. 2. To wedge a tree that is being felled in the desired direction. (P.C.F.) See wedge a tree.

trip a dam To remove the plank that closes a splash dam. (L.S.) LL

trip back To return the butt rigging to the woods. (P.C.F.)

trip boom 1. A log placed across a stream to hold the logs in a drive until the crew was ready to start them down. 2. A boom strung a safe distance above a dam, falls or fast rapids that could be opened to let logs through when the stream was clear, or closed quickly in case a jam formed below it. (L.S.) LL

trip gate A gate on an almost level flume; used to back up a head of water in order to make a current which will carry logs down to the next gate.

trip hook A self-releasing loading hook.

triple drum 1. A three-drum unit for a cat or a cat so equipped. 2. A three-drum donkey.

triple haul (yard) 1. Three donkeys in one operation, one yarding, two more either swinging or roading the logs to the landing. 2. To use three donkeys in one operation.

trip line 1. Another name for the haulback line. 2. A line used to pull the fall block down within reach of the choker setters in the O'Gorman cable logging system. 3. A light rope attached to a dog hook and used to free the hook when employed in breaking a jam, skidway or a load. (L.S.) LL Also throw line.

trip line drum The haulback drum on a donkey. The exception is the O'Gorman system, where it was a drum separate from the haulback drum.

trip line head block See haulback block.

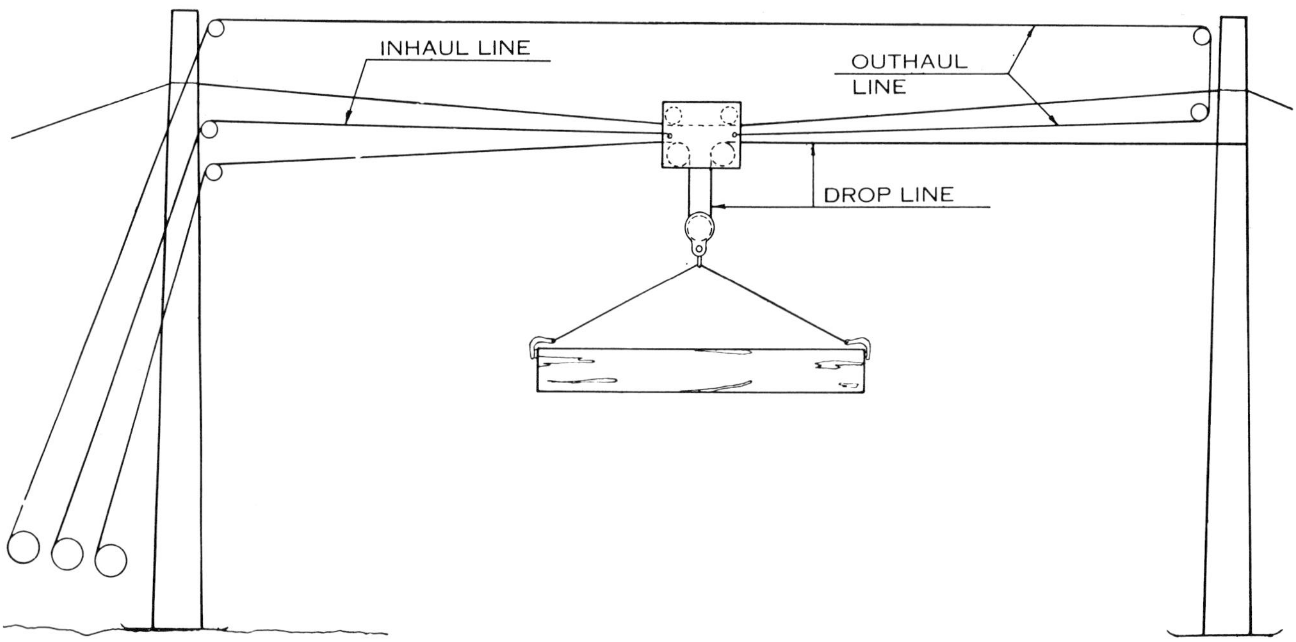

An illustration of **trolley-log loading**. Three **drums** and a **tight line** are required for this system. *From Mallory Catalog No. 52, Philip Schnell collection*

An unidentified **triple drum** donkey engine used in the early days of steam logging in the Sierra Pine Country of California. The machine is featured by its wide drums and the unusual placement of the third drum, higher above the operator's head. A special **fairlead** for the **haulback** drum, consisting of a spool running on a horizontal shaft is another unusual feature. An ordinary **block** served as a fairlead for the **mainline**. Note the exposed gears—safety engineering was yet to come into the logging industry. *From "The Timberman," 50th Anniverary Issue, October, 1949, Philip Schnell collection*

tripped'er Said of a man when quitting a job. (L.S.)

tripper 1. A light hammer or dog knocker used to knock dogs out of logs. 2. The man who used the hammer. Also pointer.

trip sill A timber placed across the bottom of the sluice way in a splash dam, against which rest the planks by which the dam is closed. (L.S.) LL

trolley See carriage.

trolley-log loading Moving logs from a roader, a yarder or a cold deck to cars by a tightline system. Loading hooks were hung from a big trolley riding on a tightline and hauled back and forth by a racking line.

trough roof A roof on a logging camp or barn, made of small logs split lengthwise, hollowed into troughs and laid from ridgepole to eaves. The joints of the lower tier are covered by inverted troughs. (N.F.) LL

truck 1. A railroad logging truck. 2. A motor truck. Trucks have gradually taken over the work done by the logging railroads. 3. A heavy wagon used to haul logs, either with animal or power traction.

truck chance An early term for a truck logging operation.

truck driver A teamster who skids logs with a bummer. (S.F.)

truck pole road An early logging road for trucks built of poles hewed off on top and flanged to keep the wheels on the track.

truck scale The board foot scale of logs loaded on a truck.

truck show A logging operation that uses motor trucks to haul logs.

tuck To lift a strand of wire rope and to push another strand under it in splicing.

tump line In the early days of packing gear in packsacks, the load on the back was eased by running an extra band from the sack around the forehead and back to the packsack. By pushing the head forward, part of the weight could be taken off the shoulders. Copied from Indian packers. (N.F.) (Eastern Canada)

tunnel loader Any log loader which straddles the track, and in which the empty cars are brought up to the loading point beneath the loader. See McGiffert loader, decker, Lidgerwood, and Surrey-Parker loader.

turkey See duffle bag, histe the turkey.

turn 1. The log or string of logs brought out of the woods in a single pull, whether it be by bull team, donkey, cat, train or truck. 2. Several thousand feet of logs held together by dogs, to be hauled along a skidway. (L.S.) 3. A single trip and return made by one team in hauling logs; for example, a four-turn road is a road the length of which will permit only four round trips each day. (L.S.) LL

turn-and-a-half road The round trip by a two-sled rig from yard (that is, from camp) to landing was a "turn." A "turn-and-a-half road" was one that the teamster traversed three times in one day, between yard and landing. It required a separate camp at the landing in which he and his horses spent the night. When new territory was to be lumbered, the cutting area was first marked out. Then the landing was selected. If these were so far apart that the teamsters could not make two (or three, or whatever) complete round trips in one day, the main camp was built as near the yard(s) as possible and a small horse-hovel was built at the landing, with a comfortable bunk for the teamster, a stove and an alarm clock. A tote-team brought in hay and grain for the horses, and the teamster who was to spend the night brought along food in pails and a wooden lunch bucket. There might be half a dozen teams hauling from the main camp, but they took turns at this night-away-from home vigil: while one was doing a turn-and-a-half, the others were making only one trip. (N.W.) (Robert E. Pike, Tall Trees, Tough Men)

turn around A cleared area in which a sleigh, truck, wagon, tractor or other skidding device could be turned around at the end of the road. (N.F.) LL

turn around the tree To move a donkey or skidder to the opposite side of a spar tree so that the other side may be yarded. (P.C.F.) Also turn the donkey.

Turney cable saver A patented device for narrowing the drum on a donkey. H. L. Turney was an engineer responsible for many improvements on donkeys. (P.C.F.)

Turney friction A type of clutch applied to donkey engine drums. (P.C.F.) See friction.

turn hooker A rigging slinger on a cat side. (P.C.F.)

turn in To go to bed.

turnip Usually a large watch tied on the pants with a leather thong. (L.S.) LL

turn of the night Midnight. (L.S.)

turnout 1. A short side road from a logging sled road to allow loaded sleds to pass. (N.F.) LL 2. A passing track on a one-way truck road. (P.C.F.) Also drive-out.

turn out 1. In turpentine orcharding, to temporarily or permanently abandon an area for turpentine purposes. (S.F.) 2. To get up in the morning. 3. To go to work. (App.)

turn-out cat A small cat used to turn the ends of bucked logs out at an angle, to face the direction of skidding and to facilitate setting the chokers. Used primarily in Pine Country. (P.C.F.) Also kitten.

turn-out hook A large, curved hook used with a turn-out cat.

turn out the crew (men) To get the lumberjacks out in the morning. (L.S.) LL

turntable 1. A section of rails on a rotating platform used to turn a locomotive around. 2. A table to turn a truck. (P.C.F.)

An unidentified horse-powered **turntable** turning a GMC truck. *From "The Timberman" collection*

turn the donkey See turn around the tree.

turn up the pops To turn up the pop valves on a steam locomotive or donkey to increase the boiler pressure.

turpentine The liquid product resulting from the distillation of crude turpentine. (S.F.) Also spirits.

turpentine ax A long-bitted ax used in cutting a box. (S.F.) Also boxing ax.

turpentine orchard A stand of pine which is being bled for crude turpentine. (S.F.) Also farm, place.

turtleback 1. A patented log bunk and car stake designed by E. B. Shields and built by the Nisqually Iron Works of Tacoma, Washington. 2. A chunky yarding block. 3. See rerailer.
tussock A turkey or a packsack. (L.S.) See duffle bag. LL
twin hooks Two hooks fastened together with a link which can be opened to take a line or a link of chain and will stay closed under a pull.
twin neck yoke A short neck yoke that goes on each end of the long neck yoke to assist in holding back a load. (L.S.) See whiffletree neck yoke. LL
twins A pair of loggers who traveled and worked together on the same jobs. If one quit or was fired, the other would leave also. (P.N.F.)
twin sleds Sleighs for hauling logs. (L.S.) LL
twister 1. Usually a stick used as a twister to tighten up on lines which are wrapped around stumps used for a tail hold. 2. A chain tightener on a load of logs at one time. 3. A tree with twisted grain. 4. See Spanish windlass.
twist the cow's tail To pass the milk. (S.F.)
twitch 1. An ox goad. 2. A short chain. 3. To skid small logs.
twitching chain A short chain having a round hook on one end and a grab hook on the other end and used in skidding logs. (N.F.) LL
twitch road A narrow road or trail for twitching logs. (N.F.) LL
two-bunk car A skeleton car; some were as long as 73 feet. (P.C.F.)
two burnt holes in a blanket Said of how one's eyes feel after a hangover. (P.C.F.)
two-cutter An early design of crosscut saw which had two cutting teeth alternating with the rakers.
two-faced Said of a double-bitted ax. (P.C.F.)
two-faced tie A pole tie with only two hewed faces. It is made from a stick of timber too small to hew four sides. (S.F.)
two-hander A large stick used with both hands to hit horses to make them pull harder and to obey orders. (N.F.) Also brolo. LL
two hundred foot to the log The lumberjack's optimum number of board feet to the log. (L.S.) LL
two-line system See Sessom's system. (P.C.F.)
two-poke unloader A jill-poke with two swinging booms for shoving logs off cars. (P.C.F.)
two-pole chute Two poles hewed to form a V on the inner surface, making a chute for small logs. (P.N.F.)
two-run cruise A ten percent estimate of standing timber. (P.N.F.)
two-sled Two logging sleds hitched one behind the other and used for hauling logs or short wood. The load is carried completely off the ground. (N.F.) See sleigh.
two-speed A yarding donkey with a low gear ratio for starting and heavy pulling, and a high gear ratio for speed once the load had gathered momentum. High gear was used on the haulback to return the butt rigging to the woods. The first two-speed was an important advance in donkey engine building. Three-speed and four-speed donkeys were built but never used widely. (P.N.F.)
two-spot See deuce.
two streaks of rust A logging railroad. (West Virginia)
two-trip road The number of round trips each day a logging team hauling logs could make. A two-trip road meant that the distance one way was about six miles. (L.S.) LL
two-up driver The driver of a two-mule team. (S.F.)
two-wheeler See high wheels.
tyee 1. The boss, meaning chief in the Chinook jargon. (P.C.F.) 2. A name of an early donkey engine. (P.N.F.)
Tyler system A tightline method of logging using a three-drum donkey. Beside the mainline and haulback there is a lifting line which is anchored to a stump at the tail tree end. A fall block rides in the bight of the lifting line and is lowered to the ground for hooking on logs, then raised for the ride to the landing. (P.N.F.) Also suspension carriage system.

A Willamette **two-speed, compound-geared yarder.** *From Willamette Iron & Steel Company Catalog, 1925, Glen Comstock collection*

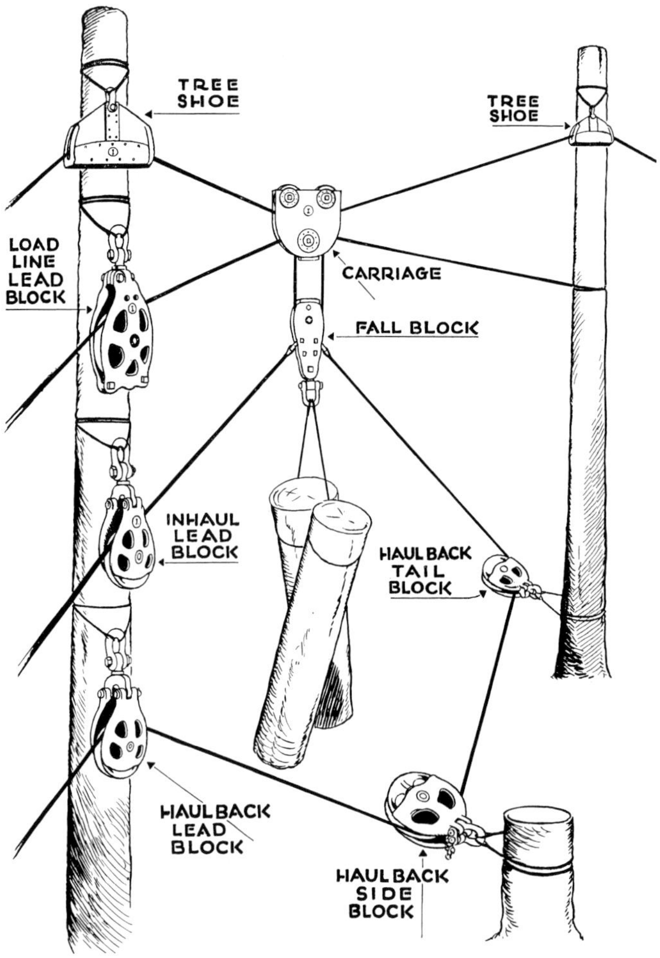

The **Tyler system.** *From Young Iron Works Catalog No. 52, 1952, Philip Schnell collection*

U

ukelele 1. A setup using two donkeys working side by side (before the donkeys with four drums were used). One donkey handled the skyline or slackline and the other handled the skidding and haulback line. The first ukelele setup on the West Coast was used in the Van Duzer area near Carlotta, Humboldt County, California about 1909 by the Holmes-Eureka Lumber Company. 2. A short handled shovel. (P.C.F.)
ukelele choker A modern choker hook. See Bardon, Wirkkala hook.
uncle The superintendent. (S.F.) (L.S.) LL
under-barker The logger who barked the log after the log was turned over on its skinned side by the top barker. Sometimes fir logs were completely barked before skidding with animals. (P.C.F.)
underbrush 1. The brush under and around a stand of timber. 2. Any kind of brush. 3. Unimportant things.
underbuck 1. To put in the undercut. 2. To cut out a break in a log while bucking. 3. To saw first from the bottom in bucking logs. (P.C.F.)
undercant To put a choker on a log to make it roll counter-clockwise.
undercut 1. The first cut in felling a tree. See notch. 2. To make an undercut. 3. To cut a wedge from below in log bucking, to relieve a bind.
undercut hold A method of arranging a choker on a log so that when a forward pull is exerted the log will roll backward. (P.C.F.)
undercutter 1. In the 1800s, an expert faller who only made the undercut and left the back cut to other loggers; a dangerous practice because heavy winds could cause the tree to split. 2. A tool used to support the back of a crosscut saw when bucking from the bottom of a log. See pocket undercutter.

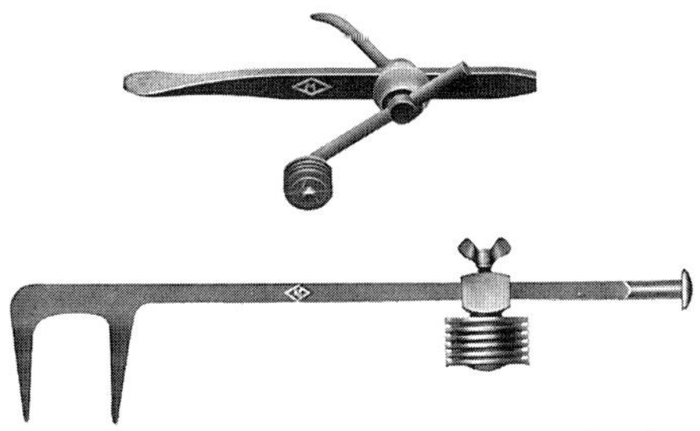

Two types of **undercutters**. *From Mallory Catalog No. 6, 1925, Glen Comstock collection*

underlay 1. To spool line onto a drum from the under side. See undershot drum. 2. To separate strands of wire rope.
undershot drum A drum on a donkey onto which wire rope is spooled from the bottom; sometimes done because on some compound-geared donkeys, the mainline and haulback drums turn in opposite directions. See underlay.
unfold To unwind a kink in a wire rope.
unhook man See chaser.
union drive A drive of logs belonging to several owners who prorate the expenses. (L.S.) LL
unit 1. In railroad logging, both the yarding and loading engines set up on one car or on one sled. A unit was used on a high lead show in B.C. as early as 1906. Also high lead unit. See car-mounted unit. 2. An area of arbitrary size designed for uniform logging or forest management treatment throughout.
unit car See skidder car.
universal lead A fairlead in all directions which is obtained by using a rotating steel tower as the head spar on a skidder operation. See swivel head.
unlay To separate strands of a wire rope.
unload To move quickly out of danger, specifically to get off a runaway.
unloader 1. The man at the dump who unloads logs off the railroad cars or trucks. 2. The machine set up to unload cars or trucks.
unravel To clear jammed pulpwood from a stream in a drive. (L.S.)
unrig To remove the lines and blocks from a yarding or loading tree. Also strip a tree, take the tree down.
up-and-down An early hand car used by the steel gang or track crew. (P.C.F.)
up a tree Said of a worker who is in a bad situation.
up-end a line To cut a line in the middle where it is beginning to wear, to splice the outer ends together, and to get the worn parts on the ends.
up-end a log To change ends with a log in order to lay the butt end in the proper direction when building a load. (P.C.F.)
up-end it To reverse or change ends. (P.C.F.)
uphill haul 1. The most satisfactory method in high lead logging because the rise of the mainline gives an extra lift to the incoming logs and reduces hang-ups. 2. A hard luck operation. 3. Said of doing things the hard way.
upright boiler The vertical boiler common on donkeys and sometimes used on geared locomotives.
upright donkey A Dolbeer-type donkey with a vertical spool instead of horizontal drums. (P.C.F.)
upright roller 1. A vertical roller with large flanges which was used

A **Willamette** Iron & Steel Company factory photo of a car-mounted **unit**. This 1920 view shows a 12x14 **two-speed compound-geared yarder** and a 9-1/4x10 **duplex loader**, mounted on a **unit car** built by Pacific Car & Foundry. This machine is destined for Kerry Timber Company, at their Sunshine Camp in the Nehalem Valley of Northwest Oregon. *Willamette Iron & Steel Co. photo, Martin E. Hansen collection*

to hold the cable on bends in the road, and to prevent it from cutting into the bank or trees. The roller was often attached to stumps or trees. See road roller, road spool. 2. The cross bar of a logging sleigh in which the tongue was set. (L.S.) LL

upset See swage.

up the pole Said of a logger on the water wagon (abstaining from alcoholic beverages). (L.S.) LL

useless as a one-legged man at a butt kicking match Said of a very useless person. (P.N.F.)

using all his horsepower Said of a man who works hard at his job. (P.C.F.)

utility drum A small drum on a donkey used for rigging up and car spotting. See haywire, straw drum, strawline.

utility engine A small steam engine on skidders used to power the strawline, spotting line and drums for tightening the heel blocks.

utility man A good logger or mill worker who can fill in on any job. (P.C.F.)

Deep River Logging Company used this rail-mounted **unloader** to unload their logs at Deep River, Washington. Deep River used **disconnected trucks** with **link-and-pin couplers**. By the 1930s, nearly all Western loggers had converted to the much safer knuckle couplers. *Courtesy of Carlton Appelo*

V

valley boys Saginaw Valley lumberjacks had legendary reputations of being rough and tough, who could outfight, outdrink or outsaw any other shanty boys in the upper or lower peninsula of Michigan.
value To estimate the amount and value of standing timber. (L.S.) See cruise. LL
valuer One who cruised timber to estimate its value. (L.S.) Also cruiser, estimator, landlooker. LL
valve oil Any dark liquid.
valves square Said of a locomotive, or any steam engine, when the valves are properly timed. (P.C.F.)
van The small store in a logging camp in which clothing, tobacco and medicine are kept to supply the crew. (N.F.) See wanigan. 2. Clothing and small ware supplied to woodsmen. (Eastern Canada) LL
van books Camp store record books. (L.S.) LL
Vanderpool A type of swivel butt rigging in which the mainline, haulback and butt hook are attached by swivels to a cloverleaf rigging plate.
Vaughn A make gas drag saw which was very popular in the West. See drag saw, Wade.

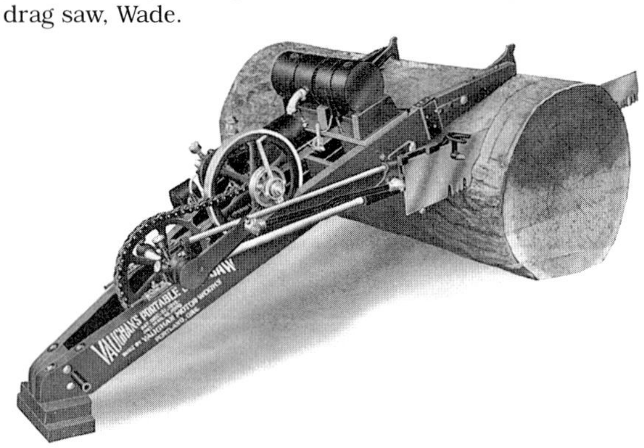

A **Vaughn drag saw**. *From Mallory Catalog No. 6, Glen Comstock collection*

v-bed A V-shaped undercut used for small timber and supposed to prevent a kickback. The tongue of the V stuck down over the stump, holding the butt on that side.
vegetable fireworks Beans. (P.C.F.)
vertical boiler The upright boiler used on donkeys and sometimes on geared locomotives. Also upright boiler.
virgin In turpentine orcharding, the crude turpentine secured during the first year a tree is bled.
virgin forest, virgin timber 1. A forest that has never been logged. Also virgin growth. 2. Old growth timber. 3. A woman difficult to seduce.
V-lead In yarding, a lead less than square lead. The line forms a V and is dangerous for the crew.
Vollans' cradle A system of bundling logs into rafts for short tows on Puget Sound; invented by B. H. Vollans of Everett, Washington. (P.N.F.)
volume The amount of timber (board feet) in a specific area, tree or log.
V-plow A plow made of two poles or planks fastened together at one end and kept apart at the other by a spreader; used in breaking out and smoothing winter roads. (L.S.) LL
Vulcan 1. An early steam donkey built by the Vulcan Iron Works of Seattle, Washington. (P.N.F.) 2. A make of locomotive put out by the Vulcan Iron Works of Wilkes Barre, Pennsylvania.

A **Vulcan** steam **donkey**. From *"The Timberman," January, 1907. Tim Anderson collection*

W

Wade A drag saw built by the R. M. Wade & Company of Portland, Oregon, which manufactured drag saws of all sizes. The company was very competitive with the Vaughn Company. The light saws made by the Wade Company were very popular, and the heavy saws made by the Vaughn Company were the preferred type for heavy work. (P.C.F.) See drag saw, Vaughn.

wagon 1. See log wagon. 2. To haul logs from the woods with a wagon.

wagon sled See sleigh. LL

The **walking dudley** consisted of a **grip wheel** mounted on a car to serve as a locomotive. The machine pulled itself along the track by means of the grip wheel and a cable fixed between the rails. Simpson Logging Company, shown here, used several to **trail** logs in the woods west of Shelton, Washington, in the 1890s. Simpson used the walking dudleys to a greater extent than any other operator and was credited with giving them their name. *U.S. Forest Service photo, from M. Koch*

wahoo 1. A kink in a choker or line. Also asshole, meow, wow. 2. Something out of line.
walk a cat To move a tractor to another area under its own power instead of transporting it by truck or train.
walkaway Burns A man who walks off the job with no excuse. Burns was a famous teamster in the early pine logging days who continually walked off the job for the slightest cause. (P.N.F.)
walk'em in To choke logs in the middle, which allowed them to come in an end at a time, thereby reducing hang-ups. (P.C.F.)
walker See walking boss.
walking anchor 1. A movable tail hold used to keep the tail hold straight behind the skyline on a skidder show. Also walking anchorage. 2. A tail hold and a line attached to a purchase block to put added tension on the skyline when changing a road.
walking appetite Said of the biggest eater in camp. (P.C.F.)
walking boss A foreman in charge of two or more operations, who spends his time going to and from the operations. Also walker.
walking dudley 1. A grip wheel mounted on a car and used like a locomotive, pulling itself back and forth along a line fixed at both ends and lying between the rails. 2. This term was applied to the grip wheel locomotives built and used at the Simpson Logging Company camps in Mason County, Washington. (P.N.F.) See Fouts' grip wheel, hill-climbing engine.
walking guy A guy line on a steel tower skidder, which was anchored to a stump and rigged so that it could be moved easily when changing roads. (P.C.F.) See walking anchor.
walking rubbers Packs with rubber feet and leather tops. (L.S.) LL
walking the cook To fire a cook because of bad food. (L.S.) LL
walking the table To walk on the table in the mess hall and to kick off all the food, in response to a poor cook. (S.F.)
walk the bugs to death To turn a shirt inside out so that the lice have to walk all the way around to the front to get in again. (P.N.F.)
wampus cat An imaginary animal to which night noises were attributed. (N.F.) LL
wanigan This word has several spellings and different meanings. 1. A boat or raft with the camp aboard, following behind a drive. (P.C.F.) 2. A float camp along the Southeastern Alaskan coast. (Alaska) 3. A place where the camp stores are kept. (N.F.) 4. The payroll charges for the camp stores. (L.S.) 5. A cook's raft which went down the river behind the logs and was tied to a tree on shore overnight. (N.F.) 6. The place where camp was made on a river drive and where the men were paid. (L.S.) 7. A locker in a camp or bateau. (Maine) Also ark, shanty-boat, van. Other spellings are "wanagan," "wanegan," "wanngan," and "wannigan." LL
wanigan car A camp commissary and office car in a railroad car camp. (P.N.F.)
wanigan man The man who prepares camp for river drives. (L.S.)
war bag A packsack. (P.C.F.)
warden The fire boss who makes the decisions on fire prevention.
war department A wife.
warming fire A small fire for warmth in cold weather; usually the responsibility of the whistle punk since he was more stationary than other members of the crew, and he also had greater need for a fire since he could not move around much.
warp To move something into position gradually. (L.S.) Also kedge.
warp it to him To make it tough for a man. (P.C.F.)
Warrington A wire rope pattern in which all wires are of equal size.
war whoops Indians. (P.N.F.)
washed his clothes Said of a river driver who fell into the water. (L.S.) LL
washer A doughnut. (L.S.) Also sinker, stovelid. LL
wash house A bath house in a logging camp. (P.N.F.)
Washington fir Douglas fir. (P.C.F.)
Washington A donkey engine built by the Washington Iron Works of Seattle, Washington.
Washington flyer A duplex aerial slackline yarder built by the Washington Iron Works of Seattle, Washington.
washout 1. An emergency stop signal. 2. Said of "washing out" the locomotive steam boilers about once a month, at which time repairs were made while the boilers were cool. (P.C.F.)

A 12 x 17 **Washington flyer** on display in Scotia, California. *Merv Johnson photo*

waste That portion of the tree which has merchantable value but is not utilized.
watch fob An early name for a small logging block. (P.N.F.)
watchman 1. A man who tends the steam donkeys or locomotives at night and has the steam up ready for work in the morning. 2. The man who patrols the lumber mill when the plant is not operating at night. 3. A man who watches for fires on a logging operation after a day's work or when shut down during the fire season.
water and chalk A poor grade of milk. (P.C.F.)
water-bar See cross-ditch.
water box See sprinkler.
water boy See water slinger, sugler.
water buck Anyone who carries water to a crew or for a donkey engine. (Red.)
water cart See sprinkler.
water content The water in a piece of lumber or a log. There is considerable water in green redwood. A piece one inch thick, one foot wide and 20 feet long may contain 70 pounds of moisture, or about 8.4 gallons of water.
waterfall A donkey or tractor winch with one drum higher than the other so that the lines would not foul. (P.N.F.) See double drum.
water hole 1. Usually a small creek that is dammed to provide a water supply for drinking, for tank trucks, for filling tanks to cool truck brakes or for supplying donkeys. (P.C.F.) 2. A saloon.
Waterhouse hook A butt hook locked by a spring latch; invented by Rollie Waterhouse, an Oregon logger.
water ladder Pole guides, up and down on which a barrel slides in filling a sprinkler by horse power. (N.F.) LL
water the logs To bring logs to a pond or booming ground. (P.N.F.)
water mark The chipped out part of the bark on a log which serves as a log brand.
water packer The man or boy who packed water in canvas bags on horses or mules up the hillsides to the buckets on the skidroad. (P.C.F.)
water scale The scale of logs in a boom or raft. LL
watershed A region or area bounded peripherally by a water parting and draining ultimately to a particular watercourse or body of water.
water slide A log chute or slide, lubricated with a small stream of water. See flume.
water slinger The man who threw water ahead of the logs so that they would slide smoothly. The water was kept in five-gallon buckets or cans at intervals along the skidroad. (P.C.F.) Also water boy. See sugler.
water tank See sprinkler.
water the brakes To drip water on the brake drums of a logging truck going down a long steep hill.

A **Cat** D6 equipped with a **Hyster** D6N **double drum**, or **waterfall** winch, working near Molalla, Oregon, circa 1945. *Philip Schnell collection*

A **Waterhouse hook**. *From Mallory Catalog No. 12, 1925, Glen Comstock collection*

water up 1. To fill a tank with water. 2. To stop for an alcoholic drink. 3. To thin the soup.
water wagon A sprinkler to ice the roads. It had a tongue at each end so that it could be pulled both ways. (L.S.) See sprinkler. LL
water yoke A wooden yoke to fit the shoulders of a man. Used in a lumber camp when a man was carrying two pails of water or two boxes of grub. (L.S.) LL
way car A railroad car which is used as a caboose and supply car. (P.C.F.)
wear a choke strap To wear a tie. (S.F.)
wear a (white) feather 1. Said of a steaming locomotive with the safety valve blowing off steam. 2. Said of any steam boiler with enough pressure to lift the pops slightly and allow a feather of steam to escape. (P.C.F.)
weather-beat Said of a logger who is kept off the job by wind, rain, low humidity or snow. (P.C.F.)
weathered Said of a logger who has been toughened by many years out in the open. (P.C.F.)

weaver's bind A method for attaching logs to the bunk of a yarding sled with chains. (N.W.)

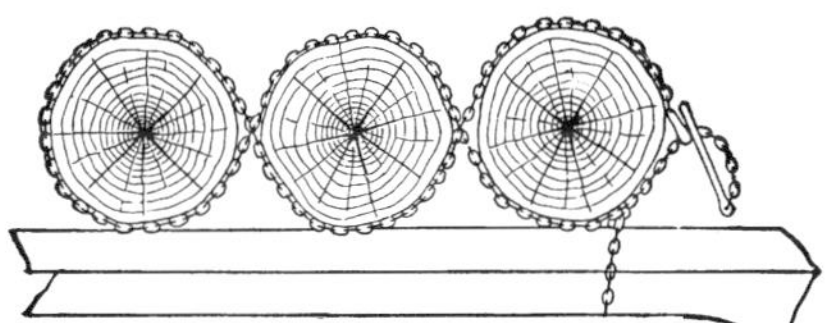

The **weaver's bind**. *From "Logging," Bryant, Wylie & Sons, 1923, Philip Schnell collection*

web 1. The solid or spoked part of a sheave between the hub and the outer groove. 2. That part of a rail connecting the running surface and the base. The running surface of the rail is the ball.
webfoot 1. An Oregon logger who is a poor worker in the redwoods. (Red.) 2. A pond worker. (P.C.F.)
wedge 1. An iron or wooden piece sharpened on one end which is driven into a saw cut with a large hammer to prevent the saw from binding and to direct the fall of the tree. 2. In river driving, the crotched wedge made out of hardwood which looked like a flattened clothespin about five inches long. The slot in the wedge was a quarter-inch wide which allowed it to fit snugly over the quarter-inch rafting line. (N.W.)

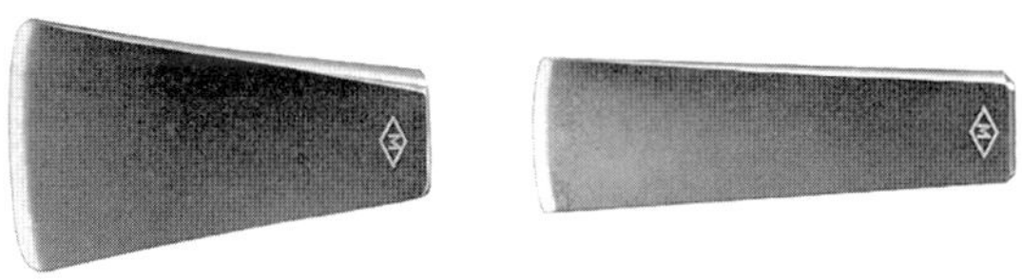

Wedges. Bucking wedge, left; **falling** wedge, right. *Mallory Catalog No. 12, 1925, Glen Comstock collection*

wedge a tree To force wedges into a saw cut, forcing the tree to fall in the desired direction. Also throw, trip.
wee-waw Said of something crooked or out of plumb. (Maine)
Weigant swivel A three-way butt rigging swivel, designed by Andy Weigant of Consolidated Timber Company, near Glenwood, Oregon.

A **Weigant swivel**. *From Mallory Catalog No. 48, Philip Schnell collection*

weighs heavy Said of a man who thinks he is very important. (P.C.F.)
well A hole dug in the snow around a tree so that the faller may cut the tree at the required height. (R.M.F.)
went out Said of a woodsman who had gone to town to get drunk. (L.S.) LL
West Coast The West Coast Lumbermen's Association of Operators and lumbermen in the Douglas fir region. (P.N.F.)
western hemlock (tsuga heterophylla) The tallest species in the genus, sometimes reaching a height of 200 feet. It grows along the Pacific Coast from Southeastern Alaska to Central California. Like the eastern hemlock, it is an important softwood timber tree. About 90 percent of the hemlock used for lumber is western hemlock. Also Alaska pine.
western penitentiary See old folks home.
western pine Ponderosa pine.
Western Pine The Western Pine Association of lumbermen in the Pine Country of the Western States.
western soft pine See ponderosa pine.
western white pine (pinus monticola) Like its eastern relative, western white pine is a very important timber tree. It develops its greatest size and highest economic values in deep porous soils on gentle north slopes and flats in Northern Idaho and Montana. Western white pine grows from the lower part of British Columbia southward into Western Montana and along the Cascade and Sierra Nevada Ranges through Washington and Oregon to Central California. Practically all of the western white pine taken from the forest is sawn into lumber.
west side West of the Cascade Mountains. (P.N.F.)
wet ass 1. Said of a steamer with a stern paddle wheel which splashed water over the stern while pulling logs from the log dumps to the mill. (P.C.F.) 2. A tug boat used to tow log booms on the Columbia River and Puget Sound. (P.N.F.) 3. A passenger steamer which carried loggers to and from camp or the nearest town. (P.N.F.) 4. Said of one who falls in a river or log pond.
wet ass drive In the early days, a steamer with a stern wheel that splashed water over the stern while pulling logs from the log dumps to the mills. (P.C.F.)
wet'er down To sprinkle with water to reduce the fire hazard, as around a donkey setting. (P.C.F.)
wetland A bog, marsh or swamp. A general term for any poorly-drained, uncultivated tract, whatever its vegetative cover and soil.
wet rag law The fire law in the Pacific Northwest which requires operators to shut down their logging when the humidity drops under 30 percent. Two thermometers (hygrometers) are used in determining the danger point, one kept damp with a cloth, thus the wet rag. The wet and dry thermometers on the hygrometer measured the evaporation rate of the wet rag, and thus determined the humidity.
wet slide A V-shaped runway made of lumber to bring logs out of the mountains or steep hills. See flume, water slide. LL
whacker A bull puncher. (S.F.) See bull puncher.
what loggers do all winter Said of the loggers who wonder what happened to the money they made the preceding summer. (P.C.F.)
wheel 1. A block. 2. The boss or super. 3. To haul logs by big wheels in the Pine Country. 4. To make a fast run.
wheel camp In the Pine Country of California, a logging camp in which big wheels are used to haul logs.
wheel cat A cat yarding with an arch or hauling big wheels.
wheel'er 1. To quit or leave camp. (L.S.) LL 2. To run fast.
Wheeler raft A flat ocean-going raft which was first built with a hole in each log through which a toggle chain was passed to attach the log to a tow chain. Later only the center line of logs was solidly coupled, and lines were woven over and under the logs parallel to the center line. The raft was designed by C. H. Wheeler of the Wheeler Lumber Company of Wheeler, Oregon. The rafts were towed to the Columbia River market. See ocean-going raft.
wheelers The team closest to the log in a log haul. (App.) (S.F.) (L.S.) (Eastern Canada) See butt team. LL
wheel loader In the early days, the man who fixed the chains around loads of logs to be hauled by big wheels.
wheel of fortune A grindstone. (L.S.) LL
wheel road Any road over which a wagon could travel, usually a tote road. (L.S.) LL
wheel team See butt team, wheelers.
when the head hit Said when the head of water arrived at the logs to be driven.
when the head ran off Said when water released by a dam was used up, the logs stopped and the men had to wait until more water built up in back of the dam. (L.S.) LL
when the short logs come down Said to describe the tail end of a drive when the small logs have been found on the banks and pushed back into the river. (P.N.F.)
whiffletree A crosspiece made of wood that hooks to the traces of a horse harness. The whiffletree keeps the traces spread and provides an attachment to which the load being pulled is fastened.
whiffletree neck yoke A heavy logging neck yoke on the end of a pole or tongue, to which short whiffletrees are attached by rings to the ends. From the ends of the whiffletree wide straps run to the breeching, giving the team added power in holding back loads on steep slopes. (N.F.) Also twin neck yoke.
whiner A mosquito. (P.N.F.)
whip 1. The dangerous snap of a line when tightened. 2. The snap of chokers that fly in all directions when jerked from logs. 3. A young tree which springs when hit by a line. 4. The up and down movement of a crosscut saw, making an uneven cut in falling. 5. The saw used in felling and bucking.
whip and luff Two cables and two blocks used to tighten a skyline or to move an object.
whippoorwill switch A device designed to sheer logs out of a slide at a given point. (App.)

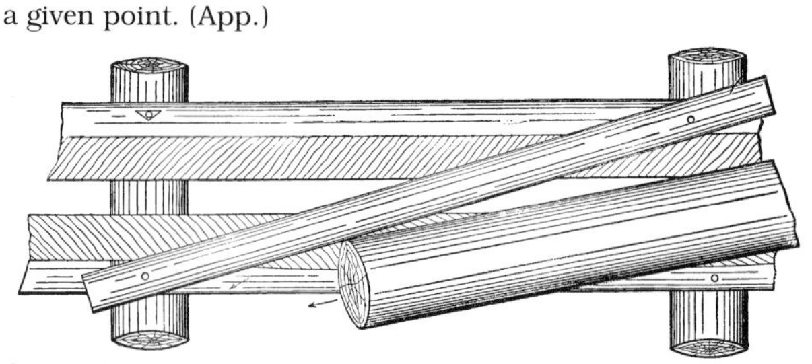

A **whippoorwill switch**. *From "Logging," Bryant, Wylie & Sons, 1923, Philip Schnell collection*

whipsaw 1. A saw, operated by two men, which is used to cut logs into lumber. 2. A bad situation. 3. To saw boards by hand with a heavy ripsaw in the early days.
whiptender An older term for a choker setter on a cable logging operation. (P.C.F.)
whirligig 1. Spokes at the end of a log unloading boom. 2. A Donovan unloader. See jill-poke.

whirling the rock Turning the grindstone while sharpening axes. (L.S.) LL
whirl lizzy (dizzy) A whirlwind caused by the draft of a forest fire. (P.N.F.)
whirly-bird logging Yarding by helicopter where other methods are impractical or barred.
whiskers 1. The jagged wires on a cable. See jaggers. 2. Seniority on a job: "He showed his whiskers." 3. The standing splinters on the stumps which pulled out when the tree fell.
whiskey jack 1. See camp robber. 2. A hydraulic jack filled with alcohol as a nonfreezing fluid. (N.F.) LL
whiskey jack lunch A poor lunch which should be thrown to the jays. (P.C.F.)
whiskey special The trains that used to take the loggers back to the camps from the skidroad in town. See drunkard and Sunday night train.
whistle A signal that is passed from the rigging slinger to the whistle punk to the donkey puncher. (P.C.F.)
whistle boy One who transmits orders from the foreman of a skidding crew to the engineer of a pullboat. (S.F.)
whistle bug See bug.
whistle punk The boy or man who relays signals from the choker setters to the donkey engineer, usually the beginner's job in a logging crew. In former days, he gave the line a flip which sent a ripple along the taut wire, and this ripple provided the jerk when it reached the whistle. Also bell man, punk, signal man. See skidder punk.
whistle water men Expert lumberjacks on river drives. (L.S.) LL
white Honest, fair, dependable or decent: "He's a white man to work for."
white collar job 1. A job in a camp when the weather is too bad to work in the woods. 2. An easy show.
white cypress A term used by woodsmen for cypress timber of light weight. (S.F.)
white-haired boy The boss's favorite.
white horse Fast rapids or waterfalls. (L.S.)
white line Milk. (App.)
white men Some loggers and lumberjacks called themselves "white men" and would not bunk in the same room with Poles, Finns, Russians, Greeks, Swedes, Italians and other foreigners.
white pine 1. A cone-bearing evergreen that reaches to a height of about 80 to 150 feet, the trunk measuring as much as 4 to 7 feet in diameter. The Michigan white pine was the most valuable to the lumbermen of the three native pine trees (white, red or Norway, and "Jack" pine) because of its height, straight trunks and smooth white wood. (N.F.) 2. The true western white pine is the Idaho white pine.
white pine uppers White pine boards graded as "selects"—top grade. (L.S.)
whitewash Milk. (P.C.F.)
white water See quick water.
white-water bucko A complimentary term for an excellent river driver.
white water man 1. The skilled river driver of the Northern Woods, the Lake States, Canada and the Pacific Northwest in the 1800s and early 1900s. 2. Sometimes used in a derogatory sense to describe a clumsy pond man.
whitewoods Trees other than redwood, such as hemlock, white fir or cedar. The following remark is heard: "Cut out the whitewoods first." (Red.)
whittler An old logger too old to work except to sit and whittle. (P.C.F.)
whooper A mythical creature that was used to tease greenhorns in the woods. It had the body of a kangaroo and the tail of a boa constrictor. The nearer that one got to it the farther away it seemed to be. (N.F.)
wickhegan An official document, a fine or the price of a permit. (Maine)
wickiup 1. A rough shelter. 2. A cabin. 3. A bunkhouse.
wide face An early donkey engine with very wide drums to hold a long line for a lengthy road haul. (P.C.F.)
wide-face block Usually a Tommy Moore or a similar block with a wide sheave and throat. (P.C.F.) See Tommy Moore block.

A **wide face donkey** at work for the Peninsula Lumber Company near St. Helens, Oregon. Popular in the days of **ground-lead** logging and **skidroads**, the wide drums could **spool** several thousand feet of line. *Labbe collection*

widen on'er To open up the throttle on a locomotive or donkey. (P.C.F.)
widow maker 1. A limb or bark from a fallen tree which belatedly falls after being snagged in the boughs of another tree; the cause of many deaths in logging. 2. A tight, bent small tree, whose outer end will fly up and hit the axman under the chin when cut. 3. A limb or debris blown from a tree or snag by the wind.
wigwam Two or more trees which lodge together when felling timber.
wigwam maker A faller who fells trees which wedge together. (P.N.F.)
wild ax handles Said of one who has diarrhea. (Maine)
wild-catting A term used to describe working on a gravity railroad. The trucks were hauled to the woods by horses, and after they were loaded, they were moved down the hill by crews using hand brakes. This system was known as "wild cat railroads," and the men were called "wild-catters." (N.W.)
wild center A spot where a log would catch on a rock during a river drive, causing a jam. (L.S.)
Wilder A brand of chewing and smoking tobacco. (L.S.) See Badger.
wild eye A temporary loop rolled in a cable. See Molly Hogan and rolled eye.
wild timber Trees leaning in all directions and difficult to fell to the lead. (P.C.F.)
Willamette 1. A donkey built by the Willamette Iron works of Portland, Oregon. 2. A geared locomotive built by Willamette on the Shay design.
Willamette hill climber A "walking dudley" power unit used on logging road inclines. (P.N.F.) See walking dudley, grip wheel.
William Shears A farmer or greenhorn. (P.N.F.)
winch A powered drum for paying out or reefing in cable (often pronounced "wench").
wincher A winch operator. (L.S.) LL Also winchy.

A **Willamette** electric **unit** built for the Snoqualmie Falls Lumber Company, a Weyerheuser subsidiary at Snoqualmie Falls, east of Seattle, Washington. Power was supplied by a cable from the power house at the sawmill, which was dubbed the **baloney**. It can be seen near the man standing at the rear of the machine. Crews found them a challenge, as the **puncher** was unable to feel the amount of tension on the lines, and much rigging was pulled down. *Willamette Iron & Steel Company photo, Labbe collection*

winch head 1. A spool or gypsy head. 2. A small donkey drum.
winch it To pull a load slowly along the ground. (P.C.F.)
wind The amount of winding or twisting grain in a log. Also corkscrew.
wind breaker See mackinaw.
windbroke 1. Said of an overworked animal that cannot breathe properly. 2. Said of a man who has trouble breathing. (P.C.F.) 3. Said of an engine with a leaky gasket. (P.C.F.)
windfall See blowdown.
windfall bucker A logger who first cuts up the windfall logs so that yarding is not delayed. (P.C.F.)
windfirm A strong seed tree. (P.C.F.)
wind has January in it Said of cold weather. (P.C.F.)
winding drum An older term for the main drum of a donkey on a ground lead show. (P.C.F.)
windjammer 1. An air brake. 2. A donkey used on a railroad or pole chute incline. 3. A concertina. (P.C.F.)
windmill See whirligig.
windrow A long narrow pile, usually of logging slash removed from a planting site.
wind-shook See shake.
wind slash See blowdown.
wind so strong it would blow your socks off Said of a very strong wind. (P.C.F.)
wind-splitter See peaker.
wind squeezer A large diesel tower skidder, made by the Washington Iron Works. It had five engines on the loading end which were operated by compressed air. (P.N.F.)
windthrow See blowdown.
windtimber Beans served in the camps. (L.S.) LL
windy shot A blast which contains more air than rock or dirt. (P.C.F.)
wing A small log jam intentionally caught on a rock or other obstruction for the effect of a sluiceway. (L.S.) LL
wing dam 1. A log or rock wall built out into a stream to prevent logs from hanging up or drifting away. 2. Usually a slanting downstream pier to narrow and deepen the channel, to guide logs past an obstruction, or to throw all the water on one side of an island. (N.F.) Also pier dam, side pier. LL
wing ding A drunk that a lumberjack went on when he was stakebound. (L.S.) LL
winged logs Logs piled out in the river and held in place to help form a channel through which a log drive could pass. (L.S.) LL
wing jam A jam formed against an obstacle in the stream and slanting upstream with the upper end resting solidly against one shore, and on the opposite side, an open channel for the passage of logs. Also abutment. (N.F.) LL
wing log 1. A side log in a boom. 2. An outside log on a load, above the bunk log and not on the bottom tier.
winos Habitual wine drinkers found on temporary jobs.
winter cut Poles cut during the winter because they are supposed to check less since they contain less moisture than those cut in the spring. (P.C.F.)
wintering The storing of logs over the winter in river driving. These "old logs" would be the first rafted out in the spring. (N.W.)

Medford Corporation (Medco) operated this **Willamette** locomotive along with two others out of Butte Falls, Oregon. Built in 1923 for Multnomah Lumber & Box Company at Yaquina, Oregon, the No. 2 worked for several other operators until ending up with Medco. She was scrapped in August of 1959. Medco's two other Willamettes (No.s 4 and 7) have been preserved. Note the **Clyde four-line skidder** sitting in the background. *Harold K. Vollrath photo, August, 1952, Philip Schnell collection.*

winter of the blue snow Said of unusual winter weather conditions. Blue snow fell one winter during Paul Bunyan's logging operation. LL

winter show 1. A logging operation that operates all year on dependable rock roads. 2. An extremely cold job. 3. A logging operation below the snow belt which can usually operate all winter.

wipe the clock, wipe the gauge To make an emergency stop on a locomotive, reducing the pressure on the air gauge to zero. (P.C.F.)

wire a car To secure a load on a flat car by using heavy wire to pull in the car stakes.

wire ax 1. An old double-bitted ax used in cutting wire rope. One bit was driven into a stump and the line laid across the top bit was hammered. 2. Any old ax with broken edges which was worthless.

wire chute A method of loading lumber onto lumber schooners; used primarily in the dogholes of California's Mendocino Coast. The wire chute was an improvement over the original wooden chutes. A steam donkey rigged with a skyline arrangement handled the units of lumber between ship and shore. Passengers were also transferred, riding a platform carried in the same manner as the lumber.

wire fall A fall block used with wire rope. (P.N.F.)

wire in To set up the whistle punk's wire from the woods to the donkey. (P.C.F.)

wire tender See spool tender.

Wirkkala hook The popular choker hook used today. The original patent was obtained by Oscar Wirkkala and was modified by Bardon. (P.C.F.) See Bardon hook, ukelele choker.

Wirkkala system A slackline method of yarding using two donkeys or the Wirkkala donkey. The donkeys were mounted on a single sled and drew steam from the same boiler. The power of all engines could be applied to one drum if needed. It took two punchers to operate. One donkey puncher worked from one side, while the other worked from the other side. The engine was a combination yarder and roader and could be used also for a standing line or slackline skidder. The Wirkkala family was and still is active in logging along the Lower Columbia River in Western Washington. (P.C.F.)

Wisconsin sticker A lumberjack who worked at one camp until he made his stake. (L.S.) LL

In this 1904 view, Caspar Lumber Company loads **redwood** lumber onto the sailing schooner Sailor Boy, by use of a **wire chute**. The location is Caspar, on California's rocky Mendocino Coast. *National Maritime Museum*

witch A chain and lever that bound the lumber together in a lumber raft. (L.S.) LL

witch planks The top traverses or stringers of a log raft. (L.S.) LL

witch stick A twig or branch of an alder brush used to locate water for the logging camp. (N.F.) LL

withe A tough flexible twig, usually a willow, which was used to bind things together. (L.S.) LL

witherlick A mythical animal in the lumber camps. (Maine)

witness Usually a man who stands back and lets other workers do the work. (P.C.F.)

witness tree A tree that is scribed and recorded to mark the location of a section corner. Also bearing tree.

wob, wobbly 1. A radical, a member of the Industrial Workers of the World, which flourished during World War I. This organization did not gain much strength in the Redwood Country of California because it was a misdemeanor to possess an I.W.W. membership card in Humboldt County. 2. To undermine: "Is he wobblin'?"

wobbly demands The wobblies were only partly successful in achieving their demands because of the bad publicity arising against the I.W.W. as a result of the Centralia incident in 1919 which destroyed their effectiveness as an organization in the timber industry. Listed below are some of the "radical demands" that the wobblies fought for:

Reduce the ten-hour work shift to eight hours.

Increase wages to a minimum of $4.85 a day.

Provide free transportation to the job (applied to timber logging camps).

Provide hiring on the job or from a union hall, eliminating buying a job from a job shark, also known as a man hunter.

Allow workers to join a union without fear of loss of a job.

In the logging camps provide for the following:

A drying room for wet clothes.

Shower facilities for the loggers.

Porcelain wash basins instead of enamel dishes.

Proper sanitary toilet facilities.

Care and hospitalization for the injured immediately instead of waiting until the end of the shift.

Abolition of the bindle.

Clean mattresses and bedding furnished by the company.

Abolition of double-tier bunks.

Abolition of muzzle-loading (getting as many men into as small an area as possible to eliminate building more bunkhouses).

No more than twelve men to a bunkhouse instead of the usual twenty to fifty.

(Carol J. Lind, "Big Timber, Big Men" Seattle: Handcock House Publishers Inc., 1978, p. 139.)

wobbly horrors Said of a mill superintendent or logging superintendent who had a fear of labor trouble, especially when the wobblies were around.

Wolff & Zwicker Builders of an early logging donkey. Wolff & Zwicker Iron Works was a large plant in Portland, Oregon, and donkeys were merely a sideline. They outfitted ships, rebuilt locomotives, etc.

Wolf link saw An early electric, air or gas power chain saw, developed in Portland, Oregon and subsequently acquired by Reed-Prentice Corp. of Worchester, Massachusetts.

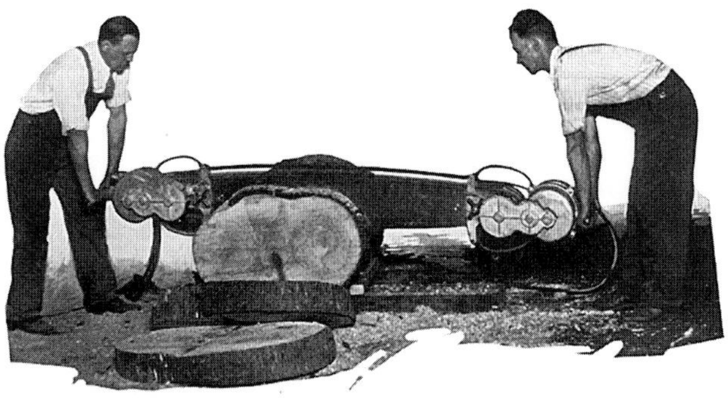

An electrically-powered **Wolf link saw**, this one having two 2-horsepower motors. *From "The Timberman," April, 1929, Philip Schnell collection*

wolf tree A large rough tree, unmerchantable, and which usually overhangs and smothers the young growth. (P.C.F.)

wonder worker A liniment that was used for everything from broken legs to toothaches. (L.S.) LL

wood 1. A log. 2. A tree. 3. A forest. 4. Any forest product.

wood boat A single sled with two skids attached by their forward ends to the bunk, and with their rear ends dragging, which was used to haul cordwood off steep or rocky slopes. (N.W.)

wood buck(er) The man who cut fuel for the wood-burning donkeys and locomotives in the early days of logging. Also wood cutter, wood split.

wood burner 1. A wood-burning locomotive. 2. A wood-burning donkey. 3. Anything old-fashioned. LL

wood butcher 1. A green bucker. 2. The logging camp carpenter. (N.F.)

wood chopper 1. A wood buck. 2. The wood buck's helper. (P.C.F.)

wood cutter See wood buck.

wooden railroad A railroad using rails of wood; usually referred to rails laminated from sawed stock, rather than poles. Although most were animal-powered, some used steam or gasoline-powered locomotives. See pole road, tram.

wood'er down To load a truck to its capacity or to overload the truck. (P.C.F.)

wood'er up To load wood cars or the wood rack on a locomotive tender with fuel wood. (P.C.F.)

wood harp See crosscut saw.

wood head, wood hick A lumberjack. (L.S.) LL

wood hook A pulp hook. (N.F.) LL

wood line A line used to pull wood logs to the side of the donkey. (P.C.F.)

wood log 1. A cull log used for fuel. 2. A log that cannot pay its way through the mill.

wood passer 1. One who transported wood fuel in a flatboat from the cutting point to a pullboat. (S.F.) 2. The man who passed wood up to the locomotive when wooding up.

woodpecker See beaver.

woodpecker porch A conk on a tree. (P.C.F.)

woods 1. The logging woods. 2. The uncut timber.

woods boss The logging superintendent.

wood split See wood buck.

wood splitter See cordwood splitter.

woods push The foreman in the logging camp. (P.C.F.)

woodsrider In turpentine orcharding, one who has charge of the operation of the whole or a part of a turpentine orchard. (S.F.)

wood up 1. To load the wood box on a locomotive. (L.S.) LL 2. To fill any wood supply: donkey, locomotive or cookstove.

wood yard The area where logs are dumped to be cut for camp fuel wood.

workala A corruption of Wirkkala, a single skyline system using two spar poles to swing logs from a yarder to the landing across a deep gulch.

work for the undertaker Said of the action of a loose choker which slips off a log or out of the pile at a landing and swings around to the great danger of anyone within reach. (P.N.F.)

working foreman Usually the boss who worked in a small crew. (L.S.) LL

work in the brotherhood notch To coast along easily in a locomotive when the engine has plenty of steam and the fireman is relaxing. (P.C.F.)

work in the company's notch 1. To work a steam engine with the throttle and cutoff notched up to give the greatest efficiency. 2. To give more than a day's work for a day's pay. (P.C.F.)

work in the corner To drop the Johnson bar "in the corner" on a locomotive to exert the greatest amount of pull. (P.C.F.)

work on the rear To bring ahead the back end of a river drive.

works (the) A logging operation.

work steam 1. To run a steam engine with the throttle partly open. 2. To make a hard pull with a steam engine. (P.C.F.)

work tandem To use two roaders to take the log output from one yarder. (P.C.F.)

work the steam dry To get the best out of a steam engine. (P.N.F.)

work under a flag To use a fictitious name when applying for work because of past troubles. (P.C.F.)

worms 1. Timber borers. 2. Marine borers. See teredo.

wormy Said of dead timber ruined by insects.

would mire a mosquito wearing snowshoes Said of very soft ground. (P.N.F.)

wouldn't that ross you? Said of a situation that is very repulsive to a person (from a rosser or debarking rig). (P.C.F.)

wow 1. A bend or curve in a piece of manufacturing. 2. A sweep in a log or cable. 3. A kink in a rail or line. See wahoo. 4. Anything out of line.

wrangler A teamster on a log-skidding job. (P.C.F.)

The Theodore Lumber Company in Perry County, Tennessee, made use of a **wooden railroad** with a small **Climax** locomotive and wooden **trucks**, about the turn of the century. Loading was done by hand with **peaveys** and **skids**. *From G.P. Vance collection, courtesy of W. Casler*

wrap 1. A turn of line around any object. 2. To anchor a guy line by wrapping it around a stump and spiking it.
wrap it up An order from the head loader to tie the logs down and to truck the load away. (P.C.F.)
wrapper 1. A cable or chain wrapped around logs on a truck to prevent them from spilling. Also binder. 2. A binding chain wrapped around the logs in the lower half of the load to hold them tightly to the sleigh. (N.F.) LL
wrapper chain See wrapper.
wrastle A dance in a lumber camp. (S.F.) (L.S.) LL
wrecker A careless worker. (P.N.F.)
wrecking frog See rerailer. LL
wrench 1. To repair equipment. 2. Occasionally used in a derogatory sense: "He wrenched the jitney."
write'er out The popular statement to the boss when a worker quits the job.
wye A track laid out in the contour of the letter Y for turning a locomotive.

X Y Z

X-tree 1. A tree marked X to be used as a spar tree; not to be cut by a felling crew. If the tree is felled, the loggers are fired immediately. (P.C.F.) 2. See broad arrow mark.
Y See wye.
YEBs Yellow-eyed beans. Pork and beans were the river driver's staple of diet, as well as the lumberman's. (Maine)
Yacolt Burn A huge forest fire which occurred between the Columbia and Lewis Rivers in Southwestern Washington state in 1902. A big fire is often called another "Yacolt." (P.N.F.)
Yakima pine (P.N.F.) See ponderosa pine.
yannigan The container or bag in which the woodsmen carried their clothes. (S.F.) (L.S.) LL
yaps Said of a person who is crazy. (L.S.) LL
yard 1. The tracks at the end of a railroad line. 2. The area within reach of a donkey engine's haul. 3. The area around a sawmill where lumber is stacked. 4. A temporary holding place for logs. (N.F.) 5. To haul logs from the woods to the landing. 6. To store loaded log cars in a yard before making up a train to go out to the dump.
yardage An older term for the yarding operations.
yarded his last turn 1. To yard the last turn in the day. 2. To quit the woods for good. 3. Said of a logger who died. (P.C.F.)
yarded wood Pulpwood that has been twitched or skidded to the yard or landing. (L.S.)
yarder 1. A donkey engine used to haul logs from the stump to the landing, or to a yarder tree. 2. A tractor used in place of a donkey engine doing the same thing. 3. A yarding block. 4. A logging arch.
yard'er over? A question, meaning "What did you say?" (P.C.F.)
yarder tree The tree to which logs are yarded and from which they are swung to the landing on a two-donkey show. (P.C.F.)
yard goat A switch engine. (P.C.F.)
yarding In the earliest West Coast logging, bull teams dragged logs over skids (hence skidding) from the logging woods to the yard, or landing, or dump. Horses were used to some extent, both pulling logs on skids, and hauling big wheels. This was still called skidding or wheeling. Oxen and horses were also used to haul logs from the woods to the dump on log carts running on a wooden tram. Though called "hauling," it was still a kind of yarding. Then Dolbeer developed the steam donkey engine which could drag logs in with a cable. A horse returned the line to the woods, or the loggers dragged it back out by hand. About this time, skidding became known as yarding, or ground lead yarding. Then the addition of the haulback did away with the line horse; the high lead replaced the ground lead; gas and diesel engines have replaced steam donkeys; and cats and arches have replaced horses and wheels. But the business of getting logs from the stump to the landing is still known as yarding.

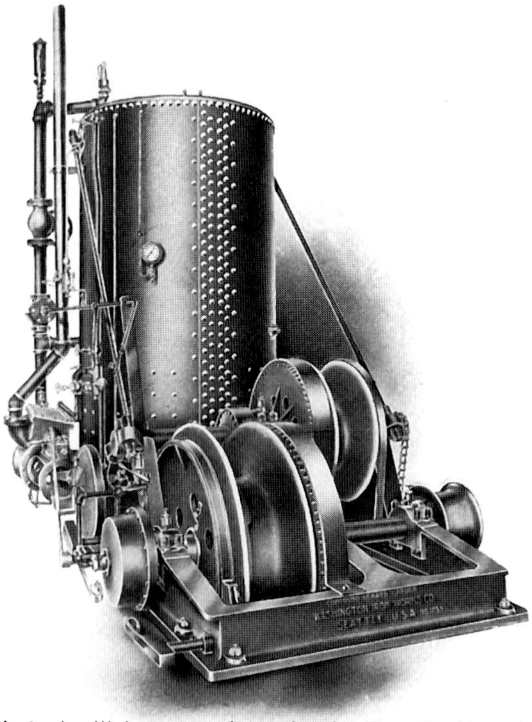

A **Washington** Iron Works **compound-geared yarder**. *From a Washington Iron Works catalog, Labbe collection*

yarding block A logging block usually linked by a swivel to a large hook which makes it easy to secure the block or hang gear on it. Also swivel block.
yarding chance See setting.
yarding donkey See yarder.
yarding grabs A pair of grabs or tongs once used in yarding.
yarding hooktender See hooktender.
yarding line The mainline on a skyline yarding system.
yarding roller A vertical spool which was used at one time to lead the rigging in a part circle back of the donkey, enabling it to yard

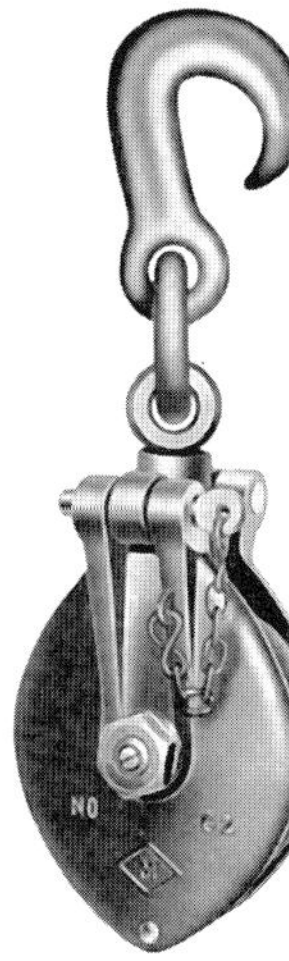

A **yarding block**. *From Mallory Catalog No. 6, Glen Comstock collection*

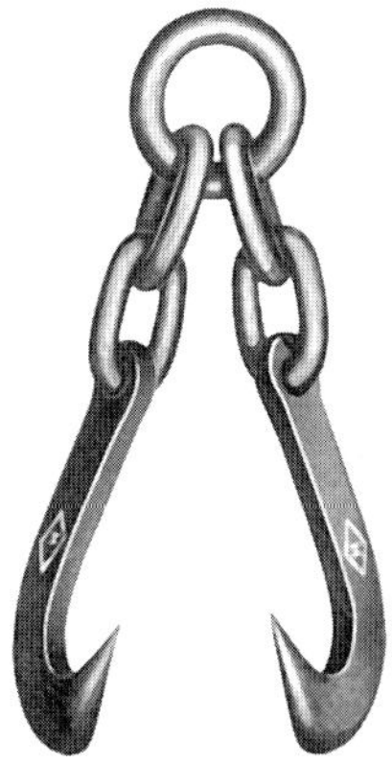

Yarding grabs. *From Mallory Catalog No. 6, Glen Comstock collection*

directly back of the landing.
yarding sled See dray.
yarding spool (roller) 1. A vertical spool on an early donkey. See Dolbeer. 2. In ground lead logging, a heavy sheave on a bracket which acted as a block for a cable to run through. It was bolted to a tree or stump. See road roller, road spool. (P.C.F.)
yarding tower 1. The big steel tower used on a steel spar skidder. See Lidgerwood. 2. A light tower built on a tractor. See lighthouse cat.
yard out To pull out finished booms to make up a tow for a tug boat. (P.N.F.)
yearling In turpentine orcharding, a term applied to a box or cup during the second season the tree is bled. (S.F.)

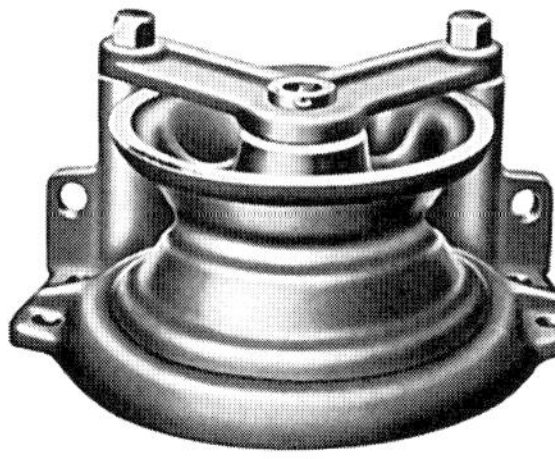

A **yarding spool**, or yarding roller. *From Mallory Catalog No. 12, 1925, Glen Comstock collection*

yeast under him Said of a man when he is fired. (App.)
yellow fir The old growth Douglas fir that grew very slowly in sheltered areas which resulted in soft wood. (P.C.F.)

A **yoke of oxen** skids a turn of logs across a crude log bridge, location unidentified. Note the heavy **yokes** on the **oxen**. A set of **carting grabs** connects the first log to the team, while **single couplers** probably connect each log together. *Coe photo from Jack Foster*

yellowjacket 1. A hornet or wasp. 2. A yellow-striped speeder or crew car on a logging railroad. (P.C.F.)
yellow pine (P.N.F.) See ponderosa pine.
YMCA whiskey A soft drink. (P.C.F.)
yoke 1. The heavy U-shaped part of a block by which it is attached to a tree, stump or other object. 2. A team of oxen in the early days. 3. The wooden frame to which the animals were hitched together.
yoke of oxen A team of oxen. (L.S.) LL
yokes See lug hooks.
you-know A tin wash basin for taking up collections. A penny was soldered to the bottom to remind the lumberjacks that they could not give less than a penny when it was passed. (L.S.)
young growth The young timber which grows up after logging or a fire. Also called second growth or third growth.
yump the yob Said of a Swedish logger who has changed camps, or who has gone to town. (P.C.F.)
Zeke The name of Paul Bunyan's bookkeeper, who used a peavey handle as a pen holder. LL
zippo Another term for gyppo, a logger who operated on a small scale. (L.S.) Also chin-whiskered jobber. LL
zoogler The pig fucker or P.F. man in the Redwood Country of California who took care of the pig. He also kept the line on the spools along the skidroad when a load was brought in and when the line was taken back to the woods. (Red.)

Bibliography

Books

Anon., *Forest Terminology, A Glossary of Technical Terms Used in Forestry*, Third ed., rev. Washingtom, D.C.: Society of American Foresters, 1958

Anon., *Safety Standards for Logging Operations*, State of Washington Department of Labor and Industries, Division of Safety, 1961

Anon., *The Heisler Locomotive, 1891-1941*. Lancaster, Pennsylvania: Benjamin F.G. Kline, Jr., 1982

Brown, Nelson Courtland, *Logging, Principles and Practices in the United States*. New York: John Wiley & Sons Inc., 1934

Brown, Nelson Courtland, *Logging, Principles and Practices of Harvesting Timber in the United States and Canada*. New York: John Wiley & Sons Inc., 1949

Bryant, Ralph Clement, *Logging: The Principles and General Methods of Operation in the United States.* New York: John Wiley & Sons, Inc., 1913

Bryant, Ralph Clement, *Logging: The Principles and General Methods of Operation in the United States.* 2nd ed. New York: John Wiley & Sons, Inc., 1923

Clarkson, Roy B., *Tumult on the Mountains*. Parsons, West Virginia: McClain Printing Co., 1964

Dean, William, and Wane, David S., editors, *Terms of the Trade - A Handbook for the Forest Products Industry*. Eugene, Oregon, Random Lengths Publications, Inc., 1978

Disque, Brice P., *History of Spruce Production Division, United States Army, and United States Spruce Production Corporation.* Undated, circa 1918

Foehl, Harold M. and Hargraves, Irene M., *Logging the White Pine in the Saginaw Valley.* Red Keg Press, 1964

Ford-Robertson, F.C., *Terminology of Forest Science, Technology, Practice and Products.* Washington, D.C.: Society of American Foresters, 1971

Gerschel, J., *Forest Terminology in English, French and German.* Nancy, France: Ecole National des Eaux et Forets, 1905

Hauff, Steve and Gertz, Jim, *The Willamette Locomotive.* Portland: Binford & Mort, 1977

Hirsimaki, Eric, *Lima, the History*. Edmonds, Washington: Hundman Publishing Co., 1986

Holbrook, Stewart H., *Holy Old Mackinaw*. New York: Macmillan, 1957.

Holbrook, Stewart H., *Tall Timber*. New York: The Macmillan Company, 1949

Syverson Brothers operated this pole road near Quincy, Oregon, circa 1893. The log cars are equpped with concave-flanged wheel to negotiate the crude poles used for rails. The pole road haulers, or locomotives, appear to be converted steam donkeys. *Labbe collection*

A Brooks-Scanlon Lumber Company **McGiffert** loads short pine logs onto **four-bunk log cars** near Bend, Oregon. This is the "swing-boom, eight-wheel" model popular in the western pine regions. This view clearly shows the traveling wheels drawn up under the machine to make room for the empy flat in the "tunnel." Horse-drawn **big wheels** bring the logs to the railhead.

Holbrook, Stewart H. *Yankee Loggers*. New York: International Paper Co., 1961.

Jackson, Francis, *Big River Dammed*. Fort Bragg, California: Mendocino County Historical Society, 1976

Koch, Michael, *Steam & Thunder in the Mountains, Saga of the Forest Railroads*. Denver: World Press, Inc., 1979

Koch, Michael, *The Shay Locomotive, Titan of the Timber*. Denver: World Press, Inc., 1971

MacKay, Donald, *The Lumberjacks*. Toronto: McGraw-Hill Ryerson Limited, 1978

McCulloch, Walter F., *Woods Words: A Comprehensive Dictionary of Loggers Terms*. Portland: Oregon Historical Society and the Champoeg Press, 1958.

Pike, Robert E., *Tall Trees, Tough Men*. New York: W.W. Norton & Company, Inc., 1967

Prentiss, Arthur M., *Pictorial Review of World's War Activities, Spruce Production Division, U.S. Army, Oregon & Washington*. Portland: A.M. Prentiss and Grant Smith-Porter Bros., 1918

Ranger, Dan Jr., *Pacific Coast Shay, Strong Man of the Woods*. San Marino, California: Golden West Books, 1964

Schenck, C.A., *Logging and Lumbering or Forest Utilization, a Textbook for Forest Schools*. Biltmore Forest School, undated

Smith, Elmer L., *Logging in the Pennsylvania North Woods*. Lebanon, Pennsylvania: Applied Arts Publishers, 1969

Sorden, L.G., *Lumberjack Lingo*, Spring Green, Wisconsin: Wisconsin House, Inc., 1969

Sorden, L.G., and Ebert, I.S., *Words of Yesteryear*. Privately printed, 1956

Sorden, L.G. and Vallier, Jacque, *Lumberjack Lingo: A Dictionary of the Logging Era*. NorthWord, Inc., 1986

Taber, Thomas T. III, Casler, Walter, *Climax, an Unusual Steam Locomotive*. River Edge, Pennsylvania: Railroadians of America, Inc., 1960

Williams, Guy, *Logger Talk*, Chapbooks, No. 41. Seattle: University of Washington, 1930

Williams, Richard L., *The Loggers*. Alexandria, Virginia: Time-Life Books, 1976

Equipment Catalogs

Associated Equipment Dealers, *Clyde Equipment Company Construction and Logging Equipment, General Catalog*. Portland, Seattle, 1952

Clyde Iron Works, *Clyde Self-Propelling Logging Machinery*. Duluth, Minnesota, 1912

Clyde Iron Works, *The Clyde Track Machine, Catalog TM*. Duluth, Minnesota, 1929

F.B. Mallory Co., *Catalog No. 6*. Portland, undated, circa 1920

F.B. Mallory Co., *Catalog No. 12*. Portland, 1925

Heisler Locomotive Works Catalog, Erie, Pennsylvania, undated, circa 1929

Kloepfer Hardware Company Limited, *Logging Supplies, Bar Iron & Steel, Blacksmiths' Tools Catalogue No. 3*. Vancouver, B.C., undated, circa 1939

Lamb Machine Company, *Logging Equipment Catalog No. 24*. Hoquiam, Washington, undated, circa 1930

Mallory Logging Equipment Co., *Catalog No. 48*. Portland, 1948

Mill and Mine Supply Company, *Logging, Mill, Mine and Cannery Supplies*, Seattle, 1916

Paul E. Keeney Co., *Rope Master Logging Blocks, Hyloy Forgings*. Portland, 1948

Seattle Car & Foundry Co., *Catalog No. 3*. Seattle, 1913

The Skookum Company, *Skookum Blocks and Forgings Catalog No. S-63*. Portland, undated, circa 1955

Smith & Watson Iron Works, *Logging Equipment Catalog No. 9*. Portland, undated, circa 1917

Washington Iron Works, *Illustrated Catalog*. Seattle, undated, circa 1921

Washington Iron Works, *Logging Equipment*. (a small catalog listing blocks, hooks, etc.) Seattle, undated, circa 1910

Willamette Iron & Steel Works, *Willamette Logging Engines Catalog*. Portland, 1913

Willamette Iron & Steel Works, *Logging Machinery*. Portland, 1920

Willamette Iron & Steel Works, *Logging Machinery*. Portland, 1925

Young Iron Works, *Catalog*. Seattle, undated, circa 1940

Young Iron Works, *Catalog No. 52*. Seattle, 1952

Periodicals

American Forests

Davis, Elrick B., *Paul Bunyon Talk*, Vol. 17 No.4, December, 1942

Morrison, J.W., *Lumberjack Rhetoric, the Terse and Picturesque Language of the Lumber Woods*. 1924

American Lumberman

Various Issues

American Speech

Carranco, Lynwood, *Logger Lingo in the Redwood Region.*, XXXI, No.2 (May 1956)

Carranco, Lynwood, *More Logger Lingo in the Redwood Region.*, XXXIV, No.1 (Feb. 1959)

Carranco, Lynwood, *Logging Railroad Language in the Redwood Region.*, XXXVIII, No.2 (May, 1962)

Clark, J.W., *Lumberjack Lingo*, No.7, October, 1931

Misfeldt, Orlo H., *Timberland Terminology*, XVI-3, October, 1941

Stevens, James, *Logger Talk*, No.3, 1924

Stevens, James, *Logger Talk*, No.1, 1925

California Folklore Quarterly

Kernan, Henry S., *Idaho Lumberjack Nicknames*, July, 1945

Dialect Notes

Carr, J.W., and Chase, G.D., *A Word-List from Aroostook*, Vol. 3, Part 5, 1909

England, George Allan, *Rural Locutions of Maine and Northern New Hampshire*, Vol. 4, Part 2, 1914

Leonard, Arthur, *Lists from Maine*, Vol. 4, Part 1, 1913

Maxfield, E.K., *Maine List*, Vol. 5, Part 9, 1926

Forest History

McColloch, Walter F., and Fry, Amelia, *The Making of "Woods Words,"* No. 16, Oct. 1972

Forestry Quarterly

Fernow, B.E., *Forest Terminology*, No.3, 1905

Lovejoy, P.S., *A Suggestion for Securing Better Professional Terminology*, No.12, 1914

Journal of Forestry

Forest Terminology: Report of Committee, Vol. XV, Jan., 1917, No.1

Forest Terminology: Terms Used in the Lumber Industry, Vol. XVI, Jan., 1918

Journal of Forestry History

Carranco, Lynwood, *Logger Language in Redwood Country*, Vol. 18, No. 3 (July, 1974)

Loggers Handbook

Various Issues from 1945-1960. Published by The Pacific Logging Congress, Portland, Oregon

The Timberman

Various Issues

West Coast Lumberman

Various Issues

Western Folklore

Davis, Wilbur A., *Logger and Splinter-Picker Talk*, April, 1950